Morphogenesis, Environmental Stress and Reverse Evolution

Jean Guex • John S. Torday
William B. Miller Jr.
Editors

Morphogenesis, Environmental Stress and Reverse Evolution

Springer

Editors
Jean Guex
Institute of Earth Sciences
Geopolis, UNIL
Lausanne, Switzerland

William B. Miller Jr.
Department of Medicine
Banner Health System
Paradise Valley, AZ, USA

John S. Torday
Department of Pediatrics, Obstetrics
and Gynecology
Evolutionary Medicine Program
David Geffen School of Medicine
University of California
Los Angeles, CA, USA

ISBN 978-3-030-47278-8 ISBN 978-3-030-47279-5 (eBook)
https://doi.org/10.1007/978-3-030-47279-5

This Springer imprint is published by the registered company Springer Nature Switzerland AG
The registered company address is: Gewerbestrasse 11, 6330 Cham, Switzerland

Contents

1 **Introduction** ... 1
Jean Guex

2 **The Controversial Cope's, Haeckel's and Dollo's Evolutionary Rules: The Role of Evolutionary Retrogradation** ... 13
Jean Guex

3 **Impact of the Environmental Stress on the Late Permian Pollen Grains from Zechstein Deposits of Poland** ... 23
Anna Fijałkowska-Mader

4 **Stress-Related Evolution in Triassic Conodonts and the Middle Norian Juvenile Mortality** ... 37
Viktor Karádi, Attila Virág, Tea Kolar-Jurkovšek, and Bogdan Jurkovšek

5 **Proteromorphosis in Early Triassic Conodonts** ... 59
Ali Murat Kiliç, Jean Guex, and Francis Hirsch

6 **Developmental Plasticity Induced by Either External or Internal Environment Co-opts Ancient Regulatory Networks** ... 97
Juan Nicolas Malagon, Sam Scanga, Ernest Ho, Armen Manoukian, and Ellen Larsen

7 **Extreme Morphological Plasticity Within *Orbulina*-"*Praeorbulina*-Like" Assemblages Related to Environmental Stress** ... 111
Ahmed Belhadji, Annachiara Bartolini, Linda Rossignol, Lahcène Belkebir, and Jean Guex

8 **Steps of Morphogenesis and Iterative Evolution of Imperforate Larger Foraminifera in Shallow Carbonate Shelves During Mesozoic Times: Possible Relations to Symbiotic and Abiotic Factors** 129
Michel Septfontaine

9 **Morphological Deformation of Foraminiferal Tests Caused by Intertidal Oil Spills (Black Tides)** 175
Marie-Thérèse Vénec-Peyré, Annachiara Bartolini, Michele Weber, and Jere H. Lipps

10 **Environmental Control on Biotic Development in Siberia (Verkhoyansk Region) and Neighbouring Areas During Permian–Triassic Large Igneous Province Activity** 197
Yuri D. Zakharov, Alexander S. Biakov, Micha Horacek, Ruslan V. Kutygin, Evgeny S. Sobolev, and David P. G. Bond

11 **Stress, Development, and Evolution in Coral Reef Communities** 233
Neil W. Blackstone and Austin P. Parrin

12 **Fossil Benthic Foraminifera Morphologic Adaptation (Kleptoplastidy) Within Low-Oxygen-Bottom Water Environments, Coupled with Geochemical Insights from the Late Cretaceous in the Levant Basin** 245
Aaron Meilijson, Sarit Ashckenazi-Polivoda, and Peter Illner

13 **Evolution as a Timeless Continuum** 289
John S. Torday and William B. Miller Jr.

14 **Chronic Disease as Reverse Evolution** 299
John S. Torday and William B. Miller Jr.

Index 311

Chapter 1
Introduction

Jean Guex

Keywords Extinctions · Atavistic development · Environmental stress

This book intends to be a continuation of a prior text entitled "Retrograde Evolution During Major Extinction Crises" (Guex 2016). That initial book was restricted to the study of catastrophic evolutionary changes during such major extinctions, which were generated by extreme and sublethal environmental stresses, leading to evolutionary reversals. This introduction will be a short discussion of usual environmental stress in geology, and a condensed and reduced presentation of Guex's 2016 book.

The concept of sublethal environmental stress denotes specific conditions that are critical to the survival or normal development of living organisms. The most common cases of environmental perturbations include pollution or poisoning by chemicals such as toxic gases, nutrient shortage, large-scale sea-level falls, major climatic changes, hydric stress, acid rain and marine anoxia. The origins of environmental stress can easily be identified and quite naturally the particular kind of stress can be hostile to some organisms and favourable to others.

The role of natural environmental stress on development and evolution is widely accepted (Badyaev 2005; Jablonka 2013). The nature of sublethal environmental stress that occurred during major extinctions such as the Permian-Triassic (PT), Triassic-Jurassic (TJ), Pliensbachian-Toarcian (Pl-To) and Cretaceous-Tertiary (KT) has been intensively investigated by means of geochemical and geochronological studies. However, most problems related to these exceptional situations belong to the realm of palaeontology, and most, if not all, such palaeontological studies are dedicated to a census of the biodiversity variations (counting how many groups survive, how many disappear, etc.) rather than to the understanding of the basic phenotypic and epigenetic variations induced by major environmental perturbations. The modes of evolution during major extinction events, the characteristics of the organisms surviving major crises and what types of transformations have affected them have not been explored. The main goal of this study is to precisely

J. Guex (✉)
Institute of Earth Sciences, Geopolis, UNIL, CH 1015 Lausanne, Switzerland
e-mail: jean.guex@unil.ch

J. Guex et al. (eds.), *Morphogenesis, Environmental Stress and Reverse Evolution*, https://doi.org/10.1007/978-3-030-47279-5_1

analyse in detail the transformations of some invertebrates during major extinction periods.

The geological record of environmental stress is naturally poor because local pollution, nutrient levels, paleotemperatures, etc. can hardly be deduced from sediments or fossils. The primary indication and proof of the sporadic sublethal environmental stress are the extinction periods themselves, which are easily recognised from the presence of major faunal turnovers reflecting the disappearance of entire phyla and their replacement by new ones.

The classical explanations of the origins of extinctions include extraterrestrial impacts, marine regressions, climatic changes, or anoxic events (Courtillot and Gaudemer 1996). The theory of giant volcanism (Courtillot 1999) has the most general explanatory power, considering the enormous potential consequences to the chemistry of seawater, global climate and temperatures. There is an almost perfect correlation between the major extinctions and periods of volcanism (Fig. 1.1, from Courtillot loc. cit.). Virtually all major extinctions are related to major ecological instability generated by giant volcanism, i.e. climatic changes and atmospheric poisoning by sulphur gases, as well as by darkening generated by fine particles and aerosols inducing major coolings

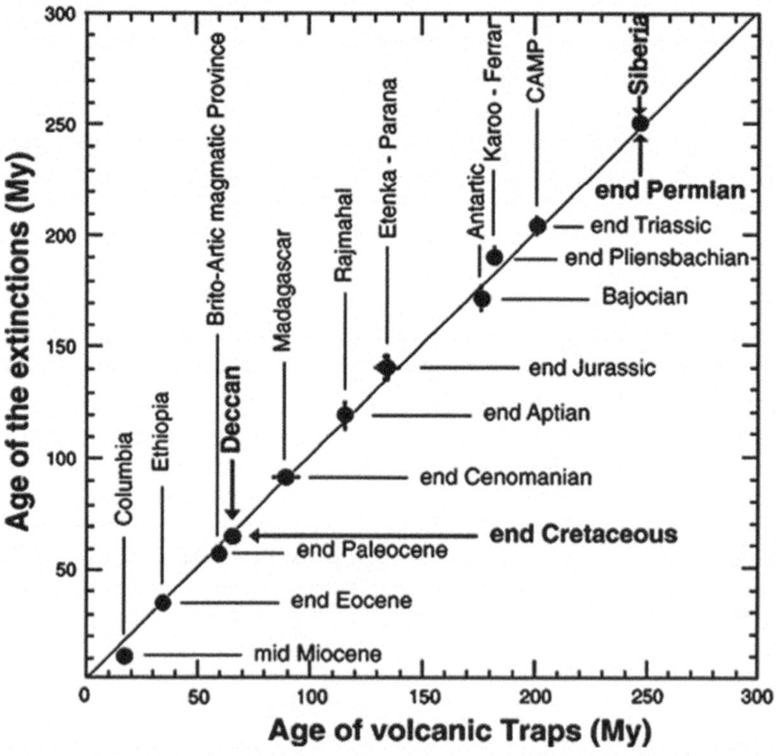

Fig. 1.1 Bivariate graph showing the correspondence between the principal mass extinctions and their geochronometrical age. From Courtillot (1999), modified

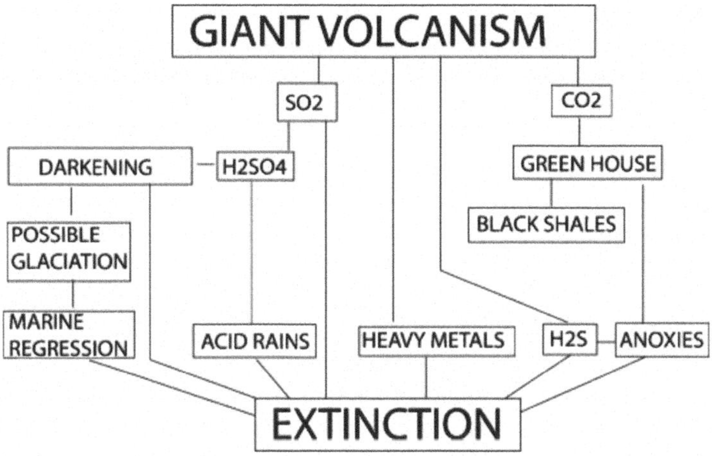

Fig. 1.2 Summary of the most obvious relationships between giant volcanism and major extinctions

and fall in the sea level (Fig. 1.2). In extremes, for example, at the Permian-Triassic boundary, the only organisms found in crisis sediments are microbialites and fern or fungus spores. Often these sediments do not contain any fossils, and hence are known as barren intervals. The post-extinction recovery is generally characterised by the explosion of simple and primitive life forms, which can be divided into two major groups:

1. Persistent opportunistic simple forms such as bacteria, fungi, ferns and some simple forms or other microfossils.
2. Primitive-looking forms derived from their immediate ancestors by retrograde evolution (a phenomenon that has been described as proteromorphosis), often associated with a reduction in body size. These organisms are not Lazarus taxa because their absence in sediments older than the ones where they are found is fully reproducible worldwide, and the duration of the intervals of time in which they are totally absent can last several millions or several tens of millions of years. In other words, a "Lazarus" explanation of such organisms cannot be reconciled with the geological record.

This introduction analyses how different groups (foraminifera, radiolaria, ammonoids, nautiloids, corals, conodonts and silicoflagellids) of invertebrates (except conodonts) survive the sublethal biotic crises during extinction episodes, and elaborates previously published works (Guex 1993, 2001, 2006). This introductory discussion also aims at developing a model explaining the heterochronous repetition of similar evolutionary lineages without invoking repetitions of identical environmental conditions.

An abundant literature has been published during the last decades concerning the Permian-Triassic boundary and the Cretaceous Paleogene boundary, which are well known for their spectacular extinctions. Two major extinctions generated by giant volcanism have been recently studied in detail by the author and his colleagues: the

Triassic-Jurassic and the Pliensbachian-Toarcian boundaries (Guex et al. 2015). These two major geological events will be discussed in light of recent geochemical and geochronological data, and will illustrate the environmental framework of the retrograde evolution observed during these events.

The End Triassic extinction has long been suspected to be related to the onset of the Central Atlantic Magmatic Province (CAMP) volcanism. However, it is only recently that U-Pb ages measured on zircons have allowed a precise correlation with the relative timescale based on the evolution of the ammonites. That correlation has been established based on very detailed stratigraphic research done in the American Cordilleras (Northern Peru and Nevada, USA) and on the discovery of ash beds deposited in the same levels as age diagnostic ammonites (Guex 1995; Schoene et al. 2010; Guex et al. 2012a). These discoveries allowed us to propose an original model explaining the precise timing of the End Triassic extinction (ETE).

One popular model to explain the ETE catastrophic event invokes super-greenhouse conditions due to extreme atmospheric CO_2 concentrations. This enrichment is often interpreted as degassing of magmatic CO_2 from huge volcanic basalt provinces (e.g. Sobolev et al. 2011 for the Permian-Triassic crisis) and/or from the degassing of carbonaceous or organic rich sediments (e.g. Svensen et al. 2009).

The second scenario invokes a short period of global icehouse conditions caused by degassing of huge volumes of volcanic SO_2, atmospheric poisoning, cooling and eustatic regression coeval with the main extinction (ETE), but probably older than the main basalt emissions. As mentioned above, this model uses the same arguments as those given below for the Late Pliensbachian cooling event.

Although both hypotheses are compatible with massive volcanic degassing related to the emission of large volumes of flood basalts, they must also be able to explain the palaeontological record in complete stratigraphic sections that display decoupling between the (marine) ammonoids and (terrestrial) plant extinctions (Guex et al. 2012a). Correlating the sedimentary and fossil records with carbon and oxygen isotope variations, and sea-level changes from the T-J and Pl-To boundaries, indicates that both boundaries are related to a regressive event followed by major sea-level rise (Guex et al. 2001, 2004, 2012a).

The data allowing us to discuss the various hypotheses of recent extinction models and the timing concerning the End Triassic extinction (ETE) and the T-J boundary have been discussed in detail in Guex (2016), and will not be repeated here. This compilation synthesises the timing of sea-level changes, and δ13Corg, δ18O and pCO_2 variations in relation with paleotemperatures, the age of the onset of the CAMP-related basaltic volcanism in the northeastern United States (Newark Supergroup) and Morocco (Argana Basin) and the ages of the two distinct End Triassic (ammonoids) and Earliest Jurassic (terrestrial plants) extinctions.

The chronology, established by ammonoids and U-Pb dating, implies that the Newark supergroup basalt postdates the ETE and the disappearance of the latest Triassic ammonite Choristoceras crickmayi (Guex et al. 2004; Schoene et al. 2010). The delay between the recovery of the Jurassic ammonites and the extinction of the very last Triassic ammonoids lasted at least 200 kyr (probably more), based on sedimentary rates in Northern Peru and Nevada. The extinction of the last Triassic ammonoids in the uppermost Rhaetian correlated with a strong negative excursion of δ13C and a marine regression (Guex et al. 2004). The δ18O record indicates a cooling episode, which could explain the regressive event recorded in the upper Rhaetian of Austria, England and Nevada. The initial regression is followed by a significant sea-level rise potentially associated with large volcanic CO_2 emissions related to the CAMP basaltic volcanism. A major plant extinction is correlated with the greenhouse conditions that postdates the ETE by at least a few hundred thousand years. The plant extinction is recorded in Greenland (McElwain et al. 1999, 2009), and is associated with a second negative δ13C recorded in the Hettangian Psiloceras planorbis beds (coeval with *P. pacificum*). The recovery of the ammonites after the End Triassic extinction calibrated with the geochronological data illustrated in Guex (2016) (Fig. 2.4), demonstrating a partial correlation between the δ13Corg curve and the diversity fluctuations, and a precise correlation between the well-known first negative excursion of the organic with the peak of the Rhaetian extinction.

The second negative excursion is restricted to the Psiloceras zone. The mid-Hettangian slowdown of the diversification is followed by an explosion of the diversity in the Upper Hettangian, and by a new positive excursion of the organic carbon. However, we note that the minimum of the ammonite diversity (D′) occurs later than the minimum of the δ13Corg curve, which is located between the *P. pacificum* and Kammerkarites beds.

A similar model has been proposed for the Pliensbachian Toarcian crisis, which is known to be correlated with the onset of the Karoo-Ferrar large igneous province (Pálfy and Smith 2000). Recent high-precision U-Pb dating on zircons of major sill intrusions in the Karoo basin can be directly correlated with the well-known Toarcian Oceanic Anoxic Event (OAE), and is concomitant with these sill intrusions into organic rich sediments of that basin (Guex et al. 2012b; Sell et al. 2014). A synthesis of the major isotopic variations through the available geochronological data and major sea-level variations allows an investigation into whether and how the geochemical and biochronological data can be correlated with the magmatic activity of the Karroo-Ferrar LIP.

The end-Pliensbachian extinction, preceding the Toarcian AOE by a few hundred kyr (Dera et al. 2010), is marked by an important diversity drop (disappearance of 90% of the ammonite taxa) associated with a generalised sedimentary gap linked to a marked regression event in NW-Europe and the Pacific area.

This regression was interpreted as being due to a major short-lived glaciation (Guex et al. 2001, 2012b) coeval with the main extinction event preceding the main basalt eruptions. Our major arguments refer to an important emersion topography observed on seismic images of the North Sea (Marjanac and Steel 1997), to the evidence of polar ice storage (Price 1999) and to the deposition of thick conglomerates (Dunlap Formation in Nevada (USA) (Muller and Ferguson 1939) and Ururoa—Kawhia area, New Zealand (Hudson 2003)). The cooling model is supported by recent δ18O data on belemnites (Gómez et al. 2008; Harazim et al. 2012), and by the discovery of glendonites in the upper part of the Pliensbachian (Suan et al. 2011). The origin of the major cooling is probably related to huge volcanogenic SO_2 degassing during the Late Pliensbachian preceding the major CO_2 emissions of the Early Toarcian (Guex et al. 2001).

The regressive phase is followed, after a few hundred thousand years, by a worldwide transgression during the Early Toarcian, with the deposition of black shale associated with the Toarcian OAE. The Toarcian OAE itself is responsible for a second extinction affecting mainly benthic foraminifera populations and brachiopods. Radiolarians were also affected, but their extinction was apparently slightly delayed in comparison with the benthos, and probably coincided with a drastic fertility drop just after the OAE.

The succession of ice house conditions immediately followed by supergreenhouse conditions can be explained thanks to a petrological model explaining the SO_2-dominated vs. CO_2-dominated degassing couplet generating the successive cold and hot conditions (Guex et al. 2016). The model invokes a thermal erosion of the cratonic lithosphere, inducing giant H_2S/SO_2 release from sulphur-bearing basal continental crust before CO_2 becomes the dominant gas associated with the giant basalt emission.

During moderate (i.e. not sublethal) environmental stress, the morphological response of many invertebrates consists often of a loss of symmetry (is this reverse evolution?), very well described by Hoffmann and Parsons (1991). Some paleontological examples will be described further in this book by Venec Peyre and Lipps.

During the major extinctions, several phyla tend to reverse their evolution and give rise to primitive-looking forms. Such reversal can be illustrated diagrammatically using one single relatively simple curve discovered by Thom (1972): the cusp catastrophe.

The catastrophe theory is a domain of the differential topology, which was invented by René Thom (1972). It aims at building the simplest continuous dynamic model, which can generate a morphology, given empirically, or a set of discontinuous phenomena.

Thom's theory concerns the phenomena where a gradual and relatively slow change produces a sudden jump of the state of the system. Such phenomena are called catastrophes. The graphical representation known under the name of "cusp catastrophe" is ideal to describe empirically the cases of the evolutionary jumps, which arise during gradual changes in the environmental stress. The surface illustrated in Fig. 1.3 represents the variable, which characterises the more or less advanced state of a taxonomic group, which varies during evolutionary space-time. This state is controlled by two parameters, which, in our case, are environmental stress and time factor.

When these parameters vary, the curve of the state of the taxonomic group under study follows a trajectory, which depends on the time and intensity of the environmental stress. When the stress gradually reaches a certain threshold, the evolutionary state of the evolving system arrives at the border of the cusp, and a jump occurs towards a previous state of a more primitive aspect. In this introduction we will use such simple diagrams to describe reverse phenomena characteristic of retrograde evolution.

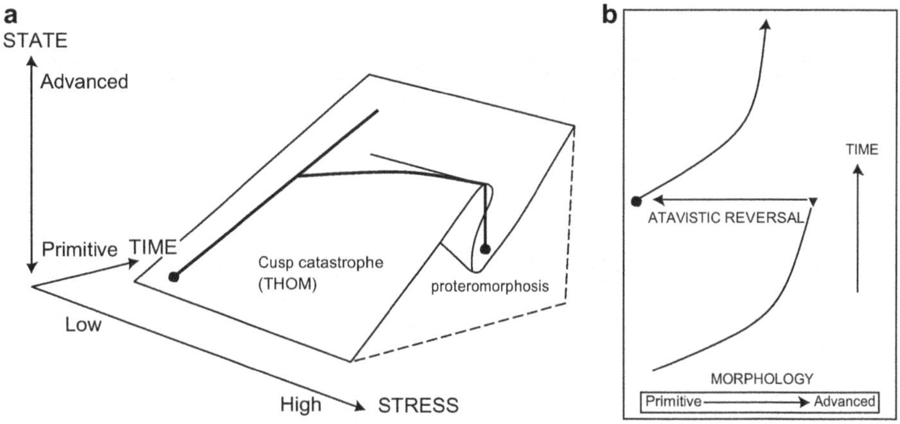

Fig. 1.3 The cusp catastrophe of Thom. (**a**) Stress and time are the parameters controlling the state of the evolving system (primitive–advanced). (**b**) Simplified graphic representation of (**a**)

One Example of Planktonic Foraminifera

A fundamental example of a Cretaceous planktonic foraminiferal anagenetic lineage starting from a very primitive form (evolute with simple rounded chambers), and giving rise to more complex forms with an involute and carinated shell, was recently identified in the lineage *Ticinella-Thalmanninella* (Desmares et al. 2008; Fig. 3.1). During the major Cenomanian oceanic anoxic event OAE2, the end forms of the lineage (group of *Thalmanninella greenhornensis*) gave rise to a very simplified atavistic group (*Thalmanninella multiloculata*, also called "*Anaticinella*"), which is a quasi-homeomorph of its ticinellid ancestor. Here we present these discoveries (Desmares et al. 2008) in light of what is known about the influence of high environmental stress on the development and variability of some other marine invertebrates.

The stratigraphic interval considered in the present section spans from the Late Albian to Late Cenomanian. There are records of two major environmental perturbations in this period: the Mid-Cenomanian Event and the Oceanic Anoxic Event called OAE2. These two anoxic events have markedly influenced the evolution of the *Ticinella-Thalmanninella* lineage, as well as two diverging lineages represented by *Rotalipora montsalvensis-praemontsalvensis* and *montsalvensis-planoconvexa* (Gonzalez-Donoso et al. 2007). The stratigraphic distribution and outline of the phylogeny of these lineages are shown in Fig. 3.1 (from Desmares et al. 2008, modified; see also Caron 1985; Robaszynski and Caron 1995 for the stratigraphic details, and Guex et al. 2012b for discussion) (Fig. 1.4).

The anagenetic *Ticinella-Thalmanninella* lineage represents an example of geometrical transformation, which is quite similar to what is known for many lineages of ammonites.

The "***Anaticinella***" plexus:

During the onset of the anoxic event OAE2, the intraspecific variability of *Thalmanninella greenhornensis* increases remarkably; this group gives rise to *Thalmanninella multiloculata*, a "species" (morphospecies) with an indistinctly

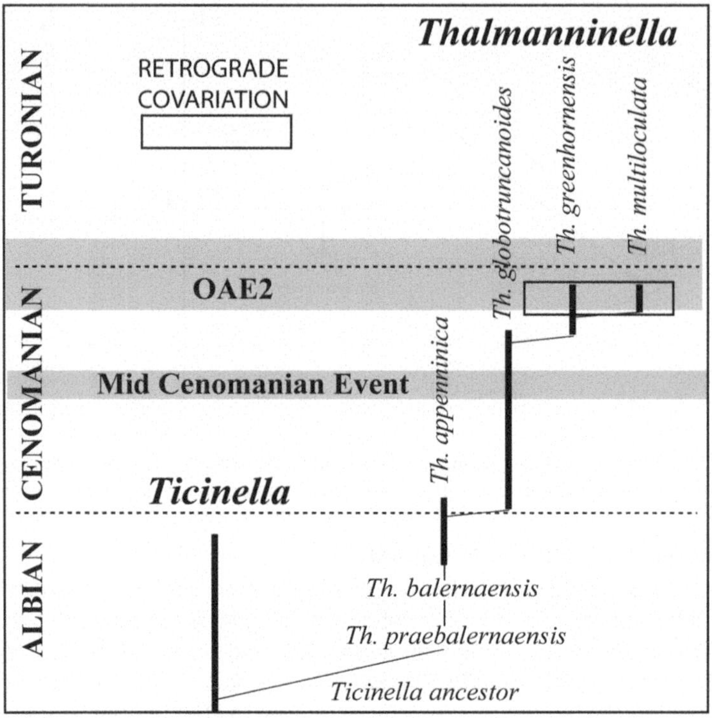

Fig. 1.4 The *Ticinella-Thalmanninella* lineage

marked or absent keel, which is the last member of the *ticinellid-thalmanninellid* lineage (Desmares et al. 2008).

Th. greenhornensis is a complex trochospiral species that displays raised sutures on the spiral side, supplementary apertures on the umbilical side and a single keel. It further presents umbilical secondary apertures and non-inflated chambers on the umbilical side.

Globular morphotypes with supplementary apertures also occur in the same assemblages as the keeled ones. They were initially referred to as *Anaticinella multiloculata* (Eicher 1972), which was first described in the North American Basin (Eicher 1972) (Fig. 1.5).

In all Upper Cenomanian Western Interior Seaway outcrops, transitional specimens between the keeled forms of *Th. greenhornensis* and the globular morphotypes of *Th. multiloculata* were observed (Desmares et al. 2008). These morphotypes do not have a keel on all chambers of their final whorls. Where present, the keel is more or less pronounced, i.e. from indistinctly marked to thick and protruding. From *Th. greenhornensis* to *Th. multiloculata*, the chambers evolve progressively from crescentic to globular shape, the raised sutures become more depressed and the periumbilical flanges disappear progressively on the umbilical side. The junction between the sutures and the periphery that is oblique in *Th. greenhornensis*

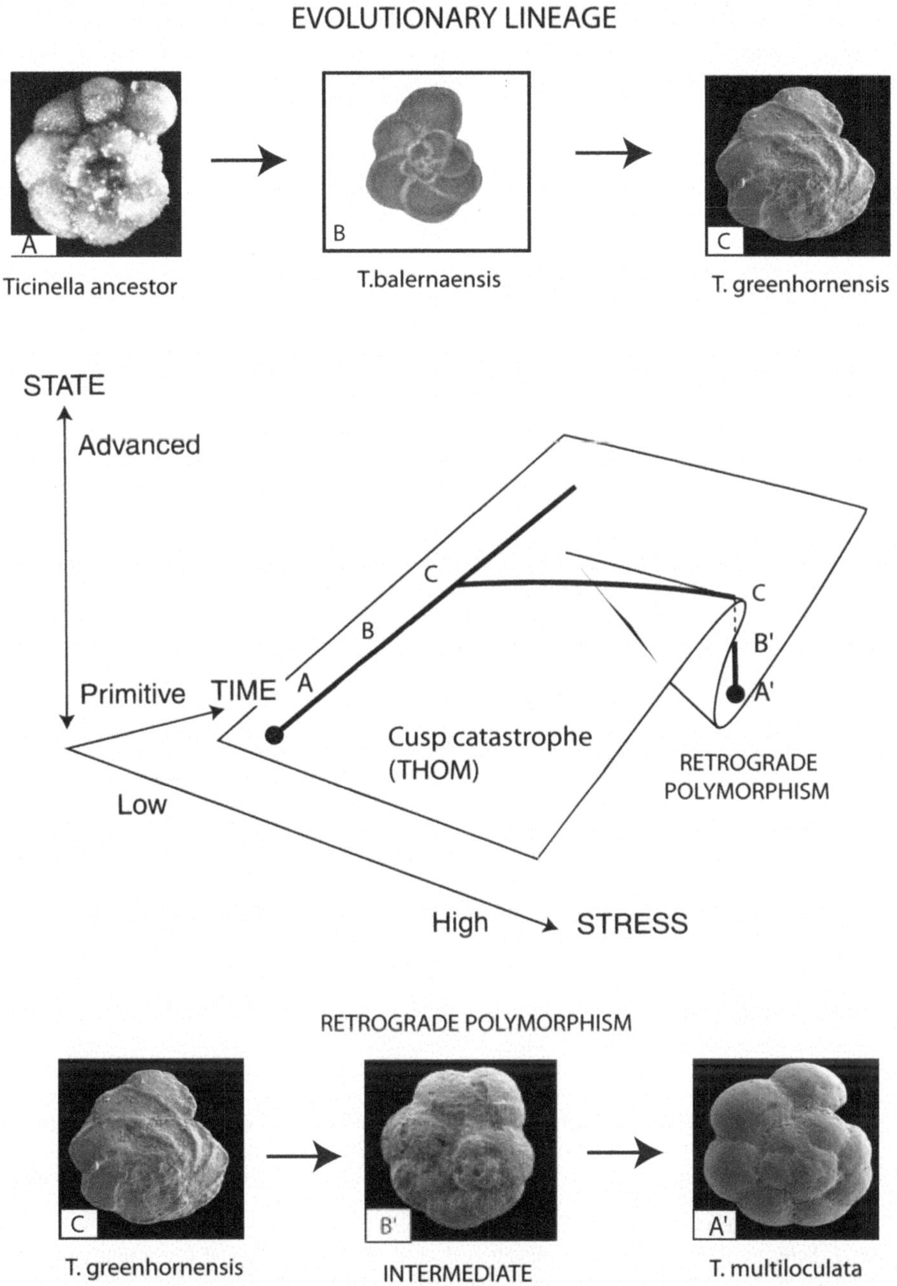

Fig. 1.5 Retrograde polymorphism affecting the *Thalmaninella* lineage during the upper Cenomanian anoxic event (data reinterpreted from Desmares et al. 2008; see Guex et al. 2012b)

becomes perpendicular in *Th. multiloculata*. In lateral view, specimens of *Th. greenhornensis* that are compressed in form gradually change to a globular shape (Desmares et al. 2008).

References

Badyaev AV (2005) Role of stress in evolution: from individual adaptability to evolutionary adaptation. In: Hallgrimson B, Hall BK (eds) Variation. Elsevier, Amsterdam, pp 277–302

Caron M (1985) Cretaceous planktic foraminifera. In Bolli HM, Saunders JB, Perch- Nielsen K (Eds), Plankton stratigraphy (pp. 17–86). Cambridge, England: Cambridge University Press

Courtillot V (1999) Evolutionary catastrophes: the science of mass extinction. Cambridge University Press, Cambridge, 173 p

Courtillot V, Gaudemer Y (1996) Effects of mass extinctions on biodiversity. Nature (London) 381:146–148

Dera G, Neige P, Dommergues J-L, Fara E, Laffont R, Pellenard P (2010) High-resolution dynamics of early Jurassic marine extinctions: the case of Pliensbachian–Toarcian ammonites (Cephalopoda). J Geol Soc London 167:21–33

Desmares D, Grosheny D, Beaudoin B (2008) Ontogeny and phylogeny of Upper Cenomanian rotaliporids (Foraminifera). Mar Micropaleontol 69:91–105

Eicher DL (1972) Phylogeny of the late Cenomanian planktonic foraminifer *Anaticinella multiloculata* (Morrow). J Foramin Res 2:184–190

Gonzalez-Donoso JM, Linares D, Robaszynski F (2007) The rotaliporids, a polyphyletic group of Albian–Cenomanian planktonic foraminifera: Emendation of genera. J Foraminifer Res 37:175–186

Gómez JJ, Goy A, Canales ML (2008) Seawater temperature and carbon isotope variations in belemnites linked to mass extinction during the Toarcian (Early Jurassic) in Central and Northern Spain, comparison with other European sections. Palaeogeogr Palaeoclimatol Palaeoecol 258:28–58

Guex J (1993) Simplifications géométriques liées au stress écologique chez certain protistes. Bulletin de la Société Vaudoise des Sciences Naturelles 82:357–368

Guex J (1995) Ammonites hettangiennes de la Gabbs Valley Range (Nevada, USA). Mémoire de Géologie Lausanne 27:1–131

Guex J (2001) Involution croissante et règle de Cope. Bulletin de la Société Vaudoise des Sciences Naturelles 87:373–379

Guex J (2006) Reinitialization of evolutionary clocks during sublethal environmental stress in some invertebrates. Earth Planet Sci Lett 240:242–253

Guex J (2016) Retrograde evolution during major extinction crises. Springer, Heidelberg, pp 1–77

Guex J, Morard A, Bartolini A, Morettini E (2001) Découverte d'une importante lacune stratigraphique à la limite Domérien-Toarcien: implications paléooceanographiques. Bulletin de la Société Vaudoise des Sciences Naturelles 87(3):277–284

Guex J, Bartolini A, Atudorei V, Taylor D (2004) High resolution ammonite and carbon isotope stratigraphy across the Triassic–Jurassic boundary at New York Canyon (Nevada). Earth Planet Sci Lett 225:29–41

Guex J, Schoene B, Bartolini A, Spangenberg J et al (2012a) Geological constraints on post-extinction recovery of the ammonites and carbon perturbation during the Early Jurassic. Palaeogeogr Paleoclimatol Palaeoecol 346–347:1–11

Guex J, O'Dogherty L, Carter ES, Dumitrica P, Bartolini A (2012b) Geometrical transformations of selected Mesozoic radiolarians. Geobios 45:541–554

Guex J, Pilet S, Muntener O, Bartolini A, Spangenberg J, Schoene B, Schaltegger U (2015) Thermal erosion of cratonic lithosphere as a trigger for mass-extinction. Scientific Reports, Nature

Guex J, Pilet S, Muntener O, Bartolini A, Spangenberg J, Schoene B, Schaltegger U (2016) Thermal erosion of cratonic lithosphere as a trigger for mass-extinction. Scientific Reports, Nature 623168

Harazim D, Van de Schootbrugge B, Sorichter K, Fiebig J, Weug A. Suan G, Oschmann W (2012) Spatial variability of watermass conditions within the European Epicontinental Seaway during the Early Jurassic (Pliensbachian–Toarcian). Sedimentology. https://doi.org/10.1111/j.1365-3091.2012.01344.x

Hoffmann AA, Parsons PA (1991) Evolutionary genetics and environmental stress. Oxford Science Publication, Oxford, pp 1–284

Hudson N (2003) Stratigraphy and correlation of the Ururoan and Temaikan stage (Lower-Middle Jurassic) sequences, New Zealand. J Roy Soc New Zeal 3:1

Jablonka E (2013) Epigenetic inheritance and plasticity: the responsive germline. Prog Biophys Mol Biol 111:99–107

Marjanac T, Steel RJ (1997) Dunlin Group sequence stratigraphy in the northern North Sea - a model for Cook sandstone deposition. AAPG Bull 81(2):276–292

McElwain JC, Beerling DJ, Woodward FI (1999) Fossil plants and global warming at the Triassic-Jurassic boundary. Science 285:1386–1390

McElwain JC, Wagner PJ, Hesselbo SP (2009) Fossil plant relative abundances indicate sudden loss of Late Triassic biodiversity in East Greenland. Science 324:1554–1556

Muller SW, Ferguson HG (1939) Mesozoic stratigraphy of the Hawthorne and Tonopah quadrangles, Nevada. Geol Soc Am Bull 50:1573–1624

Pálfy J, Smith PL (2000) Synchrony between early Jurassic extinction, oceanic anoxic event, 560 and the Karoo–Ferrar flood basalt volcanism. Geology 28:747–750

Price GD (1999) The evidence and implications of polar ice during the Mesozoic. Earth Sci Rev 48(3):183–210

Robaszynski F, Caron M (1995) Foraminifères planctoniques du Crétacé: Commentaire de la zonation Europe-Méditerranée. Bulletin de la Societe Geologique de France, 166:681–692

Schoene B, Guex J, Bartolini A, Schaltegger U, Blackburn TJ (2010) Correlating the end-Triassic mass extinction and flood basalt volcanism at the 100 ka level. Geology 38:387–390. https://doi.org/10.1130/G30683.1

Sell B, Ovtcharova M, Guex J, Bartolini A, Jourdand F, Spangenberg JE, Vicente JC, Schaltegger U (2014) Evaluating the temporal link between the Karoo LIP and climatic–biologic events of the Toarcian stage with high-precision U–Pb geochronology. Earth Planet Sci Lett 408:48–56

Sobolev SV, Sobolev AV, Kuzmin DV, Krivolutskaya NA, Petrunin AG, Arndt NT (2011) Linking mantle plumes, large igneous provinces and environmental catastrophes. Nature 477:312–316. https://doi.org/10.1038/nature10385

Suan G, Nikitenko BL, Rogov MA, Baudin F, Spangenberg JE, Knyazev VG, Glinskikh LA, Goryacheva AA, Adatte T, Riding BJ, Föllmi KB, Pittet B, Mattioli E, Lécuyer C (2011) Polar record of Early Jurassic massive carbon injection. Earth Planet Sci Lett 312:102–113

Svensen H, Planke S, Polozov AG, Schmidbauer N, Corfu F, Podladchikov Y, Jamtveit B (2009) Siberian gas venting and the end-Permian environmental crisis. Earth Planet Sci Lett 277:490–500

Thom R (1972) Stabilité structurelle et morphogenèse. Benjamin, New York, 358 p

Chapter 2
The Controversial Cope's, Haeckel's and Dollo's Evolutionary Rules: The Role of Evolutionary Retrogradation

Jean Guex

Abstract The goal of this chapter is to discuss old problems and recent polemics related to the famous Cope's, Dollo's and Haeckel's rules. The first concerns phyletic size increase: that sort of trend is observed in a multitude of phyla and shows several exceptions during periods of environmental stress. The second rule is discussed with some details because evolutionary reversions of trends are also frequent during stress episodes. The third trend, terminal addition, is very common and one can observe numerous cases where characters that are added late in phylogeny are also the first to be deleted during external stress phases. The addition of new elements at the end of ontogeny is frequently concomitant with size increases (Cope's trend).

There are many sorts of stress that are recorded in the stratigraphical data: major pollutions due to the development of massive volcanism (large igneous provinces), major climate changes due to plate tectonics, etc.

Our main examples of reverse evolution are observed in nautiloids, conodonts, planktonic foraminifera, benthic foraminifera, ammonoids, silicoflagellids, corals, trilobites, etc.

Keywords Cope's · Dollo's · Haeckel's rules · A short review

2.1 Cope's Rule

2.1.1 Some of Its Consequences

The best documented evolutionary trend in the paleontological record is the famous "rule" of size increase, also known as "Cope's rule", eponymously named for the American vertebrate specialist who described it for the first time in 1896 (Cope 1896; see also Stanley 1973). A fascinating global view of that phenomenon has

J. Guex (✉)
Institute of Earth Sciences, Geopolis, UNIL, CH 1015 Lausanne, Switzerland
e-mail: jean.guex@unil.ch

J. Guex et al. (eds.), *Morphogenesis, Environmental Stress and Reverse Evolution*, https://doi.org/10.1007/978-3-030-47279-5_2

been published recently in the Atlas of Vertebrates by Escher and Marchant (2016): almost all the vertebrate lineages show a clear tendency to increase their size over long periods of time, except during periods of major stress, when they show either a decrease in size or an extinction. The trend in size increase is more often observed at the beginning of evolutionary lineages, and much less often at the end of lineages, when organisms have attained their maximal size. They can remain unchanged for long periods, or show fluctuations in size, directed just as well towards size increase as towards size reduction, depending on environmental variations (water temperature, chemistry, etc.). The ignorance of this fact has often led to ill-founded criticisms of the Cope's rule (Gould 1996; Jablonski 1997; see Thom 1983, p. 127). The goal of this chapter is, in part, to show that this rule could be a particular case of a more general rule stating that the increases in size, surface and volume vary independently (Guex 2003). This has significant consequences for the general geometry and ornamentation of the different groups discussed in our recent book on reverse evolution concerning nautiloids, conodonts, planktonic foraminifera, benthic foramina, ammonoids, silicoflagellids and corals (Guex 2016). Major evolutionary trends observable in many groups of marine invertebrates suggest that multiple trends in the increasing involution of coiled shells (often correlated with the development of smooth forms), or of ornamental characters, are directly connected to the geometrical variations affecting the size of the organism. The evolutionary trend most frequently observed in Mesozoic ammonoids, particularly at the family level, is where the ancestral forms have an open umbilicus (evolute), and where the descendants tend to develop a narrow umbilicus (involuted and less ornamented). This trend was recognised more than 100 years ago by Hyatt (1869) in Liassic ammonites, and it appeared for the first time during the Devonian, at the beginning of the history of this group (Erben 1966). When it is completely realised, this trend starts with an open umbilicus and increases the overlap of the coiling (i.e. this is an independent consequence of the usual Cope's trend, accompanied by a surface increase of the mantle). Over the course of time, this leads to the genesis of lenticular forms (oxycones), or more or less spherical (sphaerocones) ones, or a section being depressed (cadicones). The ammonites also often show a trend to increase the sinuosity of the growth lines, and many show an increase in sutural complexity in the course of time. The major trend with increasing involution observed in the ammonites is also observed in nautiloids (Sobolev 1994) and in some gastropods (Runnegar 1987).

This trend also affects many unicellular groups like benthic and planktonic foraminifera at various stages of their development (e.g. the appearance of Orbulina (Cifelli 1969; Hottinger and Drobne 1988; Septfontaine 1988; Adams 1983)). Some benthic foraminifera develop an increase in lateral elongation (Hottinger and Drobne 1988), which geometrically corresponds to the development of cadicone coiling in ammonites. Other groups of microorganisms like the nassellarian radiolarians display a similar phenomenon, with the increase in sphericity and reduction amongst segments (Riedel and Sanfilippo 1981), and, as an ultimate result, cryptocephalisation and cryptothoracisation (Dumitrica 1970). Spherical forms like Paracannopilus also develop from spicular forms in the silicoflagellids (chrysophytes). Note in passing that these transformations can also be concomitant with an increase in the size of the organism. The specialists who have described these trends almost always provide an adaptive explanation, and an ad hoc morphofunctional origination such as improvement of the hydrodynamism. The increasing involution observed in ammonites is generally explained by a minimisation of the quantity of shell necessary to protect the animal, or by an increased hydrodynamism (Raup 1967). The discrepancy between lenticular, involuted and spherical shells seems however to show that it is not about a mechanism of optimisation of the use of shell material because the lenticular, involuted or laterally compressed forms are very frequent in the fossil record, and are far from being optimal from this point of view. Other authors explain the increasing involution of the shells and the complexification of the suture lines as an increase in the mechanical resistance to hydrostatic pressure. In benthic foraminifera, the increase in surface is supposed to facilitate the exchange of oxygen, and the increase in the elongation of the test would also increase the mobility of the animal in the loose sediment. These trends are all observed in very diverse phyla like planktonic organisms, as well as in benthic or burrowing forms. It is well known that in itself, the increase in size can be lethal, in the more or less short term, in lots of descendants. Ad hoc functional explanations are hardly satisfactory because the trends discussed above are observed in very diverse phyla, including planktonic, nektonic, benthic and burrowing organisms. Moreover, it is also well known that continuous size increase usually leads to gigantism, which can prove to be fatal within a more or less short time. The allometries observed during the geometrical/morphological evolution of shelly invertebrates show that size (i.e. diameter or length: see above), volume and surface can vary independently. Within ammonites, an increase in volumetric size, which is not accompanied by an increase in linear size (i.e. the diameter; note that the body chamber's length is often unknown for preservational reasons), will result in an increase in involution. Similarly, a decrease in linear size that is not accompanied by a decrease in volume will also lead to a drastic increase in involution. Such a process certainly accounts for the geometry of the lower Triassic small cryptogenic ammonites such as the spherical Isculitids, deriving from the serpenticone columbitids. We also note that an increase in the mantle's surface area, if not compensated for by a simultaneous increase in the volume of the animal, results in an increase in suture complexity and/or flexuosity of growth lines at the aperture. On the other hand, a decrease in volume not compensated by a decrease in the mantle surface can explain the small juvenile bulges observed in the inner whorls of the earliest Psiloceratids such as the "Knötchenstadium" of *P. spelae* (syn. *P. spelae* tirolicum). Another interesting by-product of the above-described morphogenetic rules is the "stop-and-go" growth of the ammonites, followed by an oblique reorientation of the growth lines (Guex 1967, pp. 328–329). This is obviously due to the fact that the growth's stop in shell secretion is followed by a delayed restart of the soft parts' growth, generating a rotation of the growth lines.

2.1.2 Technical Remark About Some Criticisms of Cope's Rule

In "Full House", one of the most famous books by Gould (1996), the author tries to demonstrate the absence of a global increase in complexity during the evolution of living organisms. His main argument, already discussed in Guex (2016), is that the most abundant organisms present on Earth, the bacteria, have not really evolved morphologically over the last 3 billion years. In this book, Gould uses quantitative arguments like size variations observed in the Cretaceous and Cenozoic foraminifera, and the fractal dimensions of the Ceratite suture lines during the late Palaeozoic and early Mesozoic. We will briefly examine these two quantitative arguments. His Fig. 25 (1996, p 160) shows the size variations of planktonic foraminifera during the Late Cretaceous and the Cenozoic. These diagrams are supposed to demonstrate that the rule of size increase (Cope's rule) is meaningless because it is often represented by zigzag variations, as is the case for the planktonic foraminifera observed during that period. We consider that Gould's argument is invalid because size decreases are always related to environmental stress, a phenomenon known since the beginning of the twentieth century (Shimer 1908; see also Mancini 1978). Gould's diagrams (his Fig. 25) in fact illustrate in a perfect manner that the size reductions occurred during the KT boundary, Late Eocene and Late Oligocene, which are periods of high environmental stress responsible for more or less pronounced extinctions (see Zachos et al. (2001) and Schmidt et al. (2004) for recent quantitative data). His second quantitative argument concerns the fractal dimension of some ammonoid suture lines during the late Palaeozoic and Triassic times (loc. cit Fig. 35, p 210). The measurements representing these relationships, as constructed by Gould, are distributed in a completely chaotic way, and are supposed to demonstrate that there are no relationships between sutural complexity and time during that period. The problem is that Gould overlooked the fact that he should have connected the dots representing the various measurements following two criteria. First, the ontogeny: small juvenile specimens have a suture line, which always looks less complex than an adult one. And secondly, the phylogeny within each separate lineage: the suture lines of most ceratitids become more complex during evolution. This has been illustrated in Fig. 5.3 (Guex 2016), which shows that the ceratitic suture line of the ancestor of the highly complex phylloceratids has a fractal dimension of about 1.2, whereas the resulting advanced Phylloceras has a suture line dimension of about 1.6 (see Guex 1981). The psiloceratid Neophyllites generated during the TJB extinction period also shows a drastic reduction of its sutural complexity.

2.2 Haeckel's Rule of Terminal Addition

The great Haeckel (1874) had the sad privilege of being one of the most criticised palaeontologists of his time, together with the great Lamarck. As an example, we can quote Gilbert and Barresi (2016), who wrote the following text (Internet posting): "A disastrous union of embryology and evolutionary biology was forged in the last half of the nineteenth century by the German embryologist and philosopher, Ernst Haeckel. Based on the assumption that the laws by which species arose on this planet (phylogeny) were identical to the laws by which the individuals of the species developed (ontogeny), he viewed adult organisms as the embryonic stages of more advanced organisms. This view was summarised in his "biogenetic law", that ontogeny recapitulates phylogeny. In other words, development of advanced species was seen to pass through stages represented by adult organisms of more primitive species. In this view, the creation of new phyla is a step towards the completion of human development. In earlier epochs, only the initial stages of this development occurred, producing protists and cnidarians. Later, more stages were added sequentially until a human being evolved". The absence of real fraud by Haeckel was established by the great biologist Richards (2008, 2009).

Ontogeny is the set of steps followed by the organism during the development of its structure. This can also be described as the series of steps appearing in the structure of the organism. This kind of rule would generate a great variety of forms deriving from the ontogeny of the first ancestor, and, as a by-product, the different possible evolutionary paths followed by the descendants of that hypothetical ancestor. This is of course too complex to be done experimentally nowadays, but it would obviously connect Haeckel's controversial biogenetic rule to the actual relationships between ontogeny and phylogeny.

Gilbert (loc. cit.) continues his criticism about Haeckel's *law of terminal addition*. "The embryo evolved new species by adding a step at the end of the previous ones. In such a view, humans evolved when the embryo of the next highest ape added a new stage. This provided a linear, not a branching, phylogeny. This is a critically important departure from what we usually consider as Darwinian evolution". In one of his concluding remarks, Gilbert wrote also, "Indeed, half of Stephen J. Gould's 1977 book *Ontogeny and Phylogeny* is spent exorcising the ghost of Haeckel so that we could discuss evolutionary developmental biology without having to deal with the biogenetic law". We disagree with this last statement.

Within the invertebrates, we can observe the validity of the terminal addition rule, which is very common and well recorded in the fossil record. In the modern classification of the radiolarians, based and elaborated on the ontogeny of these organisms, most of the lineages established by Dumitrica in a multitude of papers are based on the geometry of the early stages of these organisms (microsphere), which are very stable over time (see Dumitrica, in De Wever et al. 2001). The same trend can be observed in multiple foraminifera lineages. We can add that terminal addition is very common (see examples in Guex 2016), and one observes numerous cases where characters which are added late in phylogeny are also the first to be deleted during external stress phases. The addition of new elements at the end of ontogeny is frequently concomitant with size increases (Cope's trend). Examples can be found in dinosaurs (Diplodocus with an important increase in the cervical vertebrae), elephants with the defences, sabretooths, felines, etc.

Torday and Miller Jr (2018; Chap. X) have provided a mechanistic explanation for terminal addition based on the principle of cell–cell communication in evolution (Torday and Rehan 2012). Since such cellular interactions are mediated by growth factors stimulating the production of second messengers, it would be inefficient to modify this stepwise process by adding on at the beginning or in the middle, hence the "terminal" addition.

2.3 Catastrophe and Retrograde Evolution

We will come back here to a very didactic example found in ammonites (Phymatoceratinae and Hammatoceratinae of the Middle and Late Toarcian) illustrated in Fig. 2.1.

At the end of the Middle Toarcian, the Euro-Boreal province was affected by a major regional regression generated by the insurrection of the western Tethys' rift shoulder (Stampfli 1993). That regressive event began a major faunal turnover, culminating in the extinction of late Liassic ammonites such as the Hildoceratinae, the Mercaticeratinae, the Dactylioceratidae and most of the Phymatoceratinae. The disappearance of those groups was concomitant with the development of two new families, the Hammatoceratidae and the Grammoceratidae, which were dominant

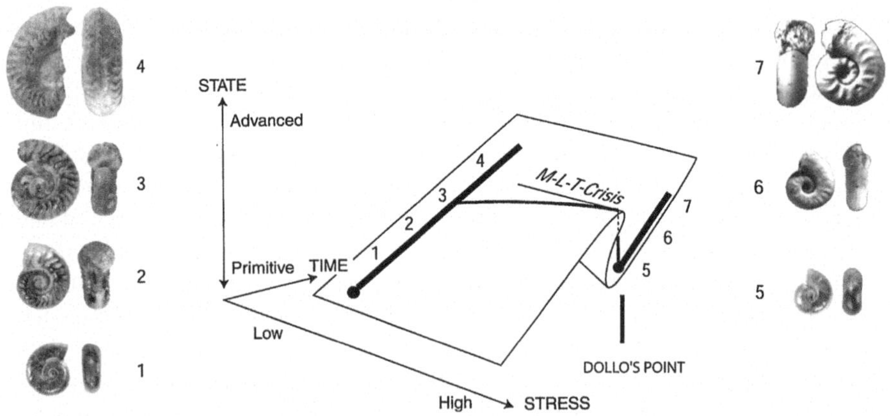

Fig. 2.1 Evolution of the lineages *Paronychoceras* (1 and 2), *Pseudobrodieia* (3) and *Brodieia* (4) in the Middle-Early Late Toarcian. Recovery of homeomorphs to extinct forms in *Onychoceras planum* (5), *O. tenue* (6) and *O. differens* (7) in the Late Toarcian. MLT = Middle-Late Toarcian Crisis. Species 5–7 appearing at the Middle-Late Toarcian crisis are quasi-homeomorphs of 1–3. Diagram not to scale. From Guex (2016), modified

during the Late Toarcian. During this stratigraphic interval, a prototypical example of atavistic reversals is observed in the microconch lineage Paronychoceras-Brodieia of the middle Toarcian Phymatoceratinae. That lineage, illustrated in Fig. 2.1, is characterised by a drastic increase in ornamental complexity, involution and size. In the Late Toarcian, that sequence is followed by the *Onychoceras* lineage, a group of unkeeled microconchs of hammatoceratids, which is atavistic and practically homeomorphic with its ancestral *Paronychoceras* (Fig. 2.1).

That case study can be analysed schematically to demonstrate the relationship between Cope's rule, terminal addition of characters and retrogradation of those characters until an atavistic point that we call "Dollo's point", beyond which the reversion would be lethal, or the shell would be lost. Figure 2.1 (1) shows the ancestor of the microconch lineage: it is smooth, with slightly evolute coiling. Its direct descendant (Fig. 2.1 (2)) is bigger and starts to develop slight ribs without a ventral keel. The next (Fig. 2.1 (3)) develops a bigger shell, and the beginning of a keel, and it is followed by forms with bifurcate ribs and a keeled venter. The end form (Fig. 2.1 (4)) is about two times bigger and develops lappets.

The Middle-Upper Toarcian crisis occurs at the boundary between the two substages; the ancestral lineage developed from Paronychoceras disappears, and is replaced by a new lineage starting with Onychoceras (Fig. 2.1 (5–7)), which follows an evolutionary morphological path identical to the ancestral one (Fig. 2.1 (1–4)).

This case study is particularly interesting because of its completeness. It shows very well the point to which the retrograde lineage cannot go further back: the Dollo's point, which will be discussed further below.

2.4 Dollo's Law of Irreversibility

Dollo's law of irreversibility of evolution was clearly explained in Dollo's (1893) seminal work on the laws of evolution in 1893. About this, he wrote "that an organism cannot return, even partially, to a previous state already realised in its ancestral series". That important sentence was rewritten by Gould 1970 as "An organism never returns exactly to a former state even if it finds itself placed in conditions of existence identical to those in which it has previously lived. But by virtue of the indestructibility of the past it always keeps some traces of the intermediate stages through which it is passed". The general opinion is that the loss of complex features in evolution is irreversible (e.g. loss of teeth in birds or loss of legs in whales). For example, Dollo had observed correctly that the protoconch of uncoiled ammonoids is fundamentally different from the initial ontogenetic stages of ancestral straight nautiloids. In fact, the existence of evolutionary inversions is extremely frequent during extinction periods (Guex 2016), but it generally does not affect the initial ontogenetic stage: there is a ratchet acting on the potential evolutionary reversal that we will discuss briefly below. Such reversals of evolution, contradicting in part Dollo's law, are generally observed during the major extinction periods, and are discussed in detail in Guex (2016: "Retrograde Evolution") (see also Marshall et al. 1994; Porter and Crandall 2003; Protas 2007).

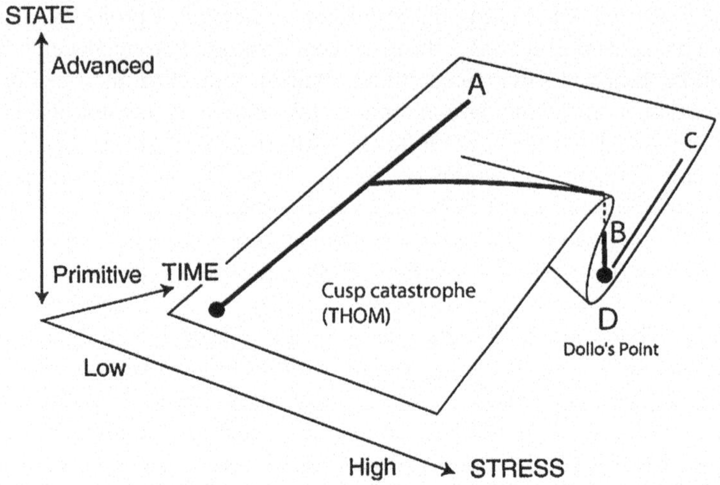

Fig. 2.2 Evolutionary trajectories of a hypothetical organism (several real examples in Guex 2016). (A) Cope's rule, terminal addition; (B) proteromorphosis, retrogradation; (C) recovery of lost characters; (D) Dollo's point

As already mentioned, we have illustrated here (Fig. 2.1) one of the cases where Cope's rule, Dollo's law and terminal addition are all observed to act in parallel in a single lineage through time. The same kinds of geometrical transformations are clearly observed in nautiloids, conodonts, planktonic foraminifera, benthic foraminа, ammonoids, silicoflagellids and corals (Guex 2016). Cases of reversal not related to environmental stress are discussed in Collin and Cipriani (2003), Collin and Miglietta (2008), Teotonio and Rose (2001), Whiting et al. (2003) and Kohlsdorf and Wagner (2006).

A diagrammatic representation of the above discussion is summarised in Fig. 2.2.

Acknowledgements I warmly thank Raphael Guex, Arthur Escher, Luis O'Dogherty, Dave Taylor, François Reuse and Annachiara Bartolini for useful discussions.

References

Adams CG (1983) Speciation, phylogenesis, tectonism, climate and eustasy: factors in the evolution of Cenozoic larger foraminiferal bioprovinces. Systematic Association Special Publication, vol. 23, p 255–289

Cifelli R (1969) Radiation of Cenozoic planktonic foraminifera. Syst Zool 18:154–168

Collin R, Cipriani R (2003) Dollo's law and the re-evolution of shell coiling. Proc Biol Sci 270:2551–2555

Collin R, Miglietta MP (2008) Reversing opinions on Dollo's law. Trends Ecol Evol 23:602–609

Cope ED (1896) The primary factors of organic evolution. Open Court Publishing Company, Chicago; 547 p

De Wever P, Dumitrica P, Caulet JP, Nigrini C, Caridroit M (2001) Radiolarians in the sedimentary record. Gordon and Breach Science, London; 553 p
Dollo L (1893) Les Lois de l'evolution. Bull Soc Belge Géol Pal Hydro 7:164–166
Dumitrica P (1970) Cryptocephalic and cryptothoracic Nassellaria in some Mesozoic deposits of Romania. Rev Roum Géol Géophys Géogr Sér Géol 14:45–124
Erben HK (1966) Über den Ursprung der Ammonoidea. Biol Rev 41:641–658
Escher A, Marchant R (2016) Atlas of vertebrates from their origin to today. Loisirs et Pedagogie, Switzerland; 40 Plates
Gilbert SF, Barresi MJF (2016) Developmental biology. In: Haeckel and the biogenetic law. Companion Website for Eleventh Edition
Gould SJ (1970) Dollo on Dollo's law and irreversibility and the status of evolutionary laws. J Hist Biol 3:189–212
Gould SJ (1977) Ontogeny and phylogeny. Belknap, Cambridge, 501 p
Gould SJ (1996) Full house. Harvard Univ. Press, Cambridge; 244 p
Guex J (1967) Contribution à l'étude des blessures chez les ammonites. Bull Soc Vaudoise Sci Nat 69:323–338
Guex J (1981) Associations virtuelles et discontinuités dans la distribution des espèces fossiles. Bull Soc Vaudoise Sci Nat 75:179–197
Guex J (2003) A generalization of Cope's rule. Bull Soc Géol 174:448–453
Guex J (2016) Retrograde evolution during major extinction crises. Springerbriefs in Evolutionary Biology, Heidelberg, pp 1–77
Haeckel E (1874) Anthropogenie oder Entwickelungsgeschichte des Menschen. Engelmann, Leipzig
Hottinger L, Drobne K (1988) Alvéolines tertiaires: Quelques problèmes liés à la conception de l'espèce. Rev Paléobiol Genève Spéc 2:665–681
Hyatt A (1869) Genesis of the Arietitidae. Museum Compar Zool Harvard Memoir 16:1–238
Jablonski D (1997) Body-size evolution in Cretaceous molluscs and the status of Cope's rule. Nature 385:250–252
Kohlsdorf T, Wagner GP (2006) Evidence for the reversibility of digit loss: a phylogenetic study of limb evolution in Bachia (Gymnophthalmidae: Squamata). Evolution 60:1896–1912
Lamark JB (1809) Philosophie Zoologique. 1809 Dentu, Paris. 235
Mancini EA (1978) Origin of micromorph faunas in the geologic record. J Paleo 52(2):311–322
Marshall CR et al (1994) Dollo's law and the death and resurrection of genes. Proc Natl Acad Sci U S A 91:12,283–12,287
Porter ML, Crandall KA (2003) Lost along the way: the significance of evolution in reverse. Trends Ecol Evol 18:541–547
Protas M (2007) Regressive evolution in the Mexican cave tetra, Astyanax mexicanus. Curr Biol 17:452–454
Raup D (1967) Geometric analysis of shell coiling: coiling in ammonoids. J Paleo 41:43–65
Richards RJ (2008) The tragic sense of life: Ernst Haeckel and the struggle over evolutionary thought. University of Chicago Press, London
Richards RJ (2009) Haeckel's embryos: fraud not proven. Biol Philos 24:147–154. https://doi.org/10.1007/s10539-008-9140-z
Riedel WR, Sanfilippo A (1981) Evolution and diversity of form in Radiolaria. In: Simpson TL, Volcani BE (eds) Silicon and siliceous structures in biological systems. Springer, New York, pp 323–346
Runnegar B (1987) Subphylum cyrtostoma, class monoplacophora. In: Cheetham, Rowell AJ (eds) Fossil invertebrates. Blackwell, Palo Alto, CA, pp 297–304
Schmidt D, Thierstein HR, Bollmann J, Schiebel R (2004) Abiotic forcing of plankton evolution in the Cenozoic. Science 303:207–210
Septfontaine M (1988) Vers une classification évolutive des Lituolidés (Foraminifères) jurassiques en milieu de plate-forme carbonatée. Rev Paléobiol Genève Spéc 2:229–256
Shimer HW (1908) Dwarf faunas. Am Nat 42:472–490

Sobolev ES (1994) Stratigraphic range of Triassic boreal Nautiloid. Mémoir Géol Lausanne 22:127–138

Stampfli GM (1993) Le Briançonnais, terrain exotique dans les Alpes? Eclogae Geol Helv 86:1–45

Stanley SM (1973) An explanation for Cope's rule. Evolution 27(1):1–26

Teotonio H, Rose MR (2001) Perspective: reverse evolution. Evolution 55:653–660

Thom R (1983) Paraboles et catastrophes. Flammarion, Paris

Torday JS, Rehan VK (2012) Evolutionary Biology, Cell–Cell Communication, and Complexity. Wiley. https://doi.org/10.1002/9781118130452

Torday JS, Miller WB Jr (2018) Terminal addition in a cellular world. Prog Biophys Mol Biol 135:1–10

Whiting MF et al (2003) Loss and recovery of wings in stick insects. Nature 421:264–267

Zachos J, Pagani M, Ioan L, Thomas E, Billups K (2001) Trends, rhythms, and aberrations in global climate 65 Ma to present. Science 292:686–693

Chapter 3
Impact of the Environmental Stress on the Late Permian Pollen Grains from Zechstein Deposits of Poland

Anna Fijałkowska-Mader

Abstract Analyzing the material from the Polish Zechstein deposits, the appearance of morphologically changed gymnosperm bisaccate pollen grains was observed. These changes concern the number of sacci, their shape and size, as well as thickening of the central body exine. Most of the malformed grains were found within the *Lueckisporites virkkiae* and *Jugasporites delasaucei* conifer species, being the dominant component of the Polish Late Permian miospore assemblages. The reason for the emergence of these abnormal forms was environmental stress caused by enhanced ultraviolet-B (UV-B) irradiation connected with Siberian Trap volcanism in the Late Permian.

Keywords Environmental stress · Pollen · Late Permian

3.1 Introduction

Late Permian gymnosperm bisaccate pollen grains are characterized by a relatively large share of malformed specimens. Properly shaped wind-dispersed bisaccate pollen grain consists of a circular central body and two half-moon-shaped, symmetric air sacci, usually lighter color than central body. Bladderlike sacci enlarge volatile surface of pollen and allow them to be transported over long distances both in air and on water (Leslie 2008). By abnormal (teratological) pollen grains the number of sacci ranges from zero (asaccate form) to a few and exine (outer part of cell wall) of the central body can be thicker. In addition a limb of exine, extending over the central body, can appear.

In the Central European Basin area (named also German Basin; Fig. 3.1a) most of the malformed grains were found within the *Lueckisporites* Potonié et Klaus, *Jugasporites* Leschik, and *Triadispora* Klaus genera (Potonié and Schweitzer 1960;

A. Fijałkowska-Mader (✉)
Polish Geological Institute—National Research Institute, Holy Cross Branch, Kielce, Poland
e-mail: anna.mader@pgi.gov.pl

J. Guex et al. (eds.), *Morphogenesis, Environmental Stress and Reverse Evolution*, https://doi.org/10.1007/978-3-030-47279-5_3

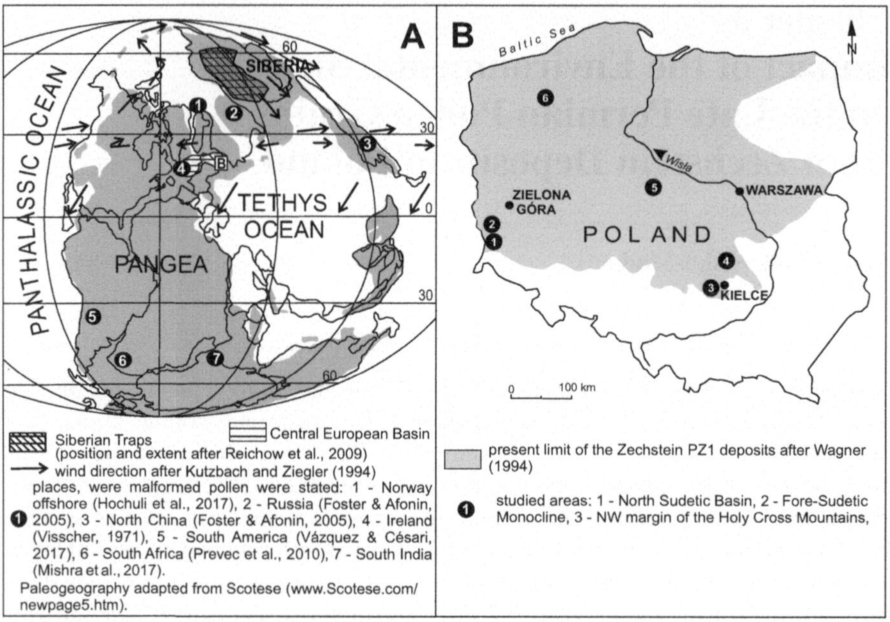

Fig. 3.1 (**a**) Location of Poland (**b**) against the paleogeographic map of the Late Permian with marked places of malformed pollen grain occurrence, (**b**) location of the studied areas in Poland

Grebe and Schweitzer 1962; Visscher 1971, 1972; Dybova-Jachowicz 1981; Fijałkowska-Mader 2012). All these taxa are the conifer pollen produced by *Ullmannia bronni* Goeppert, *U. frumentaria* (Schlotheim) Goeppert, and unspecified Voltziales, respectively (Potonié and Schweitzer 1960; Visscher 1971). To illustrate morphological variability within *Lueckisporites* and *Jugasporites* genera, Visscher (1971, 1972) introduced the "palynodeme" term, calling modified specimens morphological norms and marked with the symbolism from Aa and Ab—the unchanged forms—to E in case of the *Lueckisporites* palynodeme (Fig. 3.2), which were to arise through gradual, evolutionary changes. Although palynodeme concept was generally rejected by the palynologists (cf. Vázquez and Césari 2017), the norms of *Lueckisporites*, treated as malformed pollen, are still a useful tool, especially in palynostratigraphy of the Zechstein deposits in the Central Europe (cf. Visscher 1973; Fijałkowska 1994, 1995; Fijałkowska-Mader 1997; Fijałkowska-Mader 2013; Dybova-Jachowicz and Chłopek 2003; Fijałkowska-Mader et al. 2018).

Other Late Permian pollen do not show such strong variability. Within the *Jugasporites* genus it relates to the shape of the grain, number and shape of sacci, and shape of the tetrad mark. Based on these features, Visscher (1971) distinguished two morphological norms A and B.

Another example of malformed pollen grain is the trisaccate species *Triadispora visscheri* (Visscher) Fijałkowska. A similar specimen was described by Foster and Afonin (2005; Fig. 4f) from the Upper Permian of North China as a teratological form of the *Alisporites* sp.

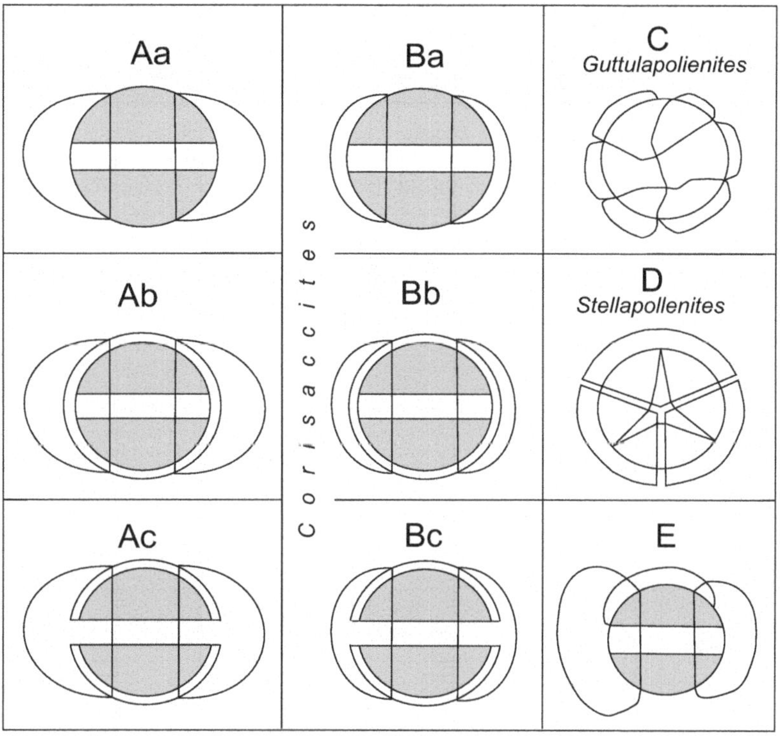

Fig. 3.2 Morphological norms within the *Lueckisporites* palynodeme after Visscher (1971, 1972)

The impulse for the author (Fijałkowska-Mader 2012) to revise the views on the causes of morphological norms within late Permian pollen grains and rejection of the theory of gradual evolutionary changes and individual variability was the work of Foster and Afonin (2005), illustrating changes in the morphology of pollen grains of the species *Klausipollenites schaubergeri* (Potonié et Klaus) Jansonius and *Alisporites* sp. from Russia and China. The authors claimed environmental stress, caused by catastrophic climate change at the end of the Permian, connected with extensive Siberian Trap volcanic activity, as the reason for these changes. Studies of Visscher et al. (2004), Hochuli et al. (2016, 2017), Black et al. (2014), Benton (2018), and Benca et al. (2018) confirmed their suggestion.

3.2 Materials and Methods

At the end of the Permian area of Poland was an eastern margin of the epicontinental Central European Basin, called also the Zechstein Basin (Fig. 3.1a), where the cyclic sedimentation of the shallow water deposits took place. Each of the four cyclothems (PZ1–PZ4; Fig. 3.3) began with marine carbonates and, through

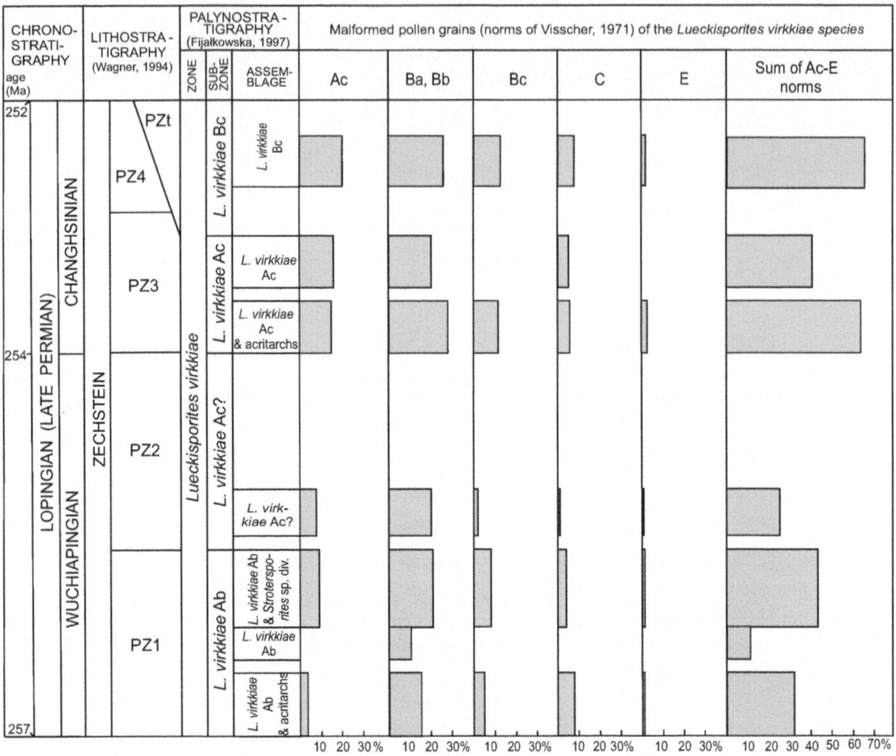

Fig. 3.3 Percent content of the malformed pollen grains of the *Lueckisporites virkkiae* species in the miospore assemblages of the Polish Zechstein deposits

evaporitic stage, passed into terrigenous deposits. In areas located near the seashore, sediments of the upper part of the PZ3 and the whole PZ4 cyclothem are replaced by the Top Terrigenous Series (PZt; Wagner 1994).

The study material came from the Zechstein deposits of Poland from 45 drill cores, 1 outcrop, and 3 mines located in the North Sudetic Basin (Fig. 3.1B1), Fore-Sudetic Monocline (Fig. 3.1B2), Nida Basin (Fig. 3.1B3), north-western margin of the Holy Cross Mts. (Fig. 3.1B4), Kuyavian Swell (Fig. 3.1B5), and Pomeranian Swell (Fig. 3.1B6). More detailed information concerning the samples is given in the Appendix Table 3.1.

Samples were treated using standard HF palynological processing technique described by Orłowska-Zwolińska (1983). Residues were sieved at 10 μm mesh size. Two slides were made of each sample. The 253 of analyzed samples yielded miospores, which were studied under Leitz Laborlux S transmitted light microscope. Generally 200 miospores were counted in each rich sample for quantitative analysis. In the case of sparser spectra, all sporomorphs have been counted.

Table 3.1 Origin of the studied material

Studied area (numbers according to the map in Fig. 3.1b)	Number of analyzed drill cores/ outcrops/ mines	Boreholes/outcrops*/mines** (number of the productive samples)
1. North Sudetic Basin	14	Czerwona Woda IG 1 (13), Kościelna Wieś IG 1 (2), I (2), II (1), III (1), IV (3), V (2), VI (4), VII (2), VIII (3), IX (2), X (3), XI (2), Niwnice* (2), "Lubichów"**(4)
2. Fore-Sudetic Monocline	2	"Rudna Północna"** (7), "Lubin Główny"** (2)
3. Nida Basin	4	Biała Wielka IG 1 (3), Brzegi IG 1 (4), Milianów IG 1 (8), Pągów IG 1 (5)
4. NW margin of the Holy Cross Mountains	23	Cierchy IG 1 (6), Goleniawy IG 1 (7), Jaworze IG 1 (20), Jaworzno IG 1 (4), Lekomin IG 1 (14), Łączna-Zaszosie IG 1 (1), Łopuszno IG 1 (30), Opoczno PIG 2 (5), Ostałów PIG 2 (3), Ostojów IG 1 (9), Podgace IG 1 (5), Radwanów IG 1 (5), Ruda Strawczyńska IG 1 (5), Siodła IG 1 (4), Sitkówka IG 1 (2), Stachura IG 1 (9), Tumlin 1 (3), Tumlin 2 (3), Tumlin-Podgrodzie IG1 (12), Tumlin-Węgle IG 1 (3), Zabłocie IG 1 (1), Zabłocie IG 2 (3), Zachełmie IG 1 (6), Zaciszowice IG 1 (3)
5. Kuyavian Swell	1	"Kłodawa"**(15)
6. Pomeranian Swell	1	Piła IG 1 (4)
Total number of productive samples		**253**

The miospore zonation of Fijałkowska-Mader (1997), based on the *Lueckisporites* palynodeme of Visscher (1971, 1972, 1973, 1978), refined by Brugman (1983) and Visscher and Brugman (1988), was applied for the determination of the biostratigraphical position of the analyzed spore-pollen assemblages.

Studies of malformed pollen grains concerned the following features: general shape of the pollen grain, number and shape of sacci, thickness of the exine of central body (in case of *Lueckisporites*), and shape of the tetrad trilete mark (in case of *Jugasporites*).

The results of studies on contemporary teratological pollen grains, carried out in the United States, Russia, Ukraine, and Slovakia (Wilson 1965; Shkarlet 1972; Tevini 1993; Kormut'ák et al. 1994; Tretyakova and Noskova 2004; Ostrolúcka et al. 1995; Kormut'ák 1996; Dzyuba 1998; Sirenko 2001; Benca et al. 2018), were applied to interpretation of the causes of morphological variability. These changes concern the occurrence of grains in unseparated tetrads, increasing the number of sacci, disappearance of terminal aperture, thickening of the exine, and reduction of grain size. The threshold amount of changed pollen grains as an indicator of environmental stress was >3% (see Foster and Afonin 2005).

3.3 Results

In total, over 10,000 specimens of the *Lueckisporites virkkiae* species, 700 specimens of the *Jugasporites delasaucei* species, and 50 specimens of the *Triadispora* sp. were analyzed.

In the case of *L. virkkiae* the sacci number varied from two (Fig. 3.4a–f) to four (Fig. 3.4h, i). Pollen without sacci were also met (Fig. 3.4g). The shape and size of the sacci varied from fully developed (Fig. 3.4a–c) to heavily reduced (Fig. 3.4d, e). Thickness of the exine of central body varied from 1–1.5 μm in normal pollen to 3 μm in malformed specimens. Similarly the thickness of exine of sacci changed from less than 1–2 μm. The author treated the Aa and Ab norms as unchanged, which strongly dominate in spectra of lower part of the *L. virkkiae* Ab subzone. In the *L. virkkiae* Ab and acritarchs assemblage the share of individual norms is as follows: Aa i Ab—67.9%, Ac—3.2%, Ba i Bb—15.8%, Bc—4.7%, C—7.9%, and E—0.5%. In the younger spectra the amount of malformed pollen increases successively up to 68.2% in the youngest *L. virkkiae* Bc assemblage (Ac—19.8%, Ba and Bb—25.1%, Bc—12.3%, C—8%, E—3%; Fig. 3.3).

Normal pollen of the *Jugasporites delasaucei* species (NA) have two half-moon or bigger sacci, thin exine (ca. 1 μm) of central body, and a characteristic tetrad mark with one longer, slightly wavy arm (Fig. 3.4j). Usually the exine of an oval area, surrounding tetrad mark, is smooth and thinner (less than 1 μm). Aberrant forms of the B norm have a variable number of sacci from one (Fig. 3.4l) or two irregular (Fig. 3.4m) to three (Fig. 3.4n, o), and sporadically four and thicker exine of central body with scarcely visible sculpture elements. There may be a trilete mark with straight, short arms of equal length, dilete or monolete mark, or no mark at all. A shape of area may become longer or area may disappear. Sometimes there is a limb of exoexine around the central body (Fig. 3.4k, n, o). The share of teratological pollen increases from 10% in the *L. virkkiae* Ab and acritarchs assemblage to over 50% in the *L. virkkiae* Bc assemblage.

Another species with malformed pollen is, according to the author, *Triadispora crassa* Klaus. The unchanged forms have round central body with a clear, trilete mark with short, straight arms and a thin, subgranulated exine. Half-moon sacci are smaller than central body (Fig. 3.4p). Abnormal pollen grains occur in literature under other species names as *Triadispora plicata* Klaus, which differs from *T. crassa* in thicker and folded exine of central body or *Triadispora visscheri* (Visscher) Fijałkowska with three sacci (Fig. 3.4q). The number of these specimens increases from a few percent in the *L. virkkiae* Ab and acritarchs assemblage to 10% in the *L. virkkiae* Bc assemblage.

Occasionally, in the analyzed material were met trisaccate pollen grains of the *Lunatisporites* genus and polysaccate, indeterminate forms (Fig. 3.4r–t).

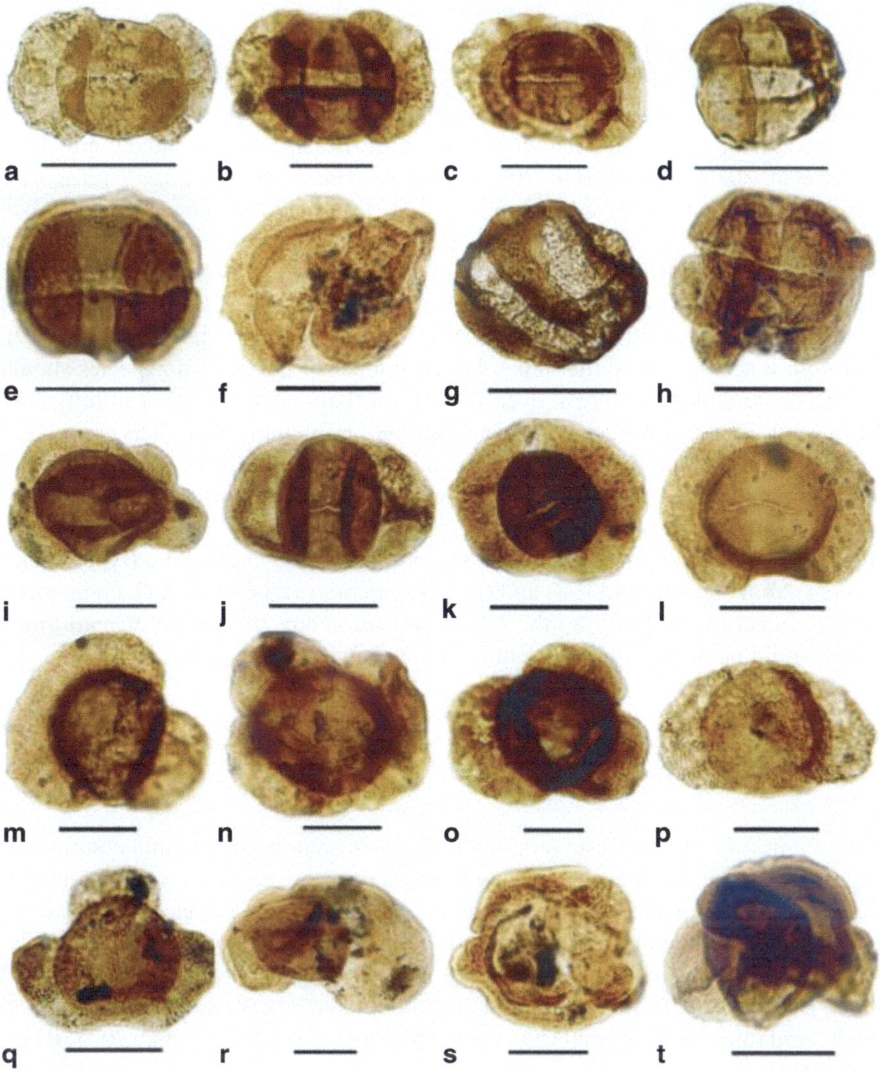

Fig. 3.4 Normal (**a**, **b**, **j**, **p**) and malformed bisaccate pollen grains occurring in the Late Permian (Lopingian) miospore assemblages of Poland. (**a–i**) *Lueckisporites virkkiae* Potonié and Klaus: (**a**) morphological norm Aa (NAa) after Visscher (1971), (**b**) Nab, (**c**) NAc, (**d**) NBa, (**e**) NBb, (**f**) NBc, (**g**) NC, (**h**, **i**) NE; (**j–o**) *Jugasporites delasaucei* (Potonié and Klaus) Leschik: (**j**) NA, (**k–o**) NB: (**k**) monosaccate grain, (**l**) bisaccate grain with a exoexine limb around the central body, (**m**) bisaccate grain with asymmetric sacci, (**n**, **o**) trisaccate grains with a limb around the central body, (**p**) and (**q**) *Triadispora crassa* Klaus: (**p**) bisaccate grain, (**q**) trisaccate grain, (**r**, **s**, **t**) indeterminate polysaccate grains
Scale bar: 30 μm. See Appendix for details

3.4 Discussion

It is puzzling that the previous reports about malformed pollen (except for the author's work), although they come from different places in the world, concern only a few taxa: *Klausipollenites schaubergeri*, *Alisporites* sp. (Foster and Afonin 2005; Prevec et al. 2010; Hochuli et al. 2017), *Klausipollenites* sp., *Hamiapollenites insolatus* Bharadwaj and Salujha (Mishra et al. 2017), and indeterminate taeniate pollen (Hochuli et al. 2017). In addition, their occurrence coincides temporarily with global crisis of forest ecosystems at the end of Permian (Visscher et al. 2004; McElwain and Punyasena 2007; Hochuli et al. 2010a, 2016; Benton and Newell 2014; Bercovici et al. 2015; Yu et al. 2015; Benca et al. 2018; Fielding et al. 2019). Therefore it seems logical that much larger number of pollen grain species should contain the teratological forms, even if we limit them only to conifer affinity. The reason may be the differences in identification and naming of pollen grains. For example the trisaccate pollen grain of *Alisporites* sp., presented in Fig. 4c of Foster and Afonin (2005), looks like the D norm of *Lueckisporites* palynodeme or *Stellapollenites* sp., whereas those in Fig. 4f resemble the *Triadispora visscheri.* Moreover, if accepting the Visscher's suggestion that the *Corisaccites* (type species *C. alutas* Venkatachala and Kar) and *Guttulapollenites* (type species *G. hannonicus* Goubin) genera belong also to the *Lueckisporites* genus, the spread of malformed pollen would be much larger especially in the southern hemisphere.

3.5 Conclusions

In the Central European Basin teratological forms occur mainly within conifer pollen. The nature of the changes, leading to an increase or reduction in the number of sacci, thickening of the central body exine, and tetrad mark shape modification, indicates environmental stress rather than evolutionary or taphonomic processes (cf. Traverse 2008). The most similar changes in modern pollen grains occur as a result of UV-B radiation (cf. Benca et al. 2018), although the others factors, such as strong winds, acid rain, and increased CO_2, cannot be excluded (Beerling et al. 2007; Black et al. 2014).

The fact that the miospore assemblages analyzed by the author are older than those containing malformed tetrads of lycopsid spores (see Visscher et al. 2004; Looy et al. 2005) indicates the multiphase of catastrophic Late Permian phenomena connected with the Siberian Traps volcanism (cf. Racki and Wignall 2005; Yin et al. 2007; Hochuli et al. 2010b).

Appendix

Accession notation: sample and lithostratigraphy.

Czerwona Woda IG 1: 674.3 m, 678.7 m, 679.4 m, 680.1 m, 681.0–2 m (PZ2), 714.9 m, 715.5 m, 716.6 m, 716.7 m, 717.8 m, 721.4 m, 721.9 m, 729.1 m (PZ1).

Kościelna Wieś IG 1: 785.4 m, 786.4 m (PZt).

I: 602.0 m, 633.9 m (PZ2).

II: 773.4 m (PZ3).

III: 988.35 m (PZ3).

IV: 755.5 m, 765.5 m (PZ2), 801.5 m (PZ2?).

V: 1044.0 m, 1067.5 m (PZ2).

VI: 1277.0 m, 1292.5 m (PZ3), 1318.3 m, 1318.6 m (PZ2).

VII: 987.3 m, 1006.5 m (PZ2).

VIII: 1075.0 m, 1093.0 m, 1103.5 m (PZ2).

IX: 893.0 m (PZt), 930.0 m (PZ2).

X: 1110.0 m, 1122.5 m (PZ2), 1139.0 m (PZ2?).

XI: 883.5 m, 898.0 m (PZ2).

Cierchy IG 1: 637.4 m, 838.2 m, 640.8 m, 641.3 m, 644.6 m, 645.5 m (PZt).

Goleniawy IG 1: 460.5 m, 463.8–9 m (PZ3), 494.6 m, 495.5 m, 499.2 m, 504.0 m, 512.0 m (PZ1).

Jaworze IG 1: 344.6 m, 346.4 m (PZ3), 369.5 m, 370.0 m, 375.0 m, 375.6 m, 412.7 m, 415.0 m, 418.7 m, 423.0 m, 423.2 m, 427.0 m, 429.7 m, 430.0 m, 433.5 m, 437.4 m, 440.0 m, 440.1 m, 445.0 m, 458.0 m (PZ1).

Jaworzno IG 1: 157.0 m, 157.1 m, 157.75 m, 158.0 m (PZ1).

Lekomin IG 1: 58.5 m, 65.7 m, 71.9 m, 72.8 m, 74.4 m, 75.8 m (PZ3), 101.0–1 m, 103.1–4 m, 111.0–1 m, 123.2 m, 124.5 m, 125.3 m, 126.2 m, 126.6–8 m (PZ1).

Łączna-Zaszosie IG 1: 435.2 m (PZ1).

Łopuszno IG 1: 1605.2 (PZ3), 1619.8 m, 1694.1 m, 1707.4 m, 1709.4 m, 1710.4 m, 1711.4 m, 1713.0 m, 1715.0 m, 1729.7 m, 1730.7 m, 1731.7 m, 1739.8 m, 1740.8 m, 1741.8 m, 1742.8 m, 1734.8 m, 1744.8 m, 1746.8 m, 1748.8 m, 1749.8 m, 1750.8 m, 1753.8 m, 1756.8 m, 1758.8 m, 1762.9 m, 1764.9 m, 1766.9 m, 1767.9 m, 1768.0 m (PZ1).

Opoczno PIG 2: 2509.2 m, 2509.6 m (PZ3), 2610.6 m, 2611.2 m (PZ2), 2972.9 m (PZ1).

Ostojów IG 1: 265.5 m, 266.3 m, 269.5 m (PZt), 318.5 m, 320.9 m, 323.4 m, 326.5 m, 330.5 m, 332.3 m (PZ1).

Ostałów PIG 2: 2207.2 m (PZt), 2252.4 m (PZ2), 2480.3 m (PZ1).

Podgace IG 1: 94.9 m, 95.9 m, 97.7 m, 103.7 m, 124.7 m (PZ1).

Radwanów IG 1: 1312.5 m (PZt), 1450.0 m (PZ3), 1513.0 m, 1603.4 m, 1645.3 m (PZ1).

Ruda Strawczyńska IG 1: 709.8–710.8 m (PZ3), 739.0–740.0 m, 743.0–744.0 m, 744.0–745.0 m, 765.0–766.0 m (PZ1).

Siodła IG 1: 181.1 m, 183.2–5 m (PZt), 212.0 m, 247.5 m (PZ1).

Sitkówka IG 1: 120.0 m, 123.0 m (PZ1).

Stachura IG 1: 601.0–4 m, 607.5 m (PZ3), 640.2 m, 655.4 m, 657.3–7 m, 670.3 m, 670.6 m, 673.5 m, 746.4 m (PZ1).

Tumlin 1: 144.5–6 m, 147.0–8 m, 158.0–8 m (PZ1).

Tumlin 2: 135.0–9 m, 146.0–7 m, 151.0–9 m (PZ1).

Tumlin-Podgrodzie IG1: 283.4 m, 283.5 m, 283.8 m, 284.6 m (PZt), 310.6 m, 313.8 m (PZ3), 367.7 m, 386.4 m, 389.5 m, 389.8 m, 395.3 m, 412.3–5 m (PZ1).

Tumlin-Węgle IG 1: 125.7 m, 129.6 m, 139.7 M (PZ1).

Zabłocie IG 1: 164.5 m (PZt).

Zabłocie IG 2: 50.6 m, 52.2 m (PZt), 54.5 m (PZ3).

Zachełmie IG 1: 145.4 m (PZ3), 185.6 m, 186.7 m, 188.3 m, 198.2 m, 227.3 m (PZ1).

Zaciszowice IG 1: 155.60 m, 160.30 m, 165.5 m (PZ1).

Biała Wielka IG 1: 1322.0 m (PZt), 1367.5 m (PZ3), 1389.8 m (PZ1).

Brzegi IG 1: 1631.0 m, 1653.0 m, 1687.5 m, 1693.4 m (PZ1).

Milianów IG 1: 1892.6 m, 1893.6 m, 1894.0 m, 1990.0 m, 1915.2 m, 1919.5 m, 1921.2 m, 1927.0 m (PZ1).

Pągów IG 1: 2474.0 m (PZt), 2586.0 m, 2589.3 m, 2600.0 m, 2658.3 m (PZ1).

Accession notation: sample slide, England Finder coordinate. (**a**) Radwanów IG1 borehole, depth 1630.0 m, slide 1 (s1), S45/2; (**b**) Podgace IG 1, 124.7 m, s2, F25; (**c**) Tumlin-Podgrodzie IG 1, 303.4 m, s1, T45/2; (**d**) Brzegi IG 1, 1639.0 m, s1, W34/1; (**e**) V, 1044.0 m, s1, S34/3; (**f**) V, 1044.0 m, s2, U38/2; (**g**) Czerwona Woda IG 1, 674.3 m, s1, K24/2; (**h**) I, 1044.0 m, s1, R37/2; (**i**) V, 1044.0 m, s1, N27/1; (**j**) Brzegi IG 1, 1689.5 m, s1, D46; (**k**) Siodła IG 1, 183.2 m, s2, O27/1; (**l**) Kościelna Wieś IG 1, 785.4 m, s2, F33/4; (**m**) Zabłocie IG 1, 164.5 m, s1, O44/2; (**n**) Zabłocie IG 2, 54.5 m, s2, M24/4; (**o**) Jaworze IG 1, 445.0 m, s1, X56; (**p**) Pągów IG 1, 2589.3 m, s2, G31/1; (**q**) Kościelna Wieś IG 1, 785.4 m, s1, S41/3; (**r**) VI, 1318.3 m, s2, V39/2; (**s**) VI, 1318.3 m, s2, R24; (**t**) Czerwona Woda IG 1, 679.4 m, s2, H38/1.

Slides are stored in the Museum Collection of the Polish Geological Institute-NRI Holy Cross Branch, Kielce

References

Beerling DJ, Harfoot M, Lomax B, Pyle JA (2007) The stability of the stratospheric ozone layer during the end-Permian eruption of the Siberian Traps. Philosophical Transactions of Royal Society A 365:1843–1866

Benca JP, Duijnstee IAP, Looy CV (2018) UV-B-induced forest sterility: implications of ozone shield failure in Earth's largest extinction. Sci Adv 4:1–10. https://doi.org/10.1126/sciadv.1700618

Benton MJ (2018) Hyperthermal-driven mass extinctions: killing models during the Permian-Triassic mass extinction. Philosophical Transactions of Royal Society A 376:1–19

Benton MJ, Newell AJ (2014) Impacts of global warming on Permo-Triassic terrestrial ecosystems. Gondw Res 25:1308–1337

Bercovici A, Cu IY, Forel M-B, Yu J, Vajda V (2015) Terrestrial paleoenvironment characterization across the Permian-Triassic boundary in South China. J Asian Earth Sci 98:225–246

Black BA, Lamarque J-F, Shields C, Elkins-Tanton LE, Kiehl JT (2014) Acid rain and ozone depletion from pulsed Siberian Traps magmatism. Geology 42:67–70
Brugman WA (1983) Permian-Triassic palynology. Laboratory of Palaeobotany and Palynology. State University of Utrecht, p 3–121
Dybova-Jachowicz S (1981) The palynological assemblage from Upper Perman of Poland. In: Pakulska Z (ed) Proceedings of international symposium central European Permian, Jabłonna, April 27–29, 1978. Instytut Geologiczny, Warszawa, pp 120–133
Dybova-Jachowicz S, Chłopek K (2003) *Perm. Palinostratygrafia* (Permian. Palynostratigraphy). In: Dybova-Jachowicz S, Sadowska A (eds) *Palinologia* (Palynology). Instytutu Botaniki PAN, Kraków, pp 120–133. (in Polish)
Dzyuba OF (1998) Paleoecological reconstructions and quality angiosperm pollen grains in stressful environments. Paleontol J 32:97–101
Fielding CR, Frank TD, McLoughlin S, Vajda V, Mays C, Tevyaw AP, Winguth A, Winguth C, Nocoll RS, Bocking M, Crowley JL (2019) Age and pattern of the southern high-latitude continental end-Permian extinction constrained by multiproxy analysis. Nat Commun 10:1–12. https://doi.org/10.1038/s41467-018-07934-z
Fijałkowska A (1994) Palynological aspects of the Permo-Triassic succession in the Holy Cross Mountains, Poland. Documenta Naturae 87:3–76
Fijałkowska A (1995) Palynostratigraphy of the Zechstein in the North Sudetic Trough. Geol Quart 39:207–228
Fijałkowska-Mader A (2012) Impact of the environmental stress on the Late Permian palynoflora from Poland. Biuletyn Państwowego Instytutu Geoleologicznego 452:23–32. (in Polish with English summary)
Fijałkowska-Mader A (1997) Correlation of the Zechstein microflora from Southern Poland. Prace Państwowego Instytutu Geologicznego 157:229–235
Fijałkowska-Mader A (2013) Palynostratigraphy, palaeoecology and palaeoclimate of the Late Permian and Triassic of the Nida Basin. Biuletyn Państwowego Instytut Geologicznego 454:5–70. (in Polish with English summary)
Fijałkowska-Mader A, Durkowski K, Sokalski D (2018) Palynological and lithological implications for Zechstein stratigraphy in the marginal part of the North-Sudetic Synclinorium (Lower Silesia, southwestern Poland). Zeitschrift der Deutschen Gesellschaft für Geowissenschaften—German J Geol 169:495–515
Foster CB, Afonin SA (2005) Abnormal pollen grains: an outcome of deteriorating atmospheric conditions around the Permian-Triassic boundary. J Geol Soc London 122:253–259
Grebe H, Schweitzer HJ (1962) Die Sporae dispersae des niederrheinischen Zechstein. Fortschritte in der Geologie von Rheinland und Westfalen 12:201–224
Hochuli PA, Hermann E, Vigran JO, Bucher H, Weissert H (2010a) Repid demise and recovery of plant ecosystems across the end-Permian. Global Planet Change 74:144–155
Hochuli PA, Vigran JO, Hermann E, Bucher H (2010b) Multiple climatic changes around the Permian-Triassic boundary event revealed by an expanded palynological record from mid-Norway. Geol Soc Am Bull 122:884–896
Hochuli PA, Sanson-Barrera A, Schneebeli-Hermann E, Bucher H (2016) Severest crisis overlooked—worst disruption of terrestrial environments postdates the Permian-Triassic mass extinction. Sci Rep 6:1–7. https://doi.org/10.1038/srep28372
Hochuli PA, Schneebeli-Hermann E, Mangerud G, Bucher H (2017) Evidence for atmospheric pollution across the Permian-Triassic transition. Geology 45:1123–1126
Kormut'ák A (1996) Development and variability of silver fir pollen in air-polluted and non-polluted habitats in Slovakia. For Genet 3:147–151
Kormut'ák A, Salaj J, Voodková B (1994) Pollen variability and seed set of silver fir (*Abies alba* Mill.) in polluted areas of Slovakia. Silvae Genetica 43:68–73
Kutzbach JE, Ziegler AM (1994) Simulation of Late Permian climate and biomes with an atmosphere-ocean model—comparisons with observation. Philos Trans R Soc B 341:327–340

Leslie AB (2008) Interpreting the function of saccate pollen in acient conifers and other seed plants. Int J Plant Sci 169:1038–1045

Looy CV, Collinsosn ME, Van konijnenburg-Van Cittert JHA, Visscher H, Brain APR (2005) The ultrastructure and botanical affinity of end-Permian spore tetrads. Int J Plant Sci 166:875–887

McElwain JC, Punyasena SW (2007) Mass extinction events and the plant fossil record. Trends Ecol Evol 22:548–557

Mishra S, Aggarwal N, Jha N (2017) Palaeoenvironmental change across the Permian-Triassic Boundary inferred from palynomorph assemblages (Godavari Graben, south India). Palaeobiodiver Palaeoenviron:1–28. https://doi.org/10.1007/s12549-017-0302-3

Orłowska-Zwolińska T (1983) Palynostratigraphy of the Upper part of Triassic epicontinental sediments in Poland. Prace Instytutu Geologicznego 104:3–89. (in Polish with English summary)

Ostrolúcka MG, Bolvanskú M, Kokár F (1995) Vitality of pine pollen (Pinus silvestris L., *Pinus nigra* Arn.) on sites with different ecological conditions. Biológia Bratislava 50:47–51

Potonié R, Schweitzer HJ (1960) Der Pollen von Ullmania frumentaria. Paläontologische Zeitschrift 34:27–39

Prevec R, Gastaldo RA, Nevelling J, Reid SB, Looy CV (2010) An autochthonous glossopterid flora with latest Permian palynomorphs and its depositional setting in the *Dicyonodon* Assemblage Zone of the southern Karoo Basin, South Africa. Palaeogeogr Palaeoclimatol Palaeoecol 292:391–408

Racki G, Wignall PB (2005) Late Permian double-phased mass extinction and volcanism: an oceanographic perspective. In: Over DJ, Morrow JR, Wignall PB (eds), Understanding Late Devonian and Permian-Triassic biotic and climatic events: towards an integrated approach. Development in Palaeontology and Stratigraphy, vol. 20, p 263–297

Reichow MK, Pringle MS, Al'Mukhamedov AI, Allen MB, Andreichev VL, Buslow MM, Davies CE, Fedoseev GS, Fitton JG, Medvedev AY, Mitchell C, Puchkov VN, Safanova IY, Scott RA, Saunders AD (2009) The timing and extent of the eruption of Siberian traps large igneous province: implications for the end-Permian environmental crisis. Earth Planet Sci Lett 277:9–20

Scotese CR (1997) Palaeogeographical map of Permian. www.Scotese.com/newpage5.htm

Shkarlet OD (1972) Influence of industrial pollution of atmosphere and soil on the size of pollen grains of the Scots pine. Ekologia 1:3–57

Sirenko EA (2001) Palynological data from studies of bottom sediments in water bodies of 30-km Chernobyl Zone. In: Belonin MD, Kirichova AI (eds) Proceedings first international seminary on the pollen as indicator of environmental state and paleoecological reconstructions. Vserossiyskiy Neftyanoy Nauchno-Issledovatel'skiy Geologorazvedochnyy Institut—VNIGRI, St. Petersburg, pp 189–190

Tevini M (1993) Effects of enhanced UV-B radiation on terrestrial plants. In: Tevini M (ed) UV-B radiation and ozone depletion: effects on humans, animals, plants, microorganisms and materials. Lewis, Boca Raton, FL, pp 125–153

Traverse A (2008) Paleopalynology. Springer, Dordrecht, pp 499–502

Tretyakova IN, Noskova NE (2004) Scotch pine pollen under conditions of environmental stress. Russ J Ecol 35:20–26

Vázquez MS, Césari SN (2017) The Permian palynological *Lueckisporites-Weylandites* Biozone in the San Rafael Block and its correlation in Western Gondwana. J South Am Earth Sci 76:165–181

Visscher H (1971) The Permian and Triassic of the Kingscourt outlier, Ireland. Geol Surv Ireland Spec Publ 1:3–115

Visscher H (1972) The *Lueckisporites* Palynodemes. Compte Rendu, seventh international congress on the stratigraphy and geology of carboniferous, Krefeld, Germany, 1971, vol. 1, p 355–358

Visscher H (1973) The Upper Permian of Western Europe—a palynological approach to chronostratigraphy. In: Logan A, Hill LV (eds) The Permian and Triassic systems and their mutual boundary, Canadian Society of Petroleum Geologists, Memoir 2, pp 200–219

Visscher H (1978) Aspects of a palynological characterization of Late Permian and Early Triassic "standard" units of chronostratigraphical classification in Europe. In: Proceedings of fourth international palynological conference, Lucknow, India, 1976, pp 238–241

Visscher H, Brugman WA (1988) The Permian-Triassic boundary in Southern Alps—a palynological approach. Mem Soc Geol Ital 34:121–128

Visscher H, Looy CV, Collinson ME, Brinkhuis H, Van Konijnenburg-Van Cittert JHA, Kürschner WM, Sephton MA (2004) Environmental mutagenesis during the end-Permian ecological crisis. Proc Natl Acad Sci U S A 101:12,952

Wagner R (1994) Stratigraphy and evolution of the Zechstein Basin in the Polish Lowland. Prace Państwowego Instytutu Geologicznego 156:3–71. (in Polish with English summary)

Wilson LR (1965) Teratological forms in pollen of *Pinus flexilis* James. J Palynol 1:106–110

Yin H, Feng Q, Lai X, Baud A, Tong J (2007) The protracted Permo-Triassic crisis and multi-episode extinction around the Permian-Triassic boundary. Global Planet Change 55:1–20

Yu J, Broutin J, Chen Z-Q, Shi X, Li H, Chu D, Huang Q (2015) Vegetation changeover across the Permian-Triassic Boundary in Southwest China. Extinction, survival, recovery and palaeoclimate: a critical review. Earth Sci Rev 149:203–224

Chapter 4
Stress-Related Evolution in Triassic Conodonts and the Middle Norian Juvenile Mortality

Viktor Karádi, Attila Virág, Tea Kolar-Jurkovšek, and Bogdan Jurkovšek

Abstract Conodonts are of great importance in Paleozoic and Early Triassic stratigraphy, which is proven by the large number of studies on their biostratigraphic relevance. Less emphasis was placed on the paleoecology of this group, even though environmental controls highly affect the distribution and applicability of conodonts. Few records are available from the Triassic, in which lethal or sublethal environmental stress is considered as the driving factor of evolutionary trends. Such cases are reviewed from the Permian-Triassic transition involving studies on the genera *Hindeodus*, *Isarcicella*, and *Neogondolella*. A summary is given on the characteristic taxa (e.g., *Pseudofurnishius murcianus*, *Nicoraella budaensis*, and *Mockina slovakensis*) of the Middle and Upper Triassic, which often formed monospecific faunas in the unstable environments of the restricted basins and marginal seas of the Tethys.

The phenomenon of the Middle Norian juvenile mortality is introduced herein, based on the faunas of four Hungarian and one Slovenian successions from the area of the western Tethys. The recovered pectiniform conodont elements were classified in growth stages and analyzed in the R programming environment in order to reveal the distribution pattern of juvenile/adult-dominated assemblages within the successions. The results show a general drop in the number of specimens and a decrease in the number of adults in the Middle Norian beds of the Danube-E blocks (Hungary), which indicates an ecologically unfavorable environment for conodont animals during this time. The faunas from the Buda Hills (Hungary) and especially from the Dovško succession (Slovenia) are rich in adult specimens and reflect that these areas were suitable habitats for conodonts in the Middle Norian. The reasons for the environmental change in the Early-Middle Norian transitional interval have yet to

---

V. Karádi (✉)
Department of Palaeontology, Eötvös Loránd University, Budapest, Hungary

A. Virág
MTA-MTM-ELTE Research Group for Paleontology, Budapest, Hungary

T. Kolar-Jurkovšek · B. Jurkovšek
Geological Survey of Slovenia, Ljubljana, Slovenia

J. Guex et al. (eds.), *Morphogenesis, Environmental Stress and Reverse Evolution*, https://doi.org/10.1007/978-3-030-47279-5_4

be determined, but the results of the present study provide a good base for future research integrating sedimentological and geochemical investigations.

Keywords Conodonts · Triassic · Middle Norian · Growth stages · Juvenile mortality

4.1 Introduction

Conodonts are the phosphatic, toothlike elements of an extinct, jawless vertebrate animal, grouped in the class Conodonta. Their rapid evolution, wide paleogeographic distribution, and high resistance to rock alteration made conodonts one of the leading microfossil groups in the biostratigraphy of Paleozoic and Triassic marine formations. Thus, it is not surprising that most studies focused on the application of conodonts. Despite the fact that the recognition of the interactions between the organisms and the environment can ameliorate the applicability of fossils, very little attention was directed to conodont paleoecology at the early stage of conodont study. Seeking to understand the mode of life of the conodont animal, two ecologic models were established based on the distribution of discrete conodont elements. The depth stratification model by Seddon and Sweet (1971) presumed a pelagic lifestyle and suggested that different depths of the water column were inhabited by different taxa. The lateral segregation model of Barnes and Fåhraeus (1975), however, considered conodonts to have been benthic or nektobenthic and indicated a change in conodont associations laterally, depending on the distance from the shoreline. Klapper and Barrick (1978) noted that neither of these models are suitable to unequivocally explain the observed distribution patterns of conodont elements. They proposed a combination of the two models, but also concluded that solely distributional data are not sufficient for analyzing the mode of life of conodonts. The discovery of the first complete conodont animal with preserved soft tissue (Briggs et al. 1983) and the development of new methods and investigative techniques (e.g., geochemistry) triggered an increase in paleobiological and paleoecological research. In the lack of recent analogues, modern lampreys and hagfish are often used as reference, since conodont animals seem to show strong affinity with these groups (e.g., Aldridge and Donoghue 1998; Goudemand et al. 2011; Iannicelli 2017; Terrill et al. 2018).

The temperature dependence of conodont animals is presumed for a long time. For example Nicoll (1976) recorded a dramatic decrease in conodont taxa in the southern hemisphere during the Late Carboniferous-Early Permian glaciation. The $\delta^{18}O$ analysis of conodont apatite revealed the stenothermal behavior of the Triassic genera *Gladigondolella* and *Norigondolella*, which represent cooler biofacies and hence deeper water environments, while other taxa seem to have been surface-dwellers (Rigo and Joachimdki 2010; Trotter et al. 2015). There is evidence proving that the distance from the coast or the platform strongly influenced the life, and consequently the distribution of conodont animals. Similar observations were made by Sandberg (1976) on Devonian and by Babcock (1976) on Permian conodonts. Both authors stated that diversity was the highest in pelagic environments, especially near the platform edge, and decreased both basinward and shoreward. According to Babcock (1976) this pattern may result from the availability of food supply. Although it is not clear what conodont animals fed on, the most recent study by Balter et al. (2019) suggests that they were zooplanktivores or primary piscivores, as concluded based on the Ca isotope composition of Late Devonian conodont elements. Evidently, if the diet of conodonts was known, it would certainly improve the investigations on the paleobiology and paleoecology of this group.

Even less research was carried out concerning the effects of environmental stress on conodonts. In the present study we aim to review how stress affected Triassic conodonts during certain events in special environments, and how it influenced their evolution. Special emphasis is placed on the juvenile faunas of the Middle Norian, based on the conodont assemblages of Hungarian and Slovenian successions.

4.2 The Permian-Triassic Crisis

The Permian-Triassic extinction is known as the largest mass extinction in the history of Earth. According to Bai et al. (2017), this crisis did not seriously affect conodonts, which provide a good biostratigraphic resolution across the Permian-Triassic boundary (PTB). Orchard (2007) gave a more accurate explanation stating that a gradual decline can be observed in the conodont families and genera from the Changhsingian, but the totality of the faunas remained stable until the late Griesbachian. Even if conodont diversity did not change significantly around the PTB, the response of this group to environmental disturbances is discernible. Provincialism had significantly reduced near the Permian-Triassic transition by the distribution of ecologically more tolerant taxa, such as *Hindeodus* (Kozur 1996; Chen et al. 2009). However, there is significant difference in the composition of the conodont faunas of the western and eastern Tethys. The PTB faunas in the western Tethys are marked by *Hindeodus* and *Isarcicella* populations in the absence or very rare representation of gondolellids and the strata are interpreted to have been deposited in the overall shallow marine environment (Kolar-Jurkovšek and Jurkovšek 2007, 2015, 2019; Kolar-Jurkovšek et al. 2011, 2018a; Aljinović et al. 2018).

Kılıç et al. (2016) presented atavistic reversals in conodont evolution in certain periods of the Late Permian-Early Triassic interval, characterized by the sudden appearance of species that are underdeveloped relative to their ancestors. They discussed the retrograde evolution that led to the development of the genus *Neospathodus* in the earliest Triassic. The reappearance of ancestral morphs in the clade is the result of sublethal environmental stress following the PTB. Hirsch (1994) stated that phylogenetic changes are related to sea-level oscillations and probably also to anoxic conditions. Kılıç et al. (2016) added that the development of atavistic features in conodonts might have been caused by chemical stress and high temperatures, which would be consistent with the first-order warming cycle of the Early Triassic (Trotter et al. 2015) in the case of the genus *Neospathodus*.

Luo et al. (2006, 2008) carried out size measurements on thousands of individual conodont P1 elements belonging to the genera *Neogondolella*, *Hindeodus*, and *Isarcicella* from the PTB interval of the Meishan section (South China), and reported abrupt size reduction of the conodonts in multiple levels of the Permian-Triassic transition. The general characteristics of a conodont element (e.g., shape of the platform, fusion of the denticles, tips of the denticles, shape of the keel, shape of the basal cavity or pit) are good indicators of the ontogenetic stage in which the element belongs. Therefore, it can be easily recognized if the element represents a juvenile individual or a small-sized adult. This was noted also in the studies by Luo et al. (2006, 2008), who considered the majority of the specimens in their collection (either *Neogondolella*, *Hindeodus*, or *Isarcicella*) to be juveniles. Incomprehensibly, Luo et al. (2006) concluded that it is not clear whether the size reduction of *Neogondolella* is resulted by the death of juvenile specimens or due to dwarfism. Luo et al. (2008) interpreted juvenile mortality as the cause for the increase in the number of small-sized specimens in the genera *Hindeodus* and *Isarcicella*. The conodonts studied by Luo et al. (2006, 2008) are indeed juveniles, but the term "size reduction" is misleading in this case according to the present authors. The phenomenon is the result of the high rate of juvenile mortality, due to environmental stress that was beyond the ecological tolerance boundary of the conodont animals, as noted also by Kozur (1996). In fact, the size of the individuals had not decreased; if these juveniles had reached the adult growth stage, they would have most likely reached the same size as those in ecologically favorable environments. These statements are supported by the study of Schaal et al. (2016) who claimed that maximum size change is inconsiderable in conodonts of the Permian-Triassic transition.

4.3 Restricted Basins of the Middle and Upper Triassic

Certain intervals during the Middle and Late Triassic were characterized by the development of such environmental niches along the southern and northern margins of the Tethys, which gave rise to monospecific conodont faunas composed of ecologically highly tolerant species. Environmental parameters (e.g., salinity, temperature, oxygen content) changed rapidly in these marginal seas, definitely causing sublethal stress to many marine organisms including conodont animals. One of the most widely known examples is the conodont species *Pseudofurnishius murcianus* of the Fassanian-Cordevolian interval. Kozur (1993) stated that *Pseudofurnishius murcianus* had its reproduction area in the pelagic environments of the southern part of the Tethys, from where it repeatedly invaded the marginal seas of the southwestern shoreline. In the adult specimens the typical morphology of the P1 elements is the unevenly developed, rudimentary platform (Fig. 4.1), which suggests that *Pseudofurnishius murcianus* had evolved under sublethal environmental stress, since this underdeveloped morphology is very similar to that of atavistic genera

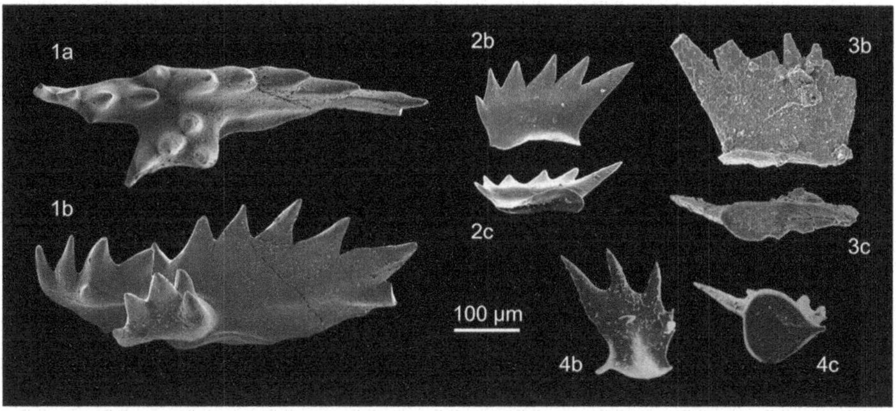

Fig. 4.1 Pectiniform elements of atavistic conodont species from the Middle and Late Triassic. The underdevelopment or the lack of the platform is indicative of sublethal environmental stress. a: upper view, b: lateral view, c: lower view. Scale bar: 100 µm. All specimens are on the same scale. (1) *Pseudofurnishius murcianus*, Ladinian, Prikrnica, Slovenia—re-figured from Kolar-Jurkovšek et al. (2018b); (2) *Nicoraella budaensis*, Julian, Helena Creek, Slovenia—re-figured from Kolar-Jurkovšek and Jurkovšek (2010); (3) *Nicoraella budaensis*, Upper Julian, Pilisvörösvár, southern part of the Pilis Hills, Hungary; (4) "*Neospathodus*" sp., Upper Julian, Zsámbék-14 borehole, Hungary—re-figured from Kristan-Tollmann et al. (1991)

often characterizing crisis events. According to Plasencia et al. (2018), *Pseudofurnishius* originated from the Late Anisian genus *Guexispathodus*, which itself is a proteromorphic taxon, too, so far only known from the eastern part of the Sephardic province (i.e., the northwestern margin of Gondwana), from Israel and Jordan, respectively. In many successions *Pseudofurnishius murcianus* is the only conodont species present in shallow-water carbonates, which demonstrates its high ecologic tolerance (Kozur 1980; Hirsch et al. 1987; Márquez-Aliaga et al. 2000; Kolar-Jurkovšek et al. 2018b; Kolar-Jurkovšek and Jurkovšek 2019).

The Late Julian humid climate pulse, termed the Carnian Pluvial Event (CPE), resulted in severe extinctions and faunal turnovers in conodont lineages (Simms and Ruffell 1989; Rigo et al. 2007). The restricted basins that existed during this time in the western part of the Tethys (e.g., in the Southern Alps and in the Transdanubian Range) were affected by an increased terrigenous input and fluctuations of the dissolved oxygen content and salinity. These environments were inhabited by different atavistic taxa. The most records are available of *Nicoraella budaensis* that was described by Kozur and Mock (1991) from the northwestern part of the Buda Hills and the southern part of the Pilis Hills (Transdanubian Range), but other occurrences are known from the Julian Alps and the Karavanke Mts. in Slovenia (Kolar-Jurkovšek et al. 2005; Kolar-Jurkovšek and Jurkovšek 2010). Kozur and Mock (1991) also mentioned the presence of this species in the pelagic limestones of Sicily, although no specimens were ever figured from that area. The P1 elements of this species consist of a single blade (Fig. 4.1), which provides a further example for the recurrence of the evolutionary trend characterized by the loss of the platform in ecologically not so favorable environments. Kovács in Kristan-Tollmann et al. (1991) described a different platformless conodont from the Zsámbék-14 borehole (Transdanubian Range) as *Neospathodus* n. sp. (Fig. 4.1), which represents approximately the same age and environment as *Nicoraella budaensis*. According to the present authors, this specimen is tentatively assigned to "*Neospathodus*", and it most likely belongs to a new genus. (Unfortunately no taxonomic description is possible, because the type material was lost.) Even if the shape of the denticles differs, the widely open basal cavity makes it homeomorphic to the Late Norian-Rhaetian genus *Misikella*, which is also a proteromorphic genus. These observations seem to support the idea that the development of the abovementioned species from Zsámbék was the result of sublethal stress, induced by the Carnian Pluvial Event.

A somewhat different trend occurs in the latest Alaunian-Sevatian in the shallow marine and restricted basin environments of the western and northern margin of the Tethys. These environments were often inhabited exclusively by the conodont species *Mockina slovakensis* (e.g., Budai and Kovács 1986; Kovács and Nagy 1989; Roghi et al. 1995; Belvedere et al. 2008). This species is frequently recovered from bituminous limestones and dolomites containing a large amount of organic matter, which suggests dysoxic to anoxic bottom water layers. Although *Mockina slovakensis* was most likely a surface dweller (Rigo and Joachimdki 2010), these restricted areas were evidently not the most suitable habitats for conodont animals. Unlike the atavistic species of the Permian-Triassic crisis, or those during the Carnian Pluvial Event, *Mockina slovakensis* retained the platform. All the specimens from the restricted basins of the Transdanubian Range have a wall-like step in their blades that is so typical of this species. However, the platform shape and ornamentation show a notably high rate of intraspecific variability (Budai and Kovács 1986; Kovács and Nagy 1989), which might be an ecologically induced diversity. Detailed research from this aspect has not yet been carried out though.

4.4 Middle Norian Juvenile Assemblages

Norian conodont faunas are often characterized by the dominance either of the genus *Norigondolella* or of the genera *Epigondolella*/*Mockina* sensu lato. This seeming competition between the genera is most likely ecologically controlled and can be related to temperature. Cooler water environments might have been favorable to the *Norigondolella* species (Trotter et al. 2015), in which they appeared in large numbers. In such assemblages the other genera are usually represented by only a few and mainly juvenile specimens. This feature is well documented throughout the Norian (Kozur 2003; Mazza et al. 2010). In the Middle Norian (Alaunian), however, another phenomenon can also be observed, which is very common but still rarely recognized. Conodont faunas of the Middle Norian often consist of a high number of juvenile and subadult specimens and a low number of adults in the majority of the Tethyan successions (e.g., Krystyn 1973; Cafiero and De Capoa-Bonardi 1981, 1982; Kolar-Jurkovšek 1982; Mao and Tian 1987; Meço 1999). According to the present authors, this hampers the Middle Norian conodont taxonomy and biostratigraphy, and repeatedly results in misinterpretations (see Karádi 2018 for details), because juveniles and subadults are not suitable for determination on the species level in most cases. It is interesting to note that the successions of western North America seem to lack this phenomenon, and in the studies from this area several adult specimens are figured (e.g., Mosher 1970; Orchard 1991, 2018). An explanation for the distribution pattern of juvenile and adult assemblages is not yet available, but it is supposedly the result of environmental stress. For better understanding of this phenomenon, the conodont faunas from Hungarian and Slovenian successions are investigated here.

4.4.1 Material and Methods

The conodont assemblages included in this study are part of a recently started research project on Middle Norian conodont biostratigraphy. The faunas were recovered from the Csővár borehole (Csv-1) and Valkó Hill in the Danube-E blocks, Mátyás Hill and the building site at Rácz Aladár Road in the Buda Hills in Hungary, and the Dovško succession in Slovenia (Fig. 4.2a). The Buda Hills and the Danube-E blocks represent the NE segment of the Transdanubian Range that was situated on the northwestern margin of the Tethys, whereas the Dovško section can be found in the transitional area between the External and Internal Dinarides, and was located at the eastern part of the Slovenian Basin during the Late Triassic (Kolar-Jurkovšek and Jurkovšek 2019) (Fig. 4.2b). The Dovško section (~65 m) covers the interval from the Lacian (Lower Norian) to the lower Alaunian (Fig. 4.3). The sections of Mátyás Hill (~19 m), Rácz Aladár Road (~3.5 m), and Valkó Hill (~2 m) are

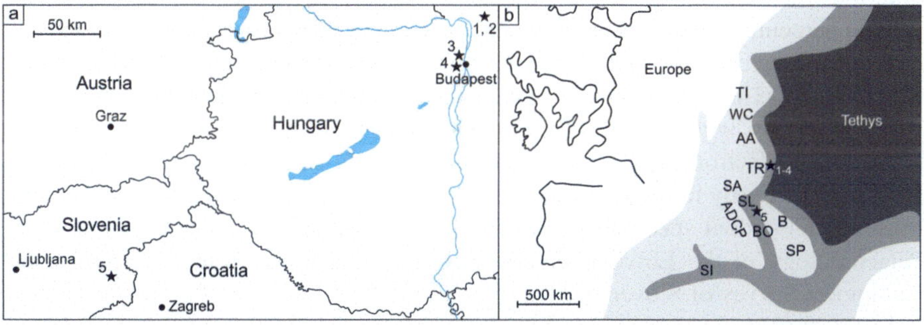

Fig. 4.2 Present-day locations (**a**) and possible paleogeographic positions (**b**) of the investigated successions within the broader region of the western Tethys. Localities are marked by stars: (1) Csővár borehole (Csv-1); (2) Valkó Hill SE; (3) Mátyás Hill; (4) Rácz Aladár Road; (5) Dovško succession. Abbreviations: *TI* Tisza unit, *WC* Western Carpathians, *AA* Austroalpine units, *TR* Transdanubian range, *SA* Southern Alps, *ADCP* Adriatic-Dinaric carbonate platform, *SL* Slovenian basin, *BO* Bosnian basin, *B* Bükk, *SP* Serbo-Pelagonian units, *SI* Sicily. The paleogeographic map is modified after Gawlick et al. (1999) and Haas et al. (2019)

composed exclusively of Alaunian beds (Fig. 4.4). The succession of the Csővár borehole (~620 m) is uppermost Tuvalian to upper Rhaetian in age, but only the upper Lacian to lower Sevatian interval is involved in this study (Fig. 4.4). Detailed geological setting and conodont biostratigraphy of the sections will be presented in upcoming articles. This study focuses only on the differences between the investigated areas in the light of the growth stages of Middle Norian conodont specimens. The authors are aware of the important study of Orchard (2018) on the Lower-Middle Norian boundary; still a simplified taxonomy is used herein; all Middle Norian conodont taxa characteristically different from the Lower Norian representatives of the genus *Epigondolella* [or *Ancyrogondolella* sensu Orchard 2018] are referred to as *Mockina* sensu lato. This decision was made due to the intergeneric variability of Tethyan associations that is not yet very well known.

Collected rock samples weighed less than 0.5 kg from Mátyás Hill, ~0.5 kg from Valkó Hill, ~2 kg from Rácz Aladár Road, 0.5–1.5 kg from the Csővár borehole depending on the available material, and ~3 kg from the Dovško section. Rocks were partly extracted in the Department of Palaeontology of the Eötvös Loránd University in Budapest (Hungary) and partly in the Geological Survey of Slovenia in Ljubljana using standard dissolution technique of dilute (10%) acetic acid. Residues of the Dovško material were enriched by density separation with bromoform that was followed by manual picking of conodonts. Pectiniform conodont elements were counted and classified in growth stages from early juvenile (GS1) to late adult (GS6) using the study of Mazza and Martínez-Pérez (2015) as reference. Among broken specimens only those with at least half-preserved platform were considered in the analysis, in which the growth stage was possible to determine. Bar plots were generated within the R programming environment (v. 3.5.2), where each bar represents the number of specimens within a sample that fell into a given growth category. The upper limit of the vertical axis on these plots was adjusted according to the maximum number of specimens that belong to a single category within a section; thus the height of the bars is comparable along each studied succession. The general percentage distribution of the growth categories within a sample was visualized behind the bars, but in this case the vertical axis was not standardized the same way as above. The maximal value of each distribution was scaled to the same height within a section, which enables the comparison of their overall shapes.

Scanning electron micrographs were taken partly in the Geological Survey of Slovenia and partly in the Department of Botany of the Hungarian Natural History Museum. Conodont specimens from Hungary are deposited in the Department of Palaeontology of the Eötvös Loránd University and those from Slovenia in the Geological Survey of Slovenia.

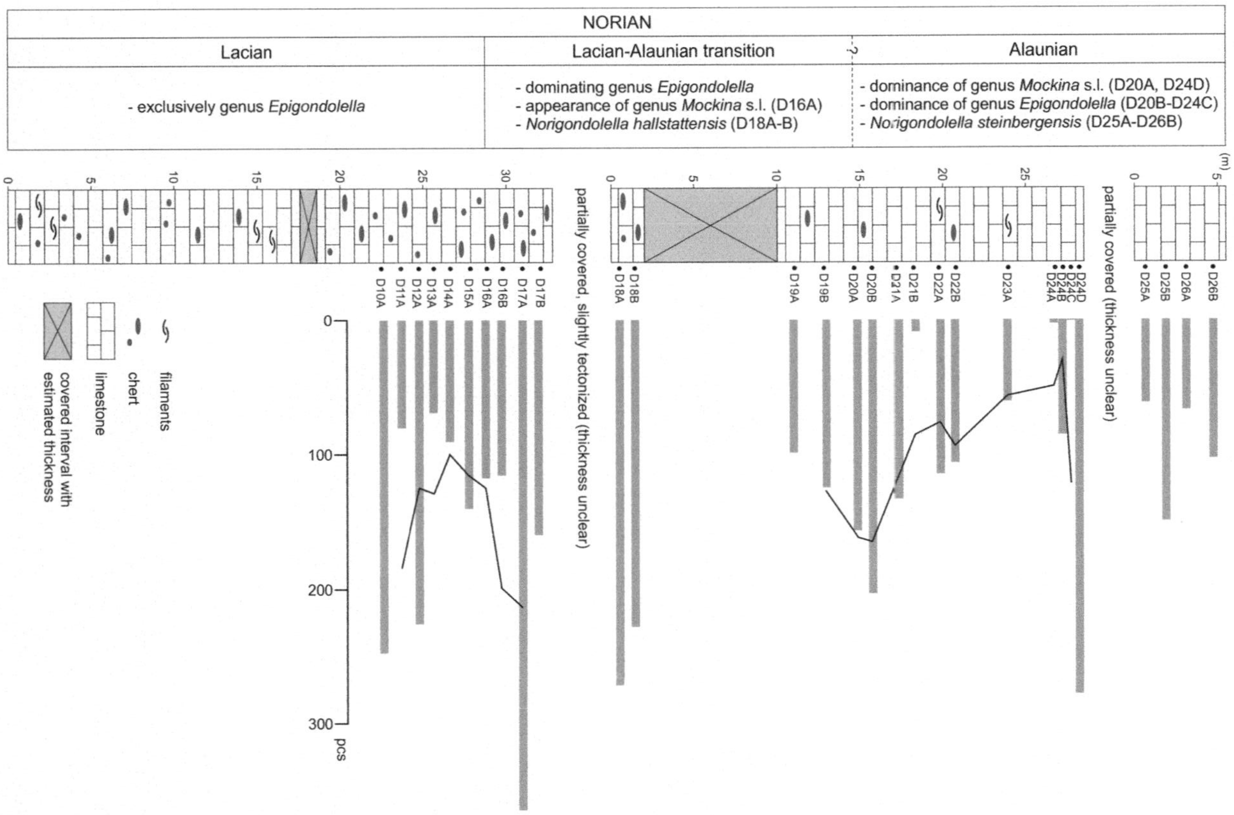

Fig. 4.3 Stratigraphic column of the Dovško succession with the position of the collected samples. The total number of specimens within each sample is represented by a grey bar adjacent to the section. A three-point moving average (depicted here as a black line) is fitted over the number of specimens

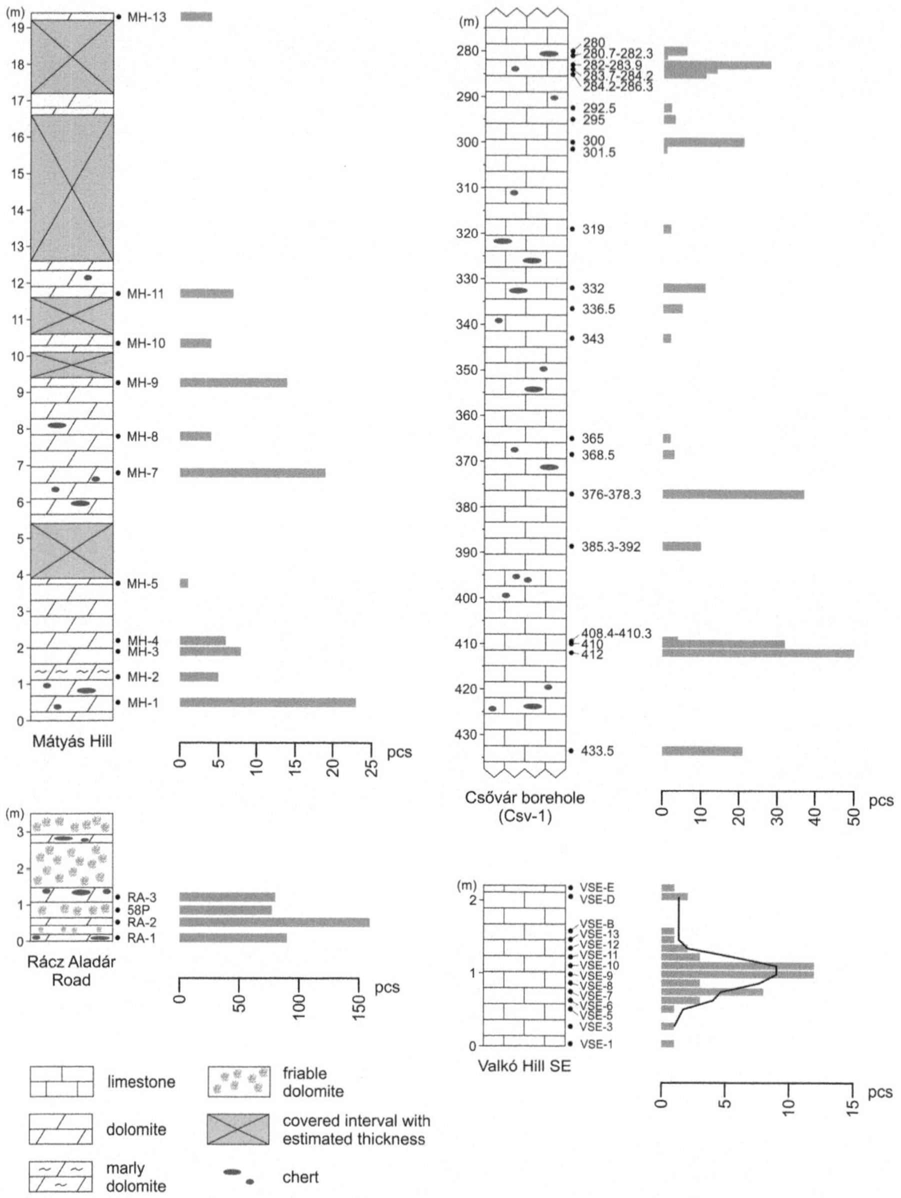

Fig. 4.4 Stratigraphic columns of the investigated successions from the Danube-E blocks and the Buda Hills with the position of the collected samples. The total number of specimens within each sample is represented by a grey bar adjacent to the sections. A three-point moving average (depicted here as a black line) is fitted over the number of specimens in the section of Valkó Hill SE, where the sampling was dense enough and nearly equidistant. Consider that the horizontal axes have different scales

4.4.2 Results

Conodonts are most abundant in the Dovško succession, even if the differences in the weight of the samples are taken into account. Most of the samples contain 100–300 specimens with the exception of the sample D17A that is extremely rich and the samples D21B, D24A, and D24C that yielded only a few specimens. A general drop in the overall yield can be observed in the intervals from D13A to D16B and from D21A to D24C (Fig. 4.3). The specimens belonging to GS3 are the most abundant in the majority of the samples (Fig. 4.5). Conodont elements of GS4 and GS5 are usually more frequent than those of GS2, whereas specimens of GS1 and GS6 turned out to be rare in most cases. The only exception is the sample D19A, in which most elements are assigned to GS1. Within the Dovško succession, four types of unimodal distributions are present: (I) quasi-symmetrical (e.g., D13A, D17B); (II) heavily right skewed (D19A); (III) heavily left skewed (e.g., D16A, D22A); and (IV) slightly right skewed, where the left shoulder is steeper than the right one (e.g., D18A). In a few cases the distributions are bimodal (e.g., D12A). Seemingly no gradual trend is present along the section regarding the shapes of the distributions.

The number of specimens and the distribution of the growth stages within the faunas of the Csővár borehole vary greatly (Figs. 4.4 and 4.6). The richest sample contains ~50 specimens. Although the sampling sights are scarce, there is a general drop in the number of recovered conodonts between 368.5 and 301.5 m. Up to 284.2 m, most frequently GS2 is the dominant category. Above this level, specimens belonging to GS3 become more abundant. It has to be outlined that the number of specimens is not sufficient to assess the distributions reliably in most samples. Nevertheless, in the richer samples distribution type I (e.g., 433.5 m, 300 m), II (e.g., 412 m), and IV (e.g., 282–283.9 m) were recognized. In addition, the sample from 410 m seems to have a bimodal distribution. Each sample from the Valkó Hill SE succession yielded fewer than 15 specimens (Fig. 4.4). The only apparent feature of the distributions is that their mode falls into the category of either GS2 or GS3 (Fig. 4.7).

Conodont elements are abundant in the section of Rácz Aladár Road, since each of the four samples yielded more than 70 specimens (Fig. 4.4). All samples are characterized by the dominance of specimens of GS3 (Fig. 4.8). Sample RA1 has a type I distribution, whereas the other samples show a type IV distribution. Due to the low number of the samples, the authors do not intend to evaluate trends. Each sample from the Mátyás Hill succession contains less than 25 conodont elements (Fig. 4.4). In almost all of the distributions, category GS2 or GS3 contains the highest number of specimens, apart from MH-11, which has an additional mode at category GS5 (Fig. 4.8). The shape of a distribution was only assessed if sufficient number of specimens were available. Distribution type II can be observed in samples MH-7 and MH-9, but the latter has a less prominent peak at GS2. Sample MH-1 has a slightly left-skewed shape, which is basically the mirrored version of distribution type IV.

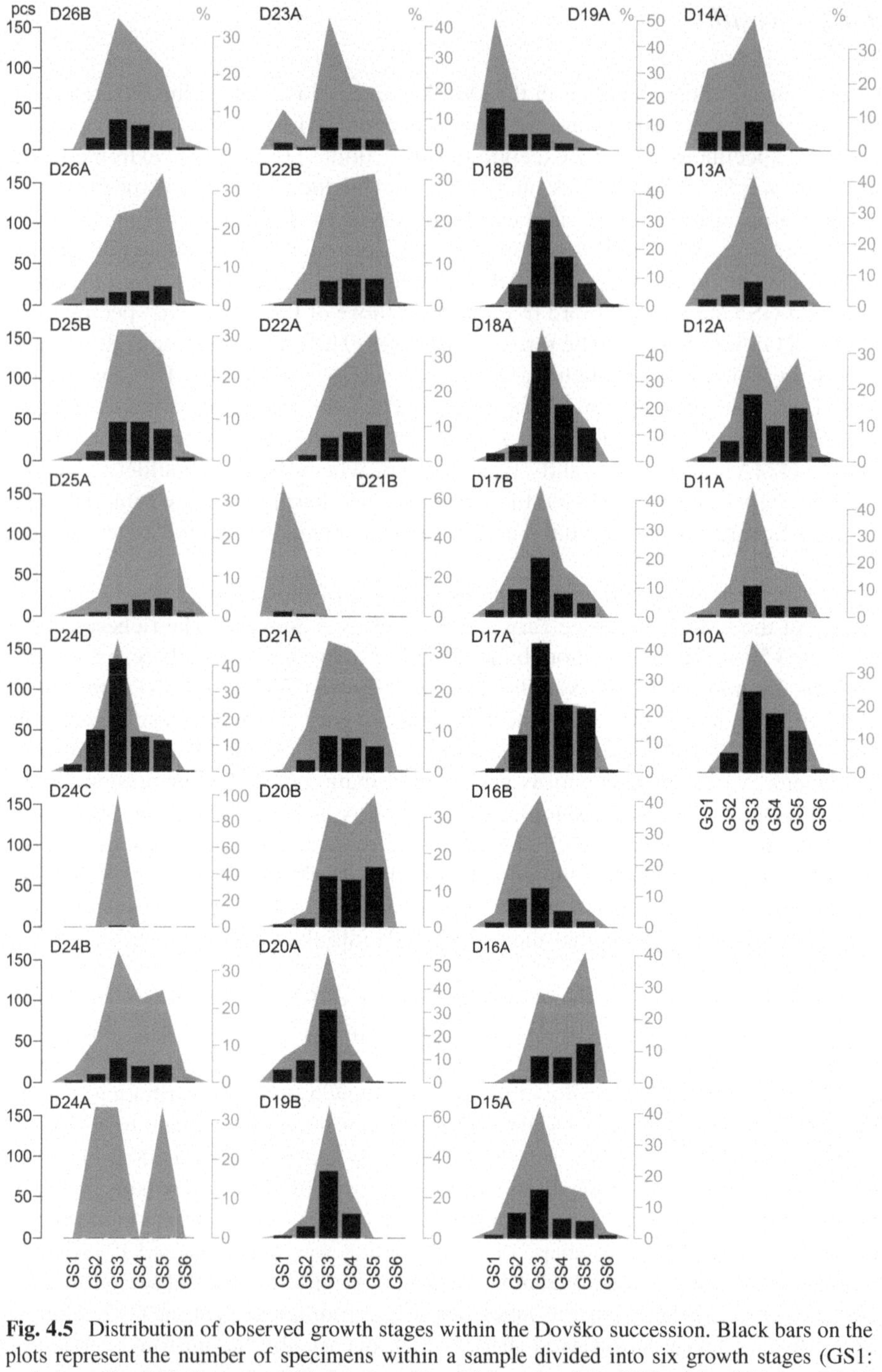

Fig. 4.5 Distribution of observed growth stages within the Dovško succession. Black bars on the plots represent the number of specimens within a sample divided into six growth stages (GS1: early juvenile, GS2: juvenile, GS3: late juvenile, GS4: early adult, GS5: adult, GS6: late adult) based on the study of Mazza and Martínez-Pérez (2015). General percentage distributions are visualized behind the bars with grey. Consider that the vertical axes for the percentage distributions have different scales, even within the same section

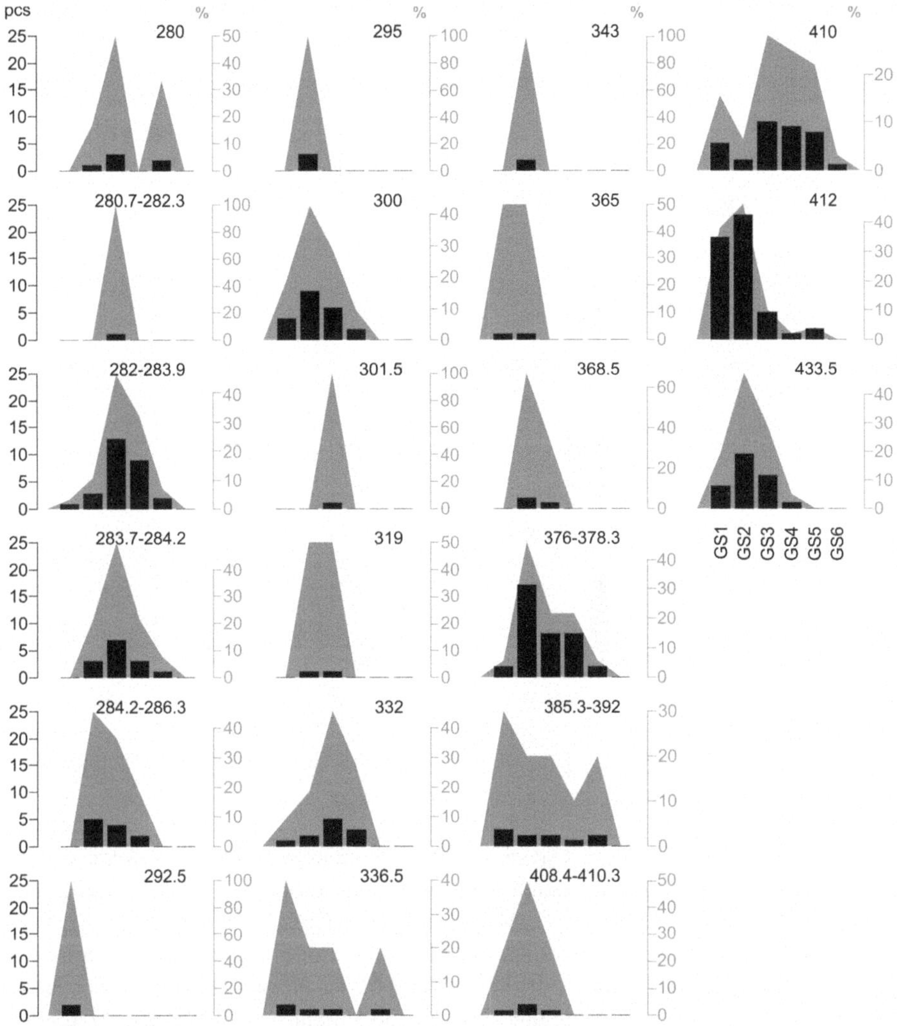

Fig. 4.6 Distribution of observed growth stages within the succession of the Csővár borehole. See details in the caption of Fig. 4.5

4.4.3 Discussion

Kılıç et al. (2017, p. 359) claim that from the Alaunian on, all genera are atavistic. This statement is only partly correct, and it reflects an incomplete knowledge on Middle Norian conodonts, due to the abundance of juvenile-dominated faunas. Atavism of "*Mockina*" *multidentata* and *Mockina postera* can be accepted as referred by Kılıç et al. (2017). But the trends do not seem to be this simple (and need

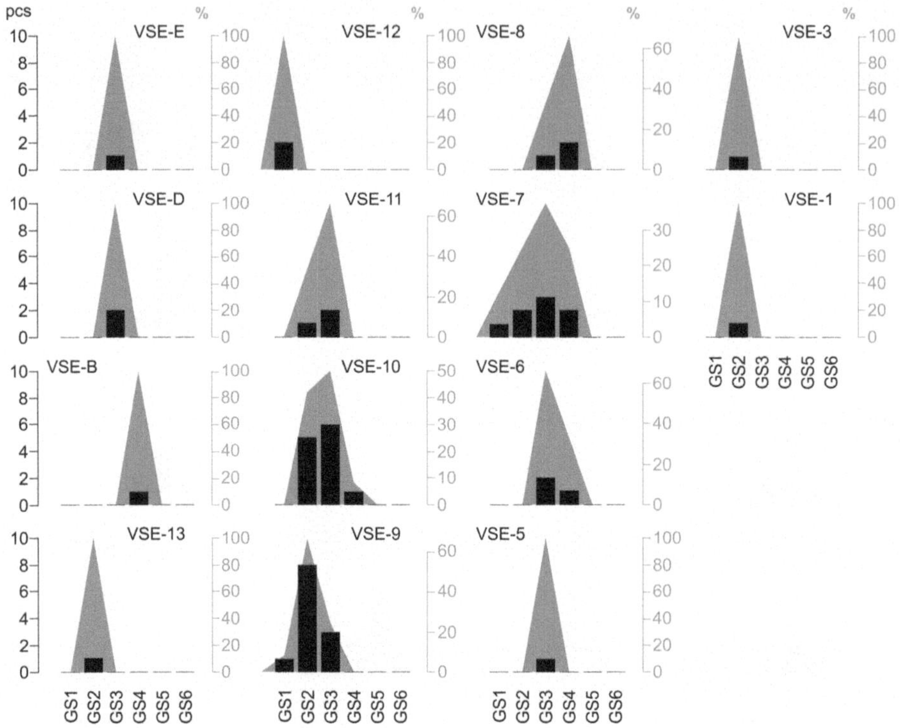

Fig. 4.7 Distribution of observed growth stages within the succession of Valkó Hill SE. See details in the caption of Fig. 4.5

to be explored), because several species of the Alaunian (e.g., *Epigondolella abneptis*, "*Mockina*" *tozeri*) do not resemble ancestral morphs of their clades, as does for example *Neospathodus* in the Early Triassic or *Misikella* in the Late Norian-Rhaetian. Nevertheless, it is not questionable that conodont lineages are characterized by retrograde evolution from the Lower-Middle Norian transition. The final stage of the trend is well documented with the appearances of the genera *Misikella* and *Parvigondolella* in the Late Norian (Karádi et al. 2019). Unfortunately, high juvenile mortality in the Alaunian hampers the research on the evolutionary patterns preceding the Sevatian records, because it reduces the chances of studying fully grown specimens that are necessary for such investigations. Therefore, understanding the causes behind the spatial distribution of juvenile and adult-bearing assemblages is of great importance.

Contrary to previous studies, in which the exact size of conodont elements was measured (e.g., Jeppsson 1976; Luo et al. 2006, 2008), in this work the investigated specimens were assigned to growth stages from early juvenile (GS1) to late adult (GS6). This was necessary, as the evolutionary trend, which initiated during the Lacian-Alaunian transition, gradually led to the development of smaller and slender conodont elements compared to those characteristic for the Lacian. Hence, size

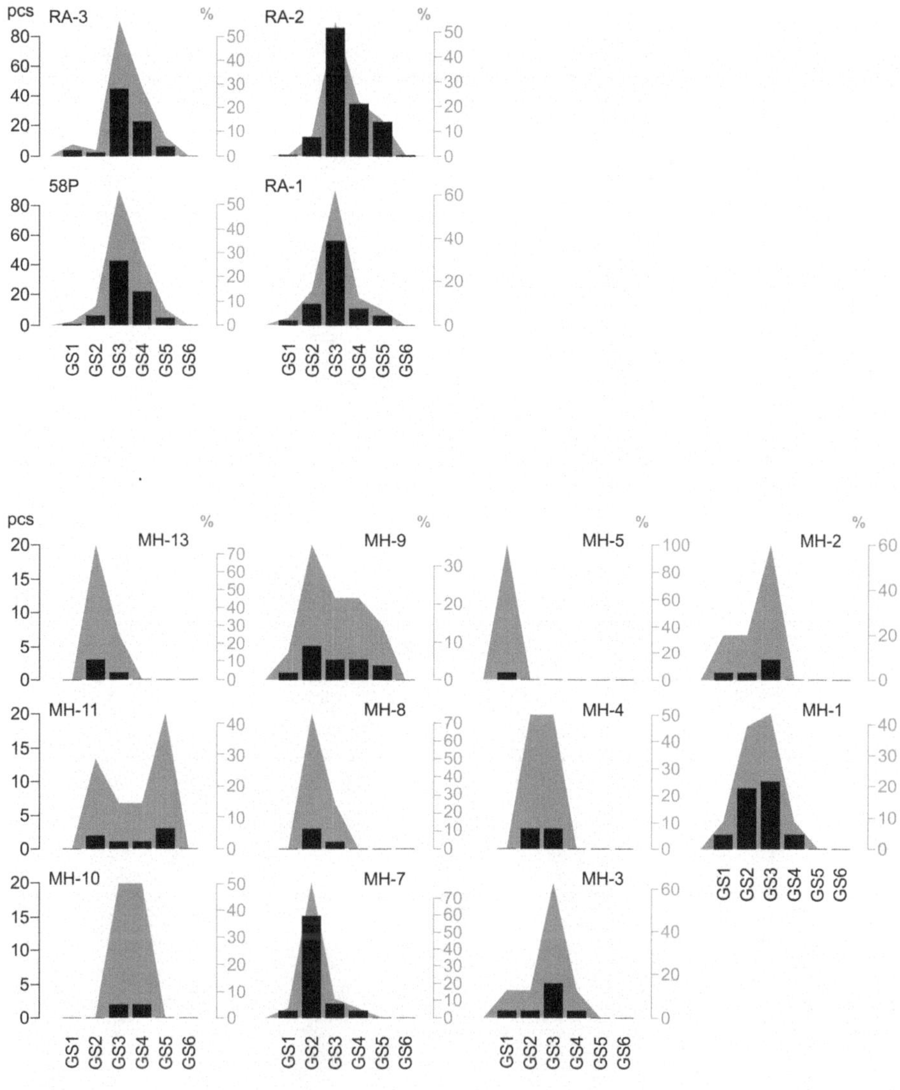

Fig. 4.8 Distribution of observed growth stages within the successions of Mátyás Hill and Rácz Aladár Road. See details in the caption of Fig. 4.5

measurement would have caused bias in the analysis, because it would have reflected not only the mass appearance of juvenile conodonts, but also retrograde evolution that is identifiable by the comparison of the dimensions of adult specimens. Characteristic conodont associations of the Middle Norian intervals of the investigated successions in the light of growth stages are shown in Fig. 4.9.

The most conspicuous feature in the succession of the Csővár borehole (Danube-E blocks) is the decrease in the number of conodonts from the sample at 368.5 m to the sample at 301.5 m. The sample below this interval (at 376–378.3 m) is

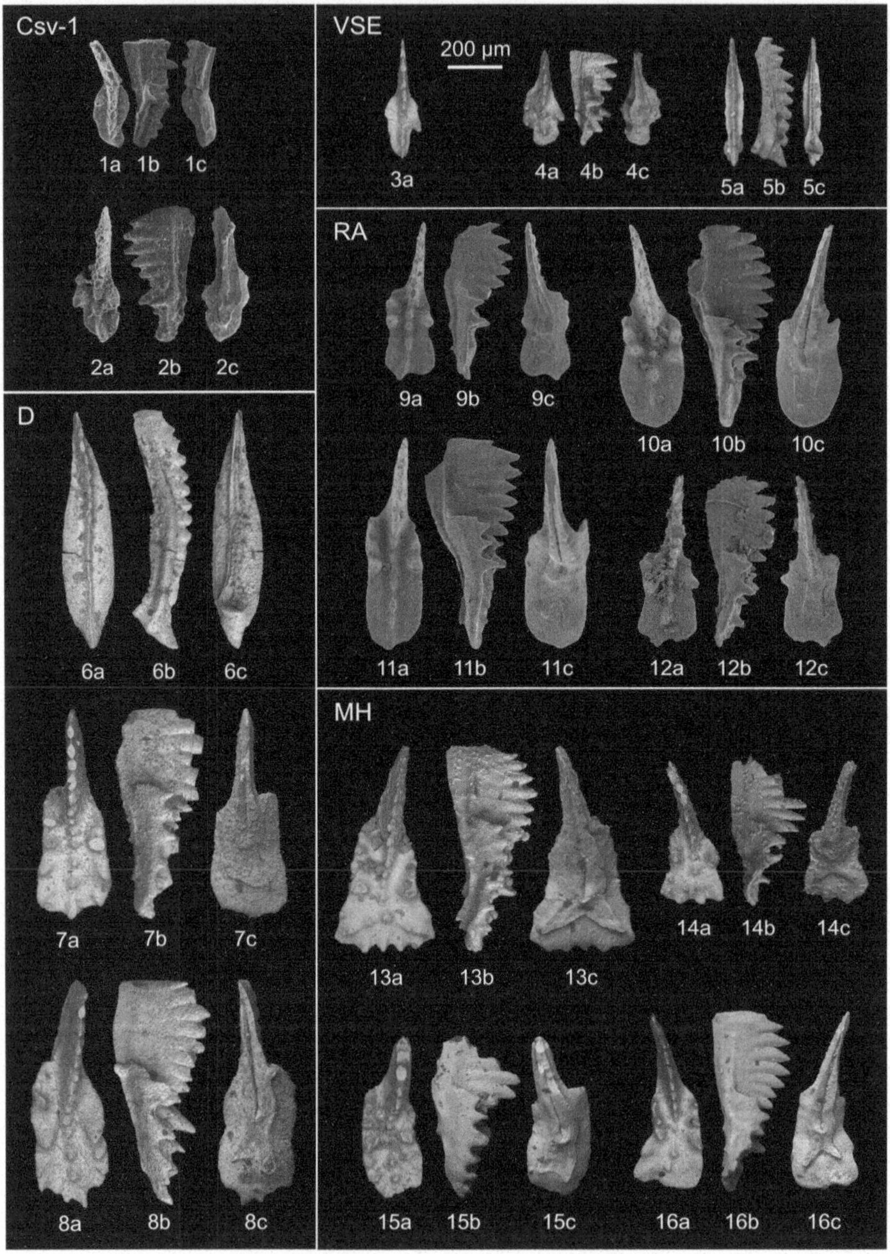

Fig. 4.9 Characteristic pectiniform conodont elements in the light of growth stages (GS) from the Middle Norian intervals of the investigated successions. (**a**) Upper view, (**b**) lateral view, (**c**) lower view. Scale bar: 200 μm. All specimens are on the same scale. Csv-1 (Csővár borehole, Danube-E blocks): (1) *Mockina* sp., GS3, sample 319 m; (2) *Mockina* sp., GS3, sample 301.5 m; VSE (Valkó Hill SE, Danube-E blocks): (3) *Mockina* sp., GS2, sample VSE-9; (4) *Mockina* sp., GS2, sample

supposedly still Lower Norian in age, whereas the first unequivocally Middle Norian sample is at 336.5 m, indicated by the appearance of specimens belonging to the genus *Mockina* sensu lato. Although the exact age of the interval between 378.3 and 336.5 m could not have been determined, it most likely includes the Lacian-Alaunian transition. The base of the Upper Norian could have been defined at the level of 300 m with the appearance of *Mockina bidentata*. Thus, it is notable that the decrease in conodont abundance coincides with the Alaunian part of the succession (Fig. 4.4). Moreover, early adult (GS4) and adult (GS5) specimens are only present in greater numbers below and above this interval, even if the most specimens in the Lacian part fall into the juvenile category (GS2) (Fig. 4.6). In the Alaunian succession of Valkó Hill SE (Danube-E blocks), the fauna is dominated by early to late juveniles (GS1–GS3). No adult (GS5) and late adult (GS6) specimens occur, and early adults (GS4) are represented by only a few conodont elements (Fig. 4.7). This indicates that the environment of the Danube-E blocks during the Alaunian was not favorable for conodont animals.

Likewise at Valkó Hill SE, juveniles (GS2) and late juveniles (GS3) are the most frequent also in the Alaunian succession of Mátyás Hill in the Buda Hills, but the total number of early adults (GS4) is higher and adults (GS5) are present, too (Fig. 4.8). It also has to be considered that, among all sections, the less amount of material was dissolved from Mátyás Hill. The number of specimens is considerably higher in the succession at Rácz Aladár Road, which makes the result and their evaluation more representative. Although the dominant category is late juvenile (GS3), the total number of early adults (GS4) and adults (GS5) by far exceeds that of early juveniles (GS1) and juveniles (GS2) (Fig. 4.8). Based on these results, it can be presumed that the environmental conditions in the area of the Buda Hills during the Middle Norian could have been more stable and more suitable for conodonts than those of the Danube-E blocks. This seems to be consistent with the paleoenvironmental model of Haas et al. (2000) and Haas (2002), which suggests a relatively restricted basin bounded by isolated platforms and connected with the open ocean only through deep channels.

The Lacian-Alaunian Dovško succession represents a similarly stable environment, favored by conodont animals. The lower part of the succession is assigned to the Lacian based on the presence of the genus *Epigondolella* without any representatives of genus *Mockina* sensu lato. The first specimens belonging to the genus *Mockina* s.l. appear in the sample D16A. This sample is considered here to mark the lowest part of the Lacian-Alaunian transition, in which *Mockina* s.l. is present, but the faunas are still dominated by the genus *Epigondolella*. From the sample D20A

Fig. 4.9 (continued) VSE-9; (5) *Norigondolella steinbergensis*, GS2, sample VSE-9; D (Dovško, transitional area between External and Internal Dinarides): (6) *Norigondolella steinbergensis*, GS4, sample D25A; (7) *Mockina* sp., GS5, sample D24D; (8) *Mockina* sp., GS5, sample D24D; RA (Rácz Aladár Road, Buda Hills): (9) *Mockina* sp., GS4, sample RA-1; (10) *Mockina* ex. gr. *matthewi*, GS5, sample RA-2; (11) *Mockina* sp., GS5, sample RA-2; (12) *Mockina* sp., GS4, sample RA-1; MH (Mátyás Hill, Buda Hills); (13) *Epigondolella triangularis*, GS5, sample MH-11; (14) *Epigondolella* cf. *triangularis*, GS3, sample MH-1; (15) *Mockina* sp., GS4, sample MH-7; (16) *Mockina* sp., GS4, sample MH-1

on, *Mockina* s.l. becomes dominant and thus this level can be supposedly assigned to the Alaunian (Fig. 4.3). Late juvenile (GS3) elements are the most abundant also here; however, generally more specimens could have been classified in early adult (GS4) and adult (GS5) categories than in the juvenile (GS2) stage all along the section. In addition, early juvenile (GS1) specimens turned out to be quite rare (Fig. 4.5). Although the high number of early juveniles in the sample D19A is apparent, this excursion is considered normal (even if it is present in the Lacian-Alaunian transition), since short-term fluctuations may happen in any environment.

Records of juvenile mortality being permanent for a longer period of time are very sporadically found in conodont literature. Kozur (1996) and Luo et al. (2006, 2008) presented and discussed such examples for the Permian-Triassic extinction and Early Triassic recovery. The geological circumstances that generated the environmental change and consequently resulted in the death of many juvenile individuals are well known. The causes for the Middle Norian phenomenon, however, still remain unexplained. Babcock (1976) stated that the key environmental gradient in the distribution pattern of Permian conodonts might have been the distance from the basin edge, which strongly affected the available food supply. This explanation should be considered for the Middle Norian. Such a sudden shortage in food would surmise an ecologic crisis in the background, most likely induced by a rapid environmental change. This is consistent with the tectonic and sedimentary disturbance traceable in many successions of the Tethys (Karádi 2018). The idea of Babcock (1976) would describe the dominance of juvenile assemblages in the more open environment of the Danube-E blocks and the abundance of adult specimens in the relatively narrow Slovenian Basin, as well as the somewhat restricted area of the Buda Hills, both basins with platforms nearby. Jeppsson (1976) used migration as an explanation for Silurian strata containing conodont specimens of similar ontogenetic stage. In this hypothesis, certain localities would represent certain generations, based on the migration routes of conodont animals. According to the present authors, this option is highly unlikely in the case of the Alaunian juvenile assemblages, because it would not explain why this pattern is not observable in Lacian and Sevatian intervals. Anyhow, the high rate of juvenile mortality of conodonts evidently characterizes the Middle Norian interval, but the causes of this phenomenon have yet to be clarified.

4.5 Conclusions

The overview of the conodont literature revealed that conodont animals suffered lethal and sublethal environmental stress several times during the Triassic. Unfavorable environmental conditions induced retrograde evolution resulting in the development of several atavistic genera, such as *Neospathodus* at the Permian-Triassic crisis, *Guexispathodus* and *Pseudofurnishius* of the Middle and early Late Triassic, or *Nicoraella budaensis* during the Carnian Pluvial Event. The extreme intraspecific variability of the Upper Norian species *Mockina slovakensis* is probably also an ecologically driven feature.

The results of the present study support the hypothesis of the authors that the high rate of juvenile assemblages with very low number of adult specimens may be the reason for the accumulation of only limited taxonomic and thus biostratigraphic knowledge on Middle Norian conodonts of the Tethyan Realm. However, even if scarce, the existence of Alaunian successions in the Tethys with conodont faunas rich in adult specimens is proven. It is not only the high number of adult conodont elements, but also their great morphological variety what highlights the importance of the Dovško succession and the sections of the Buda Hills for taxonomic and evolutionary studies. It is obvious that conodont evolution turned to retrograde phase worldwide from the Lacian-Alaunian boundary interval, supposedly due to environmental perturbations that caused sublethal stress to conodont animals. This change was gradual and led, through several transitional steps, to the development of ancestral morphs by the Late Norian, which belong to the genera *Misikella* and *Parvigondolella*. This evolutionary trend is somewhat different from that during the Permian-Triassic crisis or the Carnian Pluvial Event, when the appearance of atavistic morphs in certain intervals seems to be more sudden. The regionally (i.e., in the Tethys) characteristic juvenile conodont assemblages of the Alaunian, on the other hand, are the results of high juvenile mortality induced by lethal stress, which is in all cases related to environmental changes close to the Lacian-Alaunian transition. The effects are recognized also from a sedimentological point of view, since many successions of the Tethys show brecciation, slumping, and other sedimentary disturbance in the Middle Norian interval. The discovery of the exact causes is, however, debated and calls for studies integrating different investigation methods, which may help to explain the environmental change in the Middle Norian.

Acknowledgement This work was partially supported by the Slovenian Research Agency (program P1-0011). Further funding was provided through the PD-131536 project and the Hantken Miksa Foundation to Viktor Karádi. The mathematical statistical analyses presented in this study (conducted by co-author AV) are considered MTA-MTM-ELTE Paleo Contribution No. 323. We thank Marija Petrović for technical aid and Miloš Miler and Krisztina Buczkó for their assistance in SEM imaging.

References

Aldridge RJ, Donoghue PCJ (1998) Conodonts: a sister group to hagfishes? In: Jørgensen JM, Lomholt JP, Weber RE, Malte H (eds) The biology of hagfishes. Springer, Dordrecht, pp 15–31

Aljinović D, Horacek M, Krystyn L, Richoz S, Kolar-Jurkovšek T, Smirčić D, Jurkovšek B (2018) Western Tethyan epeiric ramp setting in the Early Triassic—an example from the central Dinarides (Croatia). J Earth Sci 29(4):806–823

Babcock LC (1976) Conodont paleoecology of the Lamar Limestone (Permian), Delaware Basin, West Texas. In: Barnes CR (ed) Conodont paleoecology. Geol. Assoc. Can. Spec. Pap. 15, p 279–294

Bai R, Dai X, Song H (2017) Conodont and ammonoid biostratigraphies around the Permian-Triassic boundary from the Jianzishan of South China. J Earth Sci China 28(4):595–613. https://doi.org/10.1007/s12583-017-0754-4

Balter V, Martin JE, Tacail T, Suan G, Renaud S, Girard C (2019) Calcium stable isotopes place Devonian conodonts as first level consumers. Geochem Perspect Lett 10:36–39. https://doi.org/10.7185/geochemlet.1912

Barnes CR, Fåhraeus LE (1975) Provinces, communities, and the proposed nekton-benthic habitat of Ordovician conodontophorids. Lethaia 8:133–149. https://doi.org/10.1111/j.1502-3931.1975.tb01308.x

Belvedere M, Avanzini M, Mietto P, Rigo M (2008) Norian dinosaur footprints from the "Strada delle Gallerie" (Monte Pasubio, NE Italy). Studi Trent Sci Nat Acta Geol 83:267–275

Briggs DEG, Clarkson ENK, Aldridge RJ (1983) The conodont animal. Lethaia 16:1–14

Budai T, Kovács S (1986) Contributions to the stratigraphy of the Rezi Dolomite Formation [*Metapolygnathus slovakensis* (Conodonta, upper Triassic) from the Keszthely Mts (W Hungary)]. Ann Rep Hung Geol Inst 1984:175–191

Cafiero B, De Capoa-Bonardi P (1981) I conodonti dei calcari ad Halobia del Trias superiore del Montenegro (Crna-Gora, Jugoslavia). Riv Ital Paleont Strat 86(3):563–576

Cafiero B, De Capoa-Bonardi P (1982) Biostratigrafia del Trias pelagico della Sicilia. Boll Soc Paleontol Ital 21(1):35–71

Chen J, Beatty TW, Henderson CM, Rowe H (2009) Conodont biostratigraphy across the Permian–Triassic boundary at the Dawen section, great Bank of Guizhou, Guizhou Province, South China: implications for the Late Permian extinction and correlation with Meishan. J Asian Earth Sci 36:442–458. https://doi.org/10.1016/j.jseaes.2008.08.002

Gawlick H-J, Frisch W, Vecsei A, Steiger T, Böhm F (1999) The change from rifting to thrusting in the Northern Calcareous Alps as recorded in Jurassic sediments. Geol Rundsch 87:644–657

Goudemand N, Orchard MJ, Urdya S, Buchera H, Tafforeauc P (2011) Synchrotron-aided reconstruction of the conodont feeding apparatus and implications for the mouth of the first vertebrates. Proc Natl Acad Sci U S A 108(21):8720–8724

Haas J (2002) Origin and evolution of Late Triassic backplatform and intraplatform basins in the Transdanubian Range, Hungary. Geol Carpath 53(3):159–178

Haas J, Korpás L, Török Á, Dosztály L, Góczán F, Hámorné Vidó M, Oraveczné Scheffer A, Tardiné Filácz E (2000) Upper Triassic basin and slope facies in the Buda Mts.—based on study of core drilling Vérhalom tér, Budapest. Bull Hungarian Geol Soc 130(3):371–421

Haas J, Jovanović D, Görög Á, Sudar MN, Józsa S, Ozsvárt P, Pelikán P (2019) Upper Triassic–Middle Jurassic resedimented toe-of-slope and hemipelagic basin deposits in the Dinaridic Ophiolite Belt, Zlatar Mountain, SW Serbia. Facies 65:23. https://doi.org/10.1007/s10347-019-0566-3

Hirsch F (1994) Triassic conodonts as ecological and eustatic sensors. In: Embry AF, Beauchamp B, Glass DJ (eds). Pangea: global environments and resources. Canadian Society of Petroleum Geologists, memoir 17, p 949–959

Hirsch F, Márquez-Aliaga A, Santisteban C (1987) Distribución de moluscos y conodontos del Tramo Superior del Muschelkalk en el sector occidental de la Provincia Sefardí. Cuad Geol Ibérica 11:799–814

Iannicelli M (2017) Solving the mystery of endless life between conodonts and lampreys, plus a reason for final extinction of the conodonts. J Oceanogr Mar Res S1:1. https://doi.org/10.4172/2572-3103.S1-001

Jeppsson L (1976) Autecology of Late Silurian conodonts. In: Barnes CR (ed) Conodont paleoecology. Geol Assoc Can Spec Pap 15, p 105–118

Karádi V (2018) Middle Norian conodonts from the Buda Hills, Hungary: an exceptional record from the western Tethys. J Iber Geol 44(1):155–174. https://doi.org/10.1007/s41513-017-0009-3

Karádi V, Cau A, Mazza M, Rigo M (2019) The last phase of conodont evolution during the Late Triassic: integrating biostratigraphic and phylogenetic approaches. Palaeogeogr Palaeoclimatol Palaeoecol. https://doi.org/10.1016/j.palaeo.2019.03.045

Kılıç AM, Plasencia P, Ishida K, Guex J, Hirsch F (2016) Proteromorphosis of Neospathodus (Conodonta) during the Permian–Triassic crisis and recovery. Rev Micropaleontol 59:33–39

Kılıç AM, Plasencia P, Guex J, Hirsch F (2017) Challenging Darwin: evolution of Triassic Conodonts and their struggle for life in a changing world. Stratigraphy Timescales 2:333–389

Klapper G, Barrick JE (1978) Conodont ecology: pelagic versus benthic. Lethaia 11:15–23

Kolar-Jurkovšek T (1982) Konodonti iz amfiklinskih skladov in baškega dolomita (Conodonts from Amphiclina beds and Bača dolomite). Geologija 25(1):167–188

Kolar-Jurkovšek T, Jurkovšek B (2007) First record of *Hindeodus-Isarcicella* population in Lower Triassic of Slovenia. Palaeogeogr Palaeoclimatol Palaeoecol 252:72–81. https://doi.org/10.1016/j.palaeo.2006.11.036

Kolar-Jurkovšek T, Jurkovšek B (2010) New paleontological evidence of the Carnian strata in the Mežica area (Karavanke Mts., Slovenia): conodont data for the Carnian Pluvial Event. Palaeogeogr Palaeoclimatol Palaeoecol 290(1–4):81–88. https://doi.org/10.1016/j.palaeo.2009.06.015

Kolar-Jurkovšek T, Jurkovšek B (2015) Conodont zonation of lower Triassic strata of Slovenia. Geologija 58(2):155–174. https://doi.org/10.5474/geologija.2015.012

Kolar-Jurkovšek T, Jurkovšek B (2019) Konodonti Slovenije/Conodonts of Slovenia. Geološki zavod Slovenije, Ljubljana, 260 p

Kolar-Jurkovšek T, Gaździcki A, Jurkovšek B (2005) Conodonts and foraminifera from the "Raibl beds" (Carnian) of the Karavanke Mountains, Slovenia: stratigraphical and paleobiological implications. Geol Q 49(4):429–438

Kolar-Jurkovšek T, Jurkovšek B, Aljinović D (2011) Conodont biostratigraphy and lithostratigraphy across the Permian-Triassic boundary at the Lukač section in western Slovenia. Riv Ital Paleont Strat 117(1):115–133

Kolar-Jurkovšek T, Jurkovšek B, Nestell GP, Aljinović D (2018a) Biostratigraphy and sedimentology of Upper Permian and Lower Triassic strata at Masore, Western Slovenia. Palaeogeogr Palaeoclimatol Palaeoecol 490:38–54. https://doi.org/10.1016/j.palaeo.2017.09.013

Kolar-Jurkovšek T, Martínez-Pérez C, Jurkovšek B, Aljinović D (2018b) New conodont clusters of *Pseudofurnishius murcianus* from the middle Triassic of Slovenia (Dinarides). Bull Am Paleont 395–396:149–163

Kovács S, Nagy G (1989) Contributions to the age of the Avicula- and Halobia-limestones (Feketehegy limestone formation) in Pilis Mts (NE Transdanubian Central Range, Hungary). Ann Rep Hung Geol Inst 1987:95–129

Kozur H (1980) Revision der Conodontenzonierung der Mittel-und Obertrias des Tethyalen Faunenreichs. Geol Paläont Mitt Innsbruck 10(3–4):79–172

Kozur H (1993) First evidence of *Pseudofurnishius* (Conodonta) in the Triassic of Hungary. Jb Geol B-A 136(4):783–793

Kozur H (1996) The conodonts *Hindeodus*, *Isarcicella* and *Sweetohindeodus* in the uppermost Permian and lowermost Triassic. Geol Croat 49(1):81–115

Kozur HW (2003) Integrated ammonoid-, conodont and radiolarian zonation of the Triassic. Hallesches Jb Geowiss B25:49–79

Kozur H, Mock R (1991) New Middle Carnian and Rhaetian conodonts from Hungary and the Alps. Stratigraphic importance and tectonic implications for the Buda mountains and adjacent areas. Jb Geol B-A 134(2):271–297

Kristan-Tollmann E, Haas J, Kovács S (1991) Karnische Ostracoden und Conodonten der Bohrung Zsámbék-14 im Transdanubischen Mittelgebirge (Ungarn). Jubileumsschrift 20 Jahre Geologische Zusammenarbeit Österreich-Ungarn 1:193–219

Krystyn L (1973) Zur Ammoniten- und Conodonten-Stratigraphie der Hallstätter Obertrias (Salzkammergut, Österreich). Verh Geol B-A 1973(1):113–153

Luo G, Lai X, Jiang H, Zhang K (2006) Size variation of the end Permian conodont *Neogondolella* at Meishan section, Changxing, Zhejiang and its significance. Sci China Ser D Earth Sci 49(4):337–347. https://doi.org/10.1007/s11430-006-0337-1

Luo G, Lai X, Shi GR, Jiang H, Yin H, Xie S, Tong J, Zhang K, He W, Wignall PB (2008) Size variation of conodont elements of the *Hindeodus–Isarcicella* clade during the Permian–Triassic transition in South China and its implication for mass extinction. Palaeogeogr Palaeoclimatol Palaeoecol 264:176–187

Mao L, Tian C (1987) Late Triassic conodonts from the uppermost Mailonggang formation in Mailonggang village of Lhünzhub County, Xizang (Tibet), China. Bull Chin Acad Geol Sci 17:159–168

Márquez-Aliaga A, Valenzuela-Rios JI, Calvet F, Budurov K (2000) Middle Triassic conodonts from northeastern Spain. Terra Nova 12:77–83

Mazza M, Martínez-Pérez C (2015) Unravelling conodont (Conodonta) ontogenetic processes in the Late Triassic through growth series reconstructions and X-ray microtomography. Boll Soc Paleontol Ital 54(3):161–186

Mazza M, Furin S, Spötl C, Rigo M (2010) Generic turnovers of Carnian/Norian conodonts: climatic control or competition? Palaeogeogr Palaeoclimatol Palaeoecol 290:120–137

Meço S (1999) Conodont biostratigraphy of Triassic pelagic strata, Albania. Riv Ital Paleont Strat 105(2):251–266

Mosher LC (1970) New conodont species as Triassic guide fossils. J Paleontol 44(4):737–742

Nicoll RS (1976) The effect of Late Carboniferous–Early Permian glaciation on the distribution of conodonts in Australia. In: Barnes CR (ed) Conodont paleoecology. Geol Assoc Can Spec Pap 15, p 273–278

Orchard MJ (1991) Upper Triassic conodont biochronology and new index species from the Canadian Cordillera. In: Orchard MJ, McCracken AD (eds) Ordovician to Triassic conodont paleontology of the Canadian Cordillera. Bull Geol Surv Can 417, p 299–335

Orchard MJ (2007) Conodont diversity and evolution through the latest Permian and Early Triassic upheavals. Palaeogeogr Palaeoclimatol Palaeoecol 252:93–117

Orchard MJ (2018) The Lower-Middle Norian (Upper Triassic) boundary: new conodont taxa and a refined biozonation. Bull Am Paleont 395–396:165–193

Plasencia P, Kılıç AM, Baud A, Sudar M, Hirsch F (2018) The evolutionary trend of platform denticulation in middle Triassic acuminate Gondolellidae (Conodonta). Turk J Zool 42:187–197

Rigo M, Joachimdki MM (2010) Palaeoecology of Late Triassic conodonts: constraints from oxygen isotopes in biogenic apatite. Acta Palaeontol Pol 55(3):471–478

Rigo M, Preto N, Roghi G, Tateo F, Mietto P (2007) A rise in the carbonate compensation depth of western Tethys in the Carnian (late Triassic): deep-water evidence for the Carnian Pluvial Event. Palaeogeogr Palaeoclimatol Palaeoecol 246:188–205

Roghi G, Mietto P, Dalla Vecchia FM (1995) Contribution to the Conodont biostratigraphy of the Dolomia di Forni (Upper Triassic, Carnia, NE Italy). Memoir Sci Geol 47:125–133

Sandberg CA (1976) Conodont biofacies of Late Devonian *Polygnathus styriacus* zone in western United States. In: Barnes CR (ed) Conodont paleoecology. Geol Assoc Can Spec Pap 15, p 171–186

Schaal EK, Clapham ME, Rego BL, Wang SC, Payne JL (2016) Comparative size evolution of marine clades from the Late Permian through Middle Triassic. Paleobiology 42(1):127–142. https://doi.org/10.1017/pab.2015.36

Seddon G, Sweet WC (1971) An ecological model for conodonts. J Paleontol 45:896–880

Simms MJ, Ruffell AH (1989) Synchroneity of climatic change and extinctions in the Late Triassic. Geology 17:265–268

Terrill DF, Henderson CM, Anderson JS (2018) New applications of spectroscopy techniques reveal phylogenetically significant soft tissue residue in Paleozoic conodonts. J Anal At Spectrom 33:992–1002. https://doi.org/10.1039/c7ja00386b

Trotter JA, Williams IA, Nicora A, Mazza M, Rigo M (2015) Long-term cycles of Triassic climate change: a new δ18O record from conodont apatite. Earth Planet Sci Lett 415:165–174

Chapter 5
Proteromorphosis in Early Triassic Conodonts

Ali Murat Kiliç, Jean Guex, and Francis Hirsch

Abstract Herein we emphasise how environment, palaeoecology and palaeobiogeography play key roles in the evolution of organisms. Nineteenth-century ammonoid biochronology led to the definition of the Mesozoic stages. Their beginning and end are bound by the biggest mass extinctions of Earth history. This study deals with the initial Triassic stages that needed a remarkably short biotic recovery time. The Lower Triassic stages, all named after nineteenth-century researchers of the Himalayas, are the Griesbachian, Dienerian, Smithian and Spathian.

If the twentieth century saw an increasing use of alternative biotic groups, among which were those conodonts and radiolarians that permit the relative timing of open marine sequences, one can say that twenty-first century isotopic excursion events became increasingly dominant. For the Early Triassic alone, the succession of positive and negative excursions determines the following five events:

(I) Late Permian-Early Triassic boundary
(II) Late Griesbachian
(III) Dienerian-Smithian boundary
(IV) Smithian-Spathian boundary
(V) Middle Spathian-Early Aegean interval

These events comprise three negative events (I, III and IV) and two positive events (II and V). The nature of these events, in control of environmental instabilities and system tracts (sea-level changes), is the key for understanding biological morphogenesis and evolution.

This contribution aims to analyse the phenomena during one of the five largest mass extinctions of Earth history and the fast recovery that followed in its aftermath.

A. M. Kiliç (✉)
Department of Geology, Balıkesir University, Balıkesir, Turkey
e-mail: alimurat@balikesir.edu.tr

J. Guex
Department of Geology, University of Lausanne, Lausanne, Switzerland
e-mail: Jean.Guex@unil.ch

F. Hirsch
Laboratory of Geology, Faculty of Sciences, Naruto University, Naruto, Japan

J. Guex et al. (eds.), *Morphogenesis, Environmental Stress and Reverse Evolution*, https://doi.org/10.1007/978-3-030-47279-5_5

Our study of the conodont subfamily Neogondolellinae suggests that stress-generating phenomena and events appear to have paced successive lineages, alternating from Darwinian anagenesis to atavistic reversal and rediversification.

Keywords Atavistic reversal · Darwinian anagenesis · Lower Triassic conodonts · Proteromorphosis · Rediversification

5.1 Introduction

Evolution during the Early Triassic can be viewed from a new perspective. The Triassic, one of the warmest periods of Earth history, was the theatre for the dismantling of Pangea, the widening of Tethys, the main equatorial seaway that opened up Pangea and the initial opening of the Atlantic. Following the Late Permian-Early Triassic extinction caused by the Siberian Traps (Burgess et al. 2017), conodonts radiated into new Triassic niches that opened with the transgressive action of Tethys. Panthalassan currents spread taxa worldwide and new geographic, oceanic and climatic barriers isolated new faunal provinces. Soon after, conodonts were diminished by events in the Central Atlantic Magmatic Province (CAMP) that led to their extinction at the end of the Triassic.

The multi-element apparatus of the family Gondolellidae evolved from the Permian septimembrate mesogondolellin apparatus to the Late Permian octomembrate neogondolellin apparatus that ranged to the Late Carnian, and then retrograded to an epigondolellin septimembrate apparatus during the Norian and Rhaetian. Bio-environments, from shallow shelf to deeper open marine, vary in salinity and temperature. The Spathian-Julian octomembrate apparatus of the family Gladigondolellidae is linked to the deeper equatorial provinces. The successive clades not only define the rate of evolution, but also show their dependence from environmental conditions and their response to ecological stress.

The Early Triassic is especially rich in retrograde evolutionary events during a very short duration of merely 5 Ma. This is the reason for its particular interest.

5.2 Part One

5.2.1 *Environmental Instabilities*

Environmental instabilities are controlled by the behaviour of our planet and the sun. In this regard disciplines from tectonics to atmospheric physics, astronomy and climatology all depend in large part on cyclic patterns, e.g. Milankovitch's cycles. Cycles are by definition repetitive, in opposition to catastrophes, like the collision of the Earth with an asteroid. All instabilities are reflected in the life cycles of living organisms. Some of these can be lethal with no return, such as global extinctions due to meteoritic impacts (Clutson et al. 2018) or volcanism that may be part of incompletely understood tectonic or astronomic cycles.

5.2.2 *Morphogenesis of the Bios*

Environmental instabilities control morphogenesis. The generation of anatomic shapes is the result of evolution that includes variations in size.

As we analyse the development of conodonts, a protovertebrate that lived during the first half of Phanerozoic times, we recognise that halfway through this period, a major extinction marked the limit between Paleozoic and Mesozoic eras. But conodonts, having survived several intervals of stress during the Late Permian mass extinction, somehow "knew" how to respond, and endured for another 50 Ma, through the Triassic. This was a time when abnormal morphological development under sublethal stress prevailed and this period is the subject of our present essay. Survival under such conditions was partly possible due to retrograde evolution during major extinction periods.

5.2.3 *The Triassic as a Period of Time*

It must be emphasised that since most research started in the eighteenth century, the Triassic Period was conceived in continental Europe where it consisted of the three units established by Alberti in 1864: Buntsandstein, Muschelkalk and Keuper. Soon these became equivalent to the Lower, Middle and Upper Triassic, yet, not without considerable correlation conflicts. The Lower Triassic became also known as the Scythian which in Europe remains dominated by variegated sandstones of continental origin, the so-called Bunter facies. In the course of the nineteenth century, ammonoid research in the Himalayas resulted in the subdivision of the Lower Triassic or Scythian into the Brahmanian and Jakutian stages. During the twentieth century the detailed ammonoid studies in Arctic Canada led to the distinction of four stages, named after nineteenth-century researchers in the Himalayas: the Griesbachian, Dienerian, Smithian and Spathian. This geological chapter is unfortunately not yet fully settled, since the twofold subdivision of the Induan and Olenekian stages appears rather inadequate. Half a century ago, the geological timescale of Harland et al. (1964) admitted a duration of 47 Ma for the Triassic. At that time Tozer (1967) estimated that the probable duration of an average ammonite species was 1.5 Ma. This led to the overestimation of the Early Triassic (Scythian) time span that, having the highest speciation rate of all Triassic subdivisions, only represents 10% of the total Triassic 52 Ma long time span (e.g. Gradstein et al. 2005). Moreover, following the so-called Late Smithian event that took place less than 1.5 Ma after the PTB mass extinction, Brayard et al. (2017) found that the so-called Paris Biota fulfilled all conditions of full recovery.

After long deliberations in the frame of UNESCO, scientists of many specialities selected the single fossil that determines the base of the Triassic in one chosen locality. They subsequently marked this Global Stratotype Section and Point (GSSP) where this fossil appears in the field with a so-called golden spike, of which Lucas (2018) recently accurately analysed the modalities. The 150-year-long quest by the most eminent specialists to determine that rare zonal beast, supposed to mark the dawn of the Triassic Period, and in this case of the Mesozoic era as well, did not end in the discovery of a large healthy-looking ammonoid from a solid family in full radiation. Instead, the choice fell on a tiny conodont, which was not from the most promising family, Gondolellidae, but a poorly defined variation within a species of the family Anchignathodontidae. Long acquainted to hostile conditions, it also survived the PTB. But the Griesbachian environment, quickly regaining a certain normality of temperature, salinity and sea level, became hostile for this family that soon became extinct. Knell (2012), in "The Great Fossil Enigma, the Search of the Conodont Animal", does not spare his criticism of the metaphorical "Golden Spikes", calling it an artificial scheme.

The passage from Paleozoic to Mesozoic was surrounded by extremely life-hostile palaeoenvironments of lower than normal sea levels and abnormal temperatures. The basins that hosted the transitory sequence were unevenly distributed, many were desert-like and fossils were rare. First came the Himalayas where Waagen and Diener (*in* Von Mojsisovics et al. 1895) compared their Mediterranean with the Indian provinces. They subdivided the Scythian into the Brahmanian and Jakutian stages with the former further divided into the Gangetian and Gandarian substages. These subdivisions alternating between Himalayas and Salt Range were defined by seven pelagic ammonoid zones compared to only one in the Mediterranean Werfen beds. For Shevyrev (2005) there were at least seven regions next to Himalayan Kashmir and Spiti, China (Meishan, Selong) and the high north (Canada, Greenland and Verkhoyansk). The boundary in the northern regions, accommodating an additional Otoceras concavum zone, meant that an ammonoid zone was missing in the rest of the world between the Late Permian Changsingensis zone and the Induan Fissiselatum zone. As a result, it was believed that a continuous P/T boundary section existed nowhere on Earth, not even in the boreal region, where it appeared that the youngest Permian was missing. However, since Zakharov et al. (2014), the Upper Permian Changsinghian stage in Siberia is equivalent to the range of the *Otoceras concavum* zone. Thus, the characteristic $\delta^{13}C$ isotope curve that makes a bulge around 252 Ma appears as the only reliable indication of the PTB level.

Kozur (2006) remarked that Kiparisova and Popov (1956) introduced the Induan and Olenekian stages for the Lower Triassic despite the fact that Von Mojsisovics et al. (1895) had already erected the **Brahmanian** and **Jakutian** stages for this time interval. The Olenekian was defined in the Boreal Realm of Russia, while the Induan stage, named after the Indus River (Salt Range), yields rich ammonoid and conodont faunas characteristic of the Tethyan Realm. In reality, however, the Induan was defined with Boreal *Otoceras* faunas (*O. concavum* zone), as equivalent to the Salt Range. Kiparisova and Popov (1956) assigned the largest part of the Lower Olenekian to the original Induan at the time that the Boreal ammonoids were still unknown. Later, Kiparisova and Popov (1964) removed the lower part of the Olenekian from the original Induan, bringing a drastic change to the scope of the Induan.

5.2.4 Stages

The distribution of taxa is not uniform as faunas depend on conditions of provincial and environmental nature. Zapfe (1983) was of the opinion that there were three main faunal realms during the Triassic: the tropical Tethyan, moderate Canadian and boreal Russian, with few common ammonoids and each with its particular fauna. Subdivisions were either Himalayan threefold, Canadian fourfold and Russian bifold either coexisting or separated. As an example, the Gangetian and Griesbachian were equal as stages, and the Nammalian stage comprised the Dienerian and Smithian as substages. While the Induan stage encompasses the Griesbachian and Dienerian as substages, the Olenekian contains the Smithian and Spathian as substages. For Tozer the Griesbachian, Dienerian, Smithian and Spathian stages replace all preceding subdivisions and should not be downgraded to substages of the ill-defined Induan and Olenekian stages. Thus, it is important to remember that Tozer (1994) established these four stages (Fig. 5.1) after his life-long study of the Early Triassic ammonoids in Canada, going as far as to get the nameless creeks exposing his stages officially named after the Himalayan

Fig. 5.1 Tozer's (1994) Early Triassic stages, based on ammonoids in Canada

SERIES	STAGE	SUBSTAGE	AMMONOID ZONE
LOWER TRIASSIC	SPATHIAN		*Keyserlingites subrobustus*
			Olenikites pilaticus
	SMITHIAN		*Anawasatchites tardus*
			Euflemingites romunduri
			Hedenstroemia hedenstroemi
	DIENERIAN		*Vavilovites sverdrupi*
			Proptychites candidus
	GRIESBACHIAN	U	*Bukkenites strigatus*
			Ophiceras commune
		L	*Otoceras boreale*
....			*Otoceras concavum*

stratigraphy pioneers, Griesbach, Diener, Smith and Spath, recognised as Canadian topographic map entities. While keeping ill-defined old names is obsolete, the introduction of the names "Induan" and "Olenekian" as stages was redundant. As there is no doubt that in order to accommodate facial realms, in the sense of Zapfe (1983), each realm when conflicting with a neighbour is to be defined by typical zones, proper for its conditions, so as for Tethyan versus Russian, or local Spathian zones within North America (Orchard 2007).

Since Harland et al. (1964), the calculated duration for the Lower Triassic has become shorter over the years. Lately, Cohen et al. (2018) published 251.902 ± 0.024 Ma for the PTB, which is less than the 252.6 ± 0.2 Ma for the PTB proposed in Schneebeli-Hermann et al. (2013). These authors adopted 251.22 ± 0.2 Ma for the Dienerian-Smithian boundary and 250.55 Ma for the Early Spathian (Paris Biota of Brayard et al. 2017).

5.2.5 Environmental Changes

Zakharov (1974) subdivided the palaeobiogeography of the Early Triassic using water temperature where boreal freshened waters averaged 14.5 °C versus 23–27 °C for normal Tethyan saline waters. More recently, Garbelli et al. (2015) analysed Tethyan seawater chemistry and temperature from the dawn to the end of the Permian mass extinction, whereas Petsios et al. (2017) studied the impact of temperature across the end-Permian extinction modelling extinction and recovery dynamics. Oxygen being essential for animal life, geochemical proxies are instrumental in determining the broad evolutionary history of oxygen on Earth. Showing the stepwise oxygenation of the Paleozoic atmosphere, Krause et al. (2018) produced a new model that is consistent with available proxy data and independently supports a Paleozoic Oxygenation Event, which likely contributed to the observed radiation of a complex, diverse fauna at this time. Silva-Tamayo et al. (2018) studied the global perturbation of the marine calcium cycle during the Permian-Triassic transition finding a globally distributed negative, globally distributed Changsinghian–Griesbachian $\delta^{44/40}$Ca anomaly in carbonate successions.

Payne and Van de Schootbrugge (2007) produced the geochemical tools that provide the reasons for the coming and going of organisms; among them are the crucial criteria for the appearance or absence of reef organisms. At the base of the Griesbachian, not only a substantial δ^{13}C excursion occurs, but also $^{87/86}$Sr and δ^{18}O trends suggest a catastrophic temperature rise, the impact of which inflicted small size on the conodonts. Schaal (2014) found strontium isotopic constraints on Permian-Triassic global change with a new high-resolution seawater $^{87/86}$Sr record and a numerical model of the strontium cycle. Strontium isotope data reveal that a rapid radiogenic excursion occurred during the first 2 million years of Early Triassic time. Model results show that the magnitude of CO_2 release during Siberian Traps volcanism is sufficient to account for much of the observed increase in seawater $^{87/86}$Sr through CO_2 enhancement of continental weathering rates. The small size of

Early Triassic marine organisms has important implications for the ecological and environmental pressures operating during and after the end-Permian mass extinction. Leu et al. (2018), investigating the size response in conodonts to environmental changes during the late Smithian extinction, found that body size reduction is a common evolutionary response to heavy environmental stress. These changes are not fortuitous as they obey trends that can be seen to be repetitive over time.

For Raup (in Barash 2006), the PTB caused a 52% drop in diversity. Carbon cycle perturbations in the organic carbon isotope chemo-stratigraphy are statistically indistinguishable, despite great variations of sedimentation rates (Retallack 2013). Since Baud and Magaritz (1989), isotopes have added a totally new dimension, that of the climate.

In the aftermath of the end-Permian mass extinction, some of the largest Phanerozoic carbon isotopic excursions are recorded in the Early Triassic Thaynes Group of Utah, USA (Thomazo et al. 2016). Among these a global Smithian-negative carbon isotope excursion has been identified, followed by an abrupt increase across the Smithian-Spathian boundary (SSB; ~250.8 Ma ago). This chemo-stratigraphic evolution is associated with palaeontological evidence that indicates a major collapse of terrestrial and marine ecosystems during the Late Smithian. It is commonly assumed that Smithian and Spathian isotopic variations are intimately linked to major perturbations in the exogenic carbon reservoir. Using oxygen isotope compositions of conodont apatite ($\delta^{18}O_{phos}$) Trotter et al. (2015) found three major warming cycles in the Sicani basin. The oxygen isotope ($\delta^{18}O$) record derived from conodont apatite reveals variable long-term climate trends throughout the Triassic Period. This record shows several major first-order negative shifts reflecting intense warming episodes, not only the well-known extreme PTB-Early Triassic event (~5 ‰), but also two large cycles of similar magnitude (~1.5, ~1.7 ‰) and duration (~7 Ma) during the Late Carnian and late Norian. Between the PTB-Early Triassic and Carnian major episodes, three rapid shorter term warming events of decreasing magnitude punctuate the Mid-Late Anisian, Early Ladinian and latest Ladinian with distinct cooler (i.e. favourable) intervals characterising the early Anisian and early Carnian, indicating a fluctuating but ameliorating Middle Triassic climate trend. Two long periods of sustained cooler conditions occurred during the Late Triassic for much of the Norian and Rhaetian. The five humid events previously recognised from the geological record, including the Carnian Pluvial Episode, are associated with the low $\delta^{18}O$ warming phases. The magnitudes of these first-order warming cycles together with widespread geological and palaeontological evidence suggest that they were at least Tethyan-wide events.

Korngreen and Zilberman (2017) summarised $\delta^{18}O_{carb}$excursions during the P-T transition and considered the Early Triassic as the hottest record of marine temperatures (~40 °C). Sun et al. (2012) (see also Payne and Clapham 2012) associated these with the extension of the desert belts into the mid latitudes and the palaeo-mid-high latitudes (50–55°), characterised by temperate/warm multi-exchanges of climates (Péron et al. 2005; Sellwood and Valdes 2006).

Accordingly, **two warming** and **two cooling**periods took place during the Late Permian–Early Triassic interval in intra-basin sediments analysed for $\delta^{18}O$ excursions (Sun et al. 2012).

(I) **First warming period:** Interpretation of $\delta^{18}O_{carb}$ profiles revealed during the Late Permian, PTB and Griesbachian indicated, by continued worldwide decline in values, a drop of 1.5‰ up to 5‰ (Holser et al. 1989; Heydari et al. 2000; Rampino and Eshet 2017; Haas et al. 2007; Horacek et al. 2009; Richoz et al. 2010; Joachimski et al. 2012; Chen et al. 2013a, b, 2015).

(II) **First cooling:** Using $\delta^{18}O_{carb}$ profiles, cooling developed during the uppermost Griesbachian and persisted during the entire Dienerian (Haas et al. 2007; Richoz et al. 2010).

(III) **Second warming:** Began at the Dienerian/Smithian boundary, lasted throughout the entire Smithian and reached its peak at the Smithian/Spathian boundary (Galfetti et al. 2007) that Sun et al. (2012) called the "Late Smithian Thermal Maximum".

(IV) **Second cooling period:** Recognised by Sun et al. (2012) and Romano et al. (2013) in the early Spathian. This was followed by climate stabilisation during most of the Spathian (see also Galfetti et al. 2007; Hochuli et al. 2010; Stefani et al. 2010; Hermann et al. 2011). The uppermost Spathian was considered by Sun et al. (2012) as the third cooling event, retaining a relatively cool climate through the early Anisian (Aegean-Bithynian).

In three southern Israel boreholes, Korngreen and Zilberman (2017) found the $\delta^{18}O$ profiles to exhibit values in the –5 to –7.5‰ average range in the Levant region, also typical of Western Tethys. The described changes in the major sources of terrestrial material to the three successions require climate changes that would expand the precipitation zone further to the southern hinterland and conversely. The succession was recorded in the general southward extension of the precipitation zone during the Late Permian (hinterland humidification), northward contraction (hinterland aridisation) from the lattermost Late Permian through the most of the Early Triassic, then re-expanding gradually during the late Early Triassic, reaching a peak during the mid-Anisian (hinterland re-humidification).

The tropical (~10°S) location of the studied successions attempts to attribute the changes of the precipitation zone to the long-term Intertropical Convergence Zone (ITCZ) expansion/contraction that severely affected the precipitation, erosion and vegetation type in the low-latitude hinterlands.

The $\delta^{18}O$ Profiles in the Levant Region The $\delta^{18}O$ composite profile (Fig. 5.2) is based on event correlations: Comparison of the three $\delta^{18}O$ profiles of the studied successions exhibiting similar trends of decrease/increase trends correlated well with worldwide climate change events: two warming phases at the P-T transition, the first during the Late Permian to the PTB and the second at the Dienerian/Smithian boundary and through the Smithian. An additional warming period is related to the early Middle Triassic (Early Anisian; EAE) warming towards the Middle Anisian (MAE) and after. A cooling period characterises the Late Spathian

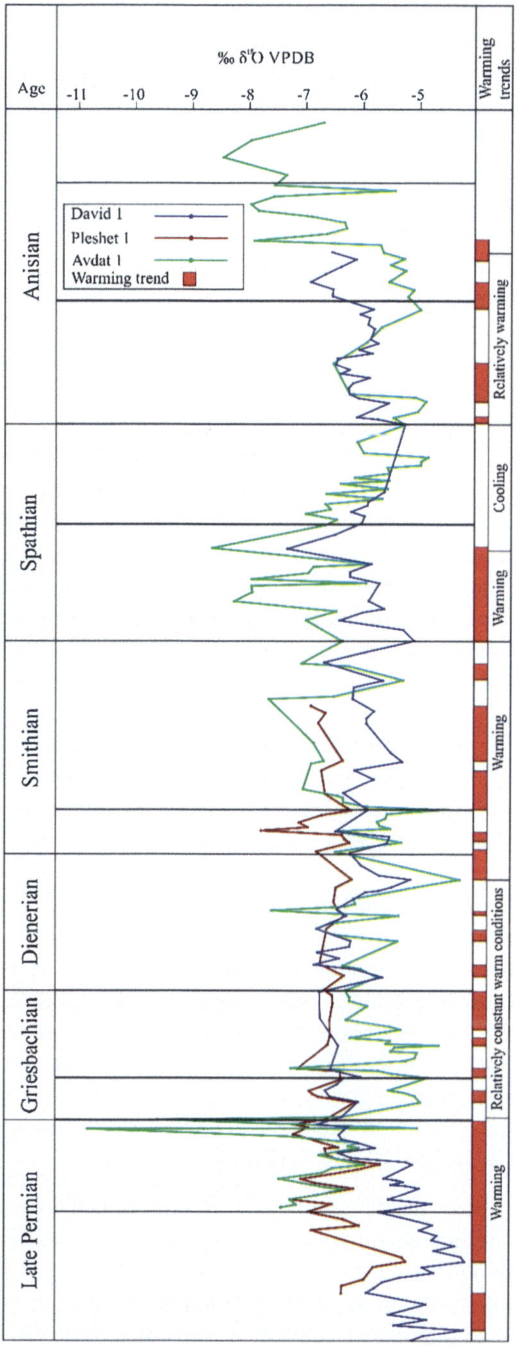

Fig. 5.2 Composite $\delta^{18}O$ profiles of three boreholes in southern Israel. The average range of values (–5 to –7.5‰) is typical of western Tethys climate trends. There are three major warming periods: (I) Late Permian to the PTB; (III) Late Dienerian—most of the Smithian; (V) Early-Middle Anisian, and the two relatively cool periods: (II) Griesbachian-Dienerian and (IV) Late Smithian-Spathian. Each of the periods may exhibit short respites with the opposite trend (after Korngreen and Zilberman 2017)

(LSE-SAB). Note the relatively extreme negative and positive values of the proximal succession and the abrupt shifts.

Late Permian-Early Triassic Negative Shifts $\delta^{18}O$ values generally vary between ~–4.0 and –8.5‰ and are much higher than those of Meishan GSSP and other sections from China with –9.5 to –13‰, 0–20°N (after Joachimski et al. 2012; Sun et al. 2012). However, the range of $\delta^{18}O$ values is highly compatible with those from Neotethyan successions (equatorial to 10° S, 25–30° S; Haas et al. 2007; Richoz et al. 2010), and is also compatible with the lighter part surrounded by land Muschelkalk sea range (–6.2 to –2‰, Korte et al. 2005). They are also lighter than marine carbonates deposited in modern equatorial range (0 to –3‰, Carpenter and Lohmann 1995). The two negative shifts of about 1–2‰ were viewed during the Late Permian, followed by a sharp and distinct low $\delta^{18}O$ peak at the PTB with an amplitude of up to –3‰, reaching a $\delta^{18}O$ value of ~–11‰. These were attributed to a significant warming trend and may be assigned to the first recognised extreme global warming of the Paleozoic-Mesozoic transition (Holser et al. 1989; Heydari et al. 2000; Haas et al. 2007; Horacek et al. 2009; Richoz et al. 2010; Joachimski et al. 2012; Sun et al. 2012). Some stabilisation of the $\delta^{18}O$ values associated with relatively short warming and cooling changes along the Griesbachian and most of the Dienerian, attributing the changes to a relatively cooling period recognised in this time interval in other parts of the world (Haas et al. 2007; Richoz et al. 2010; Sun et al. 2012), is defined as relatively constant warm conditions (Fig. 5.2).

Dienerian-Smithian Boundary The next noticeable negative shifts occurred near the Dienerian-Smithian boundary and throughout the Smithian. Considered as a warming period, the Smithian was assigned to an equivalent global warming event (Galfetti et al. 2007; Sun et al. 2012).

Smithian-Spathian Warming Event (SSB) For Chen et al. (2013a, b), recovery from the end-Permian mass extinction took several million years. This was partly due to an inhospitable environment and the three episodes of further extinction occurred in late Griesbachian near the Smithian-Spathian boundary (SSB) and in the late Spathian. A short, but **extreme warming** event precedes the Smithian-Spathian boundary. *Mercury concentrations are possibly connected to the SSB volcanism of the Siberian Traps* (after Hammer et al. 2019). However, shortly after the Late Smithian event full recovery had been achieved as shown by the Paris Biota (Brayard et al. 2017) (Fig. 5.3).

Base sea levels and $\delta^{13}C$ trends control a succession of nine conodont events: (**1**) Gradual decline from five families in the Changsinghian to three by the PTB, and to two in the Latest Griesbachian. Single neogondolellin multielement apparatus during this interval. (**2**) PTB: **Siberian Traps** caused **anoxia** and global **cooling**; Boreal stocks replaced Tethyan warm-water ones. (**3**) Late Griesbachian ***$\delta^{13}C$ minimum,*** *global* ***warming***, local regression; ***extinction*** of Anchignathodontidae; cool-water

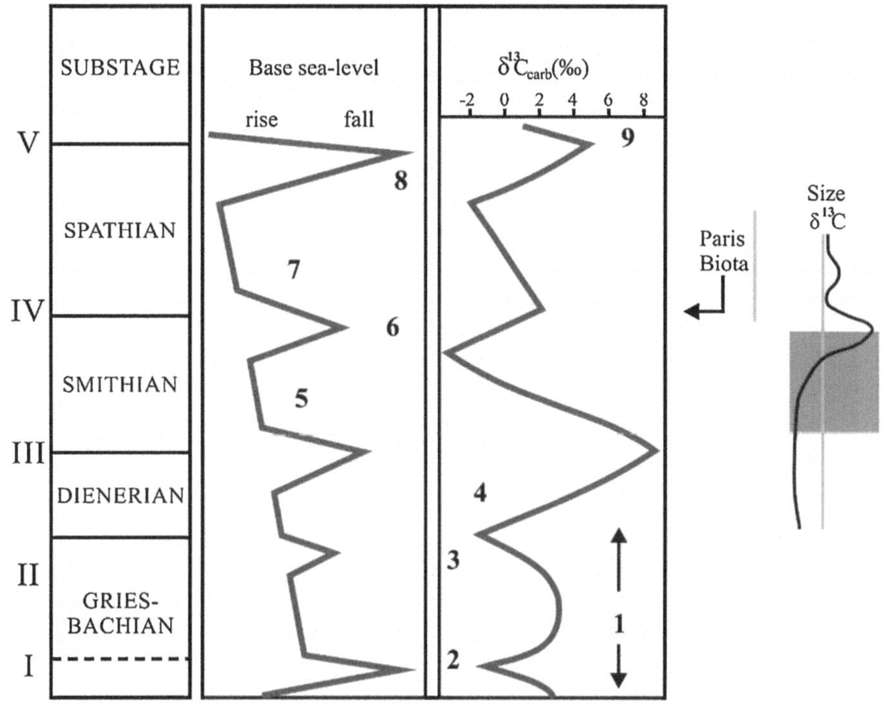

Fig. 5.3 Early Triassic Isotopic Events (modified after Orchard 2007). (I) Late Permian-Early Triassic boundary; (II) Late Griesbachian; (III) Dienerian-Smithian boundary; (IV) Smithian-Spathian boundary; (V) Middle Spathian-Early Aegean interval

segminiplanate Neogondolellinae ***retrograde*** and new ones ***initialise.*** (**4**) Dienerian ***radiation*** *of segminate proteromorphs (Neospathodus, Kashmirella)*, reinitialisations concurrent with returning normal marine conditions. (**5**) Major Basal Smithian **transgression**: Early-Middle Smithian explosive **radiation**, Triassic's strongest. (**6**) Late Smithian ***$\delta^{13}C$ minimum: low-stand****and possible renewed* ***volcanism****; biggest Triassic* ***extinction****, affecting more taxa than both PTB and final end-Triassic extinctions. Mercury concentrations are possibly connected to the SSB volcanism of the Siberian Traps* (after Hammer et al. 2019). (**7**) Early Spathian return to **normal marine** conditions, significant **radiation** and rapid **recovery** (so-called Paris Biota of Brayard et al. 2017). After this radiation, the total number of Spathian gondolellid apparatuses equalled those of the Smithian (Orchard 2007). (**8**) Late Spathian **low stand** corresponds to a gradual **decline** in diversity. Reduction and disappearance of *proteromorph* stocks with reinitialisation of segminiplanate gondolellids. (**9**) Uppermost Spathian, large **positive** carbon isotope excursion, starting in Upper Spathian and peaking with Early Anisian major **transgression.** *Conodont size variation, especially during Late Smithian* (after Leu et al. 2018).

Detailed measurements of size variation in conodonts such as *Neospathodus* show that the clade suffered significant **size reduction** during the SSB crisis, followed by gradual and steady size increase during the early Spathian.

De Wever et al. (2007) observed that radiolarian biodiversity increased constantly during the Paleozoic and decreased only slightly towards the end of the Permian. Distinctive for the PT event is not the diminishing number of taxa, but the importance of the Triassic diversification that followed. As under anoxic environments radiolarian diversity minima relate to positive $\delta^{13}C$ excursions, the end-Permian diversity drops and extinction seems linked to a decrease in $\delta^{13}C$ values. About half of the genera that disappeared at the PTB reappeared from the Spathian onwards. The Lazarus effect invoked by Wignall and Benton (1999) seems more probably to be brought by proteromorphosis (sensu Guex 2001).

The second cooling period of the Early Triassic was recognised by Sun et al. (2012) and Romano et al. (2013) in the early Spathian and was followed by climate stabilisation for the remainder of the Spathian (see also Galfetti et al. 2007; Hochuli et al. 2010; Stefani et al. 2010; Hermann et al. 2011).

Middle-Late Spathian Boundary The second positive shift during the Early Triassic appeared in the middle and late Spathian up to the earliest early Anisian. This global **cooling** event is recognised worldwide (Galfetti et al. 2007; Hochuli et al. 2010; Stefani et al. 2010; Hermann et al. 2011; Sun et al. 2012; and Romano et al. 2013).

Uppermost Spathian The uppermost Spathian is considered by Sun et al. (2012) as a **third cooling event**.

Anisian The relatively cool climate of the early Middle Triassic Anisian (Aegean-Bithynian) encompasses a period of relative warming that may be related to the Anisian aridisation events. These two noticeable negative shifts should be assigned to the Aegean warming and the mid-Pelsonian event (Korngreen and Bialik 2015), separated by the Bithynian-Early Pelsonian **humid** event (Stefani et al. 2010).

5.2.6 Stratification of Ocean Water

In the Lower Triassic $\delta^{13}O$ isotope curve of Horacek et al. (2009), the Kamura shallow-marine carbonates in Japan (Panthalassa realm) confirm the Tethys curve. These global carbon isotope curves document the profound changes in the global carbon cycle that influenced ecosystems. For these authors, an alleged delay in biological and ecological recovery from the PT extinction event is the result of the stratification of the Tethys and episodical mixing of the stratified ocean waters. These are the most likely processes to produce the observed carbon isotope signatures and seriously hamper a substantial biotic recovery.

5.2.7 Environmental Gaps

A number of gaps affect the Early Triassic.

The Chert Gap The Early Triassic Chert Gap occurred following the 30 Ma long worldwide Permian Chert Event (Murchey and Jones 1992; Beauchamp and Baud 2002). As a result, radiolarian oozes are absent in Early Triassic oceans. Radiolarian biodiversity increased constantly during the Paleozoic and decreased only slightly towards the end of the Permian. Under anoxic environments radiolarian diversity minima relate to positive $\delta^{13}C$ excursions. The end-Permian diversity drops and extinction links to a decrease in $\delta^{13}C$ values. About half of the genera that vanished with the PTB reappeared during Spathian times. Invoked as Lazarus effect (Wignall and Benton 1999), it was rather an example of proteromorphosis (sensu Guex 2001) [after De Wever et al. 2007]. A reduced chert gap is apparently present in the Mino Terrane of Japan, where deepwater chert and claystone yield the conodont genus *Hindeodus*, preserved as natural assemblages on bedding planes in claystone (Agematsu et al. 2014). These *Hindeodus* assemblages maintain the original composition and structure of an almost complete apparatus: element pairs of P_1, P_2, M, two digyrate and four bipennate of the S array, and the single alate S_0. The conodont biostratigraphy indicates that the lithological boundary between chert and claystone units in the study section corresponds exactly to the Permian-Triassic boundary. The Early Triassic chert gap seems to have caused anoxia, ocean warming and ocean acidification. One basic effect of the Traps volcanism coupled with the release of substantial amounts of hydrate methane was to slow or stop normal thermohaline circulation changing ocean bottom water from oxygenated to anoxic. Radiolarians would have been seriously affected by oceanic acidification because as ocean water became more acidic, less silica was available. Until cold, oxygenated and more alkaline conditions returned, chert production and deposition would have ceased.

The Reef Gap In the Late Permian, reefs were thriving. In a geologic blink of the eye, the Permian reefs disappeared and were not replaced immediately. The organisms that formed them went extinct, among them rugose corals, fusulinids, many echinoderms, bryozoans and brachiopods. For about 5 Ma years the only few carbonate build-ups were composed of calcareous algae with no larger organisms.

Reef ecosystems are vulnerable to even very small changes in temperature. Oxygen content of the water, its salinity, slight alkalinity (normal ocean pH is 8.1), clarity and nutrient supply are generally quite constant. Each of these conditions can be affected by protracted volcanism and hydrate methane release. The waters would have risen and become dysoxic, and the relative acidity (pH) changed by acid rain. The oceanic nutrient supply would have been enhanced by episodic volcanic ash falls and continuous weathering of continental rock by acid rain, but the upwelling of nutrients would have been much diminished in a stagnant ocean. Two major groups of corals, the rugose corals and the tabulate corals, disappeared forever. In part, this may reflect the need of corals to build calcium carbonate skeletons, which would have been difficult to fulfil as ocean waters became more acidic. In addition, many corals contain photosynthetic algal symbionts that can be ejected during periods of reefal stress. While such ejection may help a coral deal with stress for a short term, in the long term it deprives the coral of a major food source. Photo-synthesisers also contributed oxygen that corals may have been deprived of in increasingly dysoxic waters. The reef gap was apparently filled by microbial build-ups (Friesenbichler et al. 2018; Heindel et al. 2018).

The Coal Gap Extensive coal deposits extend over Gondwanian. Their absence for the 5 Ma years of the Early Triassic is known as the coal gap (Veevers 1994; Retallack et al. 1996). Peat and coal ceased completely during the Early Triassic as the result of the disappearance of peat-forming plants in the end-Permian extinction. Relatively few plants are adapted to the harsh conditions of waterlogged, acidic, dysaerobic environments in which peat is formed. The end-Permian extinctions decimated these plants and it took millions of years for other plants to become tolerant of such conditions (Retallack et al. 1996).

One considerable impact on terrestrial vegetation is acid rain. Protracted episodes of severe acidity triggered by pulses of hydrate methane release from seafloor slumps or pulses of volcanic eruption over tens to hundreds of thousands of years, coupled with increased precipitation, would have been quite sufficient to cause the extinction of many species of vegetation, including those whose peaty and humified remains became coal. More generally, mycorrhizal fungi, the root symbionts that provide most terrestrial plants with soil nutrients, may have been particularly vulnerable to this altered environment. Increased warmth, lower levels of atmospheric oxygen (hypoxia) and elevated levels of carbon dioxide may also have added to the demise of plants.

The extinction of many peat-forming plants at the end of the Permian, and their slow replacement during the Middle Triassic, has thus been cited as the reason for the coal gap (Retallack et al. 1996). The areas previously covered with forests were replaced by red beds indicating arid conditions. Early Triassic red beds followed the end of Permian throughout much of the world, such as the Moenkopi Formation in North America and the Buntsandstein in Europe.

Changes in Carbon Isotopes at and After the Permian-Triassic Boundary Following an initial move to negative values at the Permian-Triassic boundary, carbon isotopes swung wildly for roughly 5 million years indicating enormous changes in oceanic chemistry. Aside from confirming that the Early Triassic was a period of continual major instability, the findings provide no insight into the possible causes of the disturbances. As the discoverers of the carbon fluctuations note, these could represent either repeated environmental disturbances or their consequences, as ecosystems were decimated and rebuilt (Payne et al. 2004). In other words, carbon isotope changes could be telling us either of causes (for example, methane outbursts) or of consequences (for example, severe disturbances of the biosphere) [after Dorritie 2004, 2005, 2006, 2007].

5.2.8 Methane

Recently Rothman et al. (2014) noted that for millions of years prior to the Permian extinction, the world had been in one of its most biologically productive time frames. By 260 million years ago, the level of atmospheric oxygen soared allowing animals to thrive. Yet at the same time, the level of carbon dioxide was still higher than we find today and consequently plants thrived as well. All terrestrial leaf litter and organic waste and all ocean plankton that died carried enormous quantities of energy-rich compounds to the deepest ocean bottoms, and due to temperatures, which were globally very high, there was little ocean circulation. Thus, little wind, few Gulf Stream-like currents and little motion carrying cold, oxygen-rich surface water down to the deep sea took place. Thus, enormous amounts of energy-rich dead bodies from the oxygen-rich world fell onto oxygen-deficient sea bottoms. Simultaneously they became surrounded by the organic compound acetate that prevented the microbes from consuming the dead bodies that filled the ocean bottom. This resulted in deep-sea microbes that exhale methane instead of carbon dioxide, called "methanogens", to capture two acetate-processing genes from a totally different kind of microbe. That process of gene capture is an epigenetic event called "capture" by Rothman's MIT team. The results were deadly to all organisms that needed oxygen, and all organisms that died at sustained temperatures at or above about 35 °C, which are mostly multicellular (Ward 2018; p.144).

5.3 Part Two

5.3.1 Phylogenetic Reconstructions

Hirsch (1994a, b) proposed that under lethal stress conditions beginning in the Dienerian, neogondolellin segminiplanate morphs (platforms) were replaced by segminate (cavital) proteromorphs (*Neospathodus*). After the Dienerian-Smithian boundary reinitialisations of neogondolellin homeomorphs appeared as successive peramorphic trends (Fig. 5.4).

Budurov et al. (1988) introduced a parallel *Kashmirella* lineage, while Kozur et al. (1998), paying attention to the rapid change from neospathodids to gondolellids and vice versa, established the genera *Sweetospathodus* and *Triassospathodus* without significant justification. Considering the taxonomy of Permian and Triassic gondolellids in detail, Kozur (1989) considered that *Neogondolella* Bender and Stoppel had evolved from segminate *Neospathodus* Mosher through the transitional forms of *Chiosella* Kozur from the end of the Olenekian stage to the beginning of the Anisian stage and that Neogondolella was the basic group of the Middle to the beginning of the Late Triassic. Klets and Kopylova (2007) illustrated the replacement of *Clarkina* by *Neospathodus,* the latter generating up to four lineages of platform-bearing "gondolellids". These authors also wrote that endemic

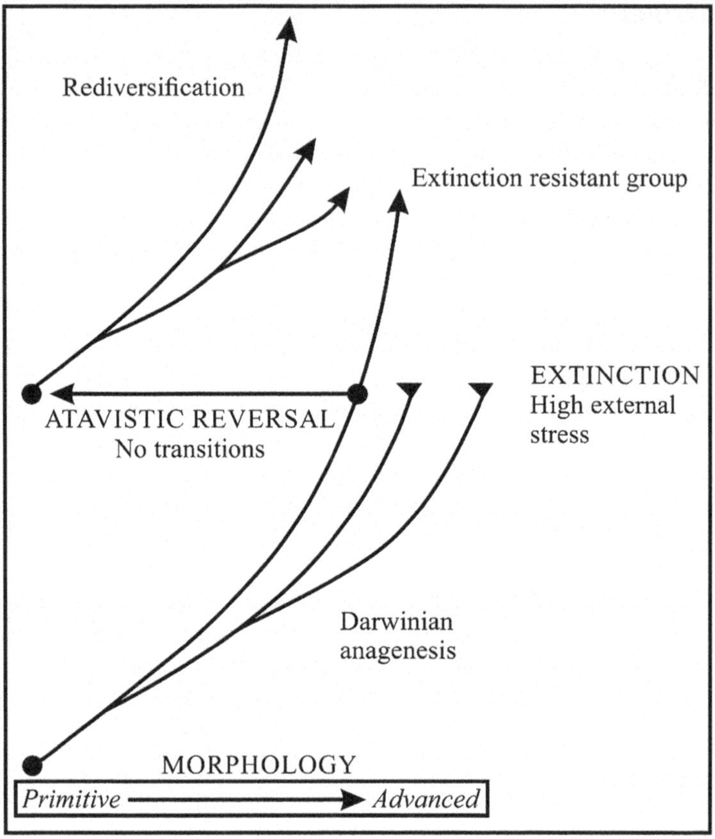

Fig. 5.4 Alternation of morphologies. Darwinian anagenesis undergoes extinction under high external stress; atavistic reversal permits rediversification (after Guex 2001)

"Neogondolella" buurensis, "N." composita, "N." jakutensis, "N." taimyrensis and *"N." sibirica*, having characteristics of *Neogondolella*, are widely distributed in the Early Olenekian of the northern latitudes.

Orchard's (2007) temporal distribution of Lower Triassic conodonts differentiated families and subfamilies, such as the subfamily Neogondolellinae that, having survived the PTB, added a number of Griesbachian species to the lineage before radiating upwards into the Spathian. Next appeared *Borinella* (Dienerian-Spathian) and the non-platform *Sweetospathodus* (Dienerian). The latter radiated into the Smithian subfamily Mullerinae that comprises the genera *Discretella, Conservatella, Guangxidella* and *Wapitiodus.* New is the appearance of subfamily Scythogondolellinae (Dienerian-Smithian). Spathian "*Neogondolella*" morphs comprised species such as "*regalis*", *shevyrevi, jubata* and derivatives like the genus *Columbitella.* Orchard (2007) also proposed *Borinella* as the origin for the genera *Gladigondolella* and *Cratognathus.* For this author, the Dienerian apparition of the segminate genus *Neospathodus* (Dienerian-Spathian) radiated into the non-platform subfamilies Cornudininae (Smithian-Spathian) and Novispathodinae (Spathian) and the genera *Eurygnathodus* (Smithian), *Icriospathodus* (Spathian), *Platyvillosus* (Spathian), *Triassospathodus* (Spathian) and *Kashmirella* (Spathian).

Family Ellisoniidae encompasses the genera *Ellisonia, Hadrodontina, Merillina, Parachirognathus, Foliella, Sweetocristatus* and *Pachycladina.*

Except Orchards' (Orchard 2007) mention of *Scythogondolella mosheri* as a paedomorphic descendant of *Scythogondolella* n. sp. in the Canadian Arctic, all phylogenies are exclusively peramorphic for him.

5.3.2 Conodont Radiation

Three peramorphic trends survived the PTB: Anchignathodontidae that became extinct by the Late Griesbachian $\delta^{13}C$ minimum; Ellisoniidae that remained dormant until the dawn of the Smithian, when they radiated "until before" the end of the Spathian; and Neogondolellinae that in the course of the Triassic secured the survival of their lineage with atavistic reversals (proteromorphosis).

(A) Family Anchignathodontidae Clark (After Brosse et al. 2015)

The genus *Hindeodus* (Rexroad and Furnish 1964), type species *H. typicalis* (Sweet), is widely variable and several species evolved. Conodonts belonging to this genus rapidly differentiated into more than ten species during latest Permian to earliest Triassic time (Kozur 1996, 2004; Angiolini et al. 1998; Orchard and Krystyn 1998; Nicoll et al. 2002; Perri and Farabegoli 2003; Orchard 2007; Chen et al. 2009; Metcalfe 2012). *H. typicalis* has an apparatus that according to Sweet (in Clark et al. 1981) consists basically of all the elements of *Ellisonia teicherti* (Sweet 1970). Recent studies on clusters of *Hindeodus* have identified an octomembrate 15-element apparatus of P_1, P_2, S_0, S_1, S_2, S_3, S_4 and M elements (Agematsu et al. 2014). For Sweet (1992), the short-living *Isarcicella isarcica* as well as *Hindeodus parvus* are morphologically different elements of a single species.

(B) Family Ellisoniidae Mueller

The genus *Ellisonia* is confined to shallow environments in the Griesbachian and Dienerian but during the Smithian, the family underwent the outburst of the genera *Furnishius, Hadrodontina* and *Pachycladina,* each with several species. Their differentiated distributions define several subprovinces from North America to the Alpine and Levantine Tethys. This Smithian conodont outburst may correspond to a peak of apatite, the mineral of the conodont elements. Smithian Ellisoniidae and Mullerinae both mark the phenomenon with large size and robust elements.

(C) Family Gondolellidae Lindstroem

It is widely agreed that classification within the family is based on the composition of the multi-element apparatus. This is critical for establishing its identity as a member within a subfamily, as in the case of *Clarkina* and *Neospathodus* within the subfamily Neogondolellinae.

We have to consider two evolutionary morphotypes: anagenetic and atavistic (retrograding).

Subfamily Neogondollelinae Hirsch
The succession of Late Permian anagenetic "gondolellid" morphs comprises the genera *Mesogondolella, Jinogondolella* and *Clarkina.* These form lineages are separated from each other by retrograded morphs. Meso- and jinogondolellin forms have a different apparatuses and they belong in different subfamilies. The first neogondolellin genus is *Clarkina.* Orchard (2007; p. 16) observed its Early Triassic explosive radiation consisting of the sudden appearance of numerous new species and genera. Most remarkable for the newly emerged conodonts was the morphological diversity of at least 12 multielement apparatuses within Lower Triassic gondolellids alone, compared with perhaps one in the Late Permian. *The following taxa occur according to the observed events.*

Since the Late Permian-Early Triassic Boundary, the Early Triassic part of the radiation of genus ***Clarkina*** (Kozur 1989) (*Neogondolella* in Orchard 2007, 2008) counts a dozen Griesbachian species, most of them Permian survivors (Orchard 2008). The evolution of this segminiplanate genus in the Meishanensis zone consists of five species ranging upward from the preceding Iranica zone and six new species that appeared, all of which—*Clarkina meishanensis* Zhang et al., *C. orchardi* Mei, *C. kazi* (Orchard), *C. nassichuki* (Orchard), *C. taylorae* (Orchard) and *C. tulongensis* Tian—ranged up into the Induan of Tibet and Spiti (Orchard 2007; Orchard et al. 1994; Orchard and Krystyn 1998) followed by Triassic ones. These Triassic *Clarkina krystyni* and *C. discreta* are believed to have radiated into the Dienerian as *Clarkina griesbachensis* at the starting point of the Dienerian, followed by *C. mongeri* and the Smithian *C. composita* that radiated into *C. altera, C. jakutensis* and *C. sibirica.* For Orchard (2008), also the appearance of the genus *Borinella* is presented as the result of Darwinian anagenesis.

The major innovation that took place in the latest Griesbachian *Strigatus*zone was that of the genus *Neospathodus* (Mosher 1968), as noted by Orchard (2008). The remarkable survival of the neogondolellin apparatus was enabled by the succession of the large number of retromorphs, known as *Neospathodus* or other generic names.

The type species of the genus ***Neospathodus*** (Mosher 1968) is *N. cristagalli* that has a multi-element apparatus like that of *Neogondolella* (Orchard 2005). The appearance of *Neospathodus* as the result of environmental stress at the end of the Griesbachian (Hirsch 1994a, b) appears as a typical example of retrograde evolution (Guex 2016; Kiliç et al. 2016). The latter authors proposed *C. krystyni* as possible origin.

Several genera have been detached from the genus *Neospathodus* based on slightly different morphologies of the P_1 as well as the apparatus, when available. Among these, Kozur et al. (1998) separated *Triassospathodus* from *Spathognathodus homeri* (Bender 1970) (alias *Neospathodus homeri*, Bender 1970). This separation from *Neospathodus* was based on the down-turning of the posterior lower margin of the P_1 element, while for Orchard (2005), the elements S_2–S_4 and S_0 significantly differ and the S_2 element is particularly distinctive. According to Shigeta et al. (2014), *Triassospathodus homeri* evolved from *T. symmetricus* in the Early Spathian. *T. homeri* extends from Chios and N. Dobrogea to Oman, Salt Range, Kashmir, China and Japan. For Budurov et al. (1988), the *Neospathodus* lineage comprises *Neospathodus cristagalli* (Huckriede), *N. dieneri* Sweet, *N. pakistanensis* Sweet, *N. waageni* Sweet, *N. conservativus* (Mueller), *N. zarnikovi* Buryi, *N. triangularis* (Bender) and *N. homeri* (Bender), *N. discretus* (Mueller), *N. conservativus* (Mueller), *N. zarnikovi* Buryi and *N. bransoni* (Mueller).

Seen as unrelated *Neospathodus*homeomorphs by Orchard (1995), Budurov et al. (1988) saw the lineage of ***Kashmirella*** Budurov, Sudar and Gupta, 1988, that ranges from latest Griesbachian to Early Aegean as prevailing during the Early Triassic next to that of the genus *Neospathodus.*

The *Kashmirella* lineage includes the Latest Griesbachian-Dienerian *K. kummeli* (Sweet 1970), the Smithian type species *K. albertii* (Budurov et al. (1988)), the Smithian *K. novaehollandiae* (McTavish 1973), the Spathian *K. spathi* (Sweet 1970), the Spathian *K. zaksi* (Buryi 1979) and the Spathian *K. timorensis* (junior synonym is *Chiosella*, Kozur 1989), often seen as the senior synonym of *K. gondolelloides* (Bender) from which the genus *Paragondolella* evolved. Characteristic in the type species *Kashmirella alberti* is a thin blade in the upper part and a lateral thickening in the lower part, similar to a platform; its broad basal cavity resembles that of *Metapolygnathus* Hayashi and its derivates, and may also be bifurcated. The genus *Kashmirella* extends to the Tethys (including Romania, Oman, Himalaya, South China, Timor), North America (Nevada, SE Alaska, W. Canada), the Arctic (Svalbard) and Australia. Budurov et al. (1988) saw the lineage of *Kashmirella* as prevailing during the Early Triassic next to that of the genus *Neospathodus.* Thus, after the several atavistic retrogradations that were initiated in the latest Griesbachian by the segminate *Neospathodus* (*cristagalli, dieneri*) and *Kashmirella* (*kummeli*), the apparatus of subfamily Neogondolellinae was preserved in the lineage as *Paragondolella* (Latest Spathian-Julian) as well as far beyond in *Norigondolella* (Norian), but not in Epigondolellinae that lacks the bifid S_3.

The genus ***Eurygnathodus*** Staesche, 1974, type species *Eurygnathodus costatus* for *Platyvillosus costatus* Staesche 1964: The morphologically very innovative taxon has an ellipsoid platform with an ornamentation of transversal ridges, and its sub-rounded anterior basal groove is narrow. *Eurygnathodus costatus* has a very large geographical distribution: North Italy (Staesche 1964); Croatia (Aljinović et al. 2006); Western Serbia (Budurov and Pantić 1973); Kashmir and Spiti, India (Krystyn et al. 2007); South Primorye, Northeastern Asia (Igo 2009); Southwest Japan (Koike 1988); Northeastern Vietnam (Maekawa and Komatsu 2014); Northwest and Western Malaysia (Koike 1982); Nevada, USA (Sweet et al. 1971); and British Columbia, Canada (Beyers and Orchard 1991). In South China, it was reported from western Hubei Province (Wang and Cao 1981; Zhao et al. 2013), Chaohu of Anhui Province (Zhao et al. 2008), Sichuan Province (Tian et al. 1983), Guizhou Province (Wang et al. 2005; Chen et al. 2015), Guangxi Province (Yang et al. 1986), Lowest Middle Smithian *Flemingites rursiradiatus* in Vietnam (Shigeta et al. 2014), South Tirol (Staesche 1964), Spiti, Kashmir, South China, Malaysia, SW Japan, BC Canada (*milleri* zone), S. Primorye (*dieneri-pakistanensis, waageni-novaehollandiae* zones (Upper Dienerian-Smithian)) (Zhang 1990).

The genus ***Borinella*** (Budurov and Sudar 1994), type species "*Neogondolella*" *buurensis* Dagys, comprises a variety of morphotypes. The Dienerian form described as *B. chowadensis* has essentially the same apparatus as *Neogondolella* (Orchard 2007). However, no apparatus was described for *Borinella nepalensis*, *B. buurensis* nor *B. megacuspa*. A number of morphologically related genera are presently combined under this genus. *Borinella* (Budurov and Sudar 1994) replaces *Kozurella* (Budurov and Sudar 1993) (*Pseudogondolella* sensu Kozur 1988, p. 244; *Chengyuania*, Kozur 1994, p. 529–530), the type species of which is "*Gondolella*" *nepalensis* (Kozur and Mostler 1976). Budurov and Sudar (1994) also assigned "*G.*" *nepalensis* to the genus *Kashmirella* (type species, *K. albertii*). For Orchard (2008), *Borinella* would be the valid name for *Kashmirella* and *Kozurella*, including the species *K. albertii*, *K. nepalensis* and *N. buurensis*. However, separation of these taxa should be based on their apparatuses, which are unknown in detail. Limited material of *Borinella buurensis* resembles elements of *Wapitiodus*. For Orchard (2008), an early *Borinella megacuspa* possibly developed in the Latest Griesbachian, followed in the Dienerian by *B. nepalensis* and *B. chowadensis*, from which the Smithian lineage of *Borinella sweeti* and *B. buurensis* derived. For Orchard (2005), *Borinella* largely disappeared at the end of the Smithian, but its characteristic anterior blade denticulation reappeared in the Spathian of Oman next to the dominant genus "*Neospathodus*" and with early Tethyan forms of the genus *Gladigondolella*. If in these collections, no apparatus of the type species of *Gladigondolella* occurs, the P_1 element of *G.* ex gr. *carinata* Bender, and perhaps of *Borinella* n. sp. do, can be thought as the first stage in the development of *Gladigondolella*. If *Borinella* was the root stock for *Gladigondolella*, then it would represent a migration of the former into lower latitude environments, as Kozur et al. (1998) postulated to have occurred among conodonts after the PTB extinction (in Orchard 2007; p. 100). The earliest occurrence of *Borinella* cf. *nepalensis* was in the uppermost Dienerian of the South Primorye Abrek Bay (Shigeta et al. 2009) and in Spiti (Krystyn et al. 2007), while its range extended to the Early Smithian in pelagic environment (Orchard 2007).

The genus ***Smithodus*** Budurov, Buryi and Sudar, 1988, type species *Smithodus longiusculus* (Buryi 1979), originally attributed to *Neospathodus*, is characterised by a sturdy blade at the end, a strong wide concave basal cavity below the main cusp. The species is found within the waageni zone of northeastern Vietnam (Shigeta et al. 2014).

The genus ***Platyvillosus*** Clark, Sincavage and Stone 1964, type species *Platyvillosus asperatus* Clark, Sincavage and Stone 1964, has a platform with a large concave basal cavity with the basal pit located in the centre. A basal furrow can be clearly observed from the basal pit to the anterior end. Protrusions on the upper surface are either low nodes or high denticles, which are either randomly arranged or in a definite pattern.

The genus ***Scythogondolella*** (Kozur 1989), after which the subfamily Scythogondolellinae (Orchard 2007) was proposed, is a Smithian genus having derived from the genus *Clarkina discreta*. However, for Orchard (2007) the type species *Scythogondolella milleri* developed from *Sc. mosheri* that would appear to descend as a paedomorph from *Scythogondolella?* n. sp. B in the Canadian Arctic. Klets and Kopylova (2007) proposed its derivation from *Neospathodus*. The genus extends from Siberia to Arctic Canada and Nevada.

From the genus ***Paragondolella*** (Mosher 1968), type species *Paragondolella excelsa* (Mosher 1968), the Spathian-Aegean *Paragondolella regale* derived from *Kashmirella timorensis*, as the first species of the *Paragondolella* lineage that ranges up to the Julian. Budurov (in Muttoni et al. 2000; p. 233), classified the species *N.* ex gr. *regalis* Mosher as *Paragondolella regale* (Mosher). Deriving from *Kashmirella timorensis*, it was the first species of the anagenetic *Paragondolella* lineage. Supporting this attribution is the mention in Golding (2018) of the suggestion by Henderson (2006) and Henderson and Mei (2007) that the S_0 element morphology found in the species *regale* should refer the species to *Clarkina* or *Neoclarkina* instead of *Neogondolella*. For Chen (2015), the segminate derivation of *Paragondolella*, i.e. the genus *Triassospathodus* as proposed by Mosher (1968), has been challenged by an apparatus comparison of *Paragondolella* and *Triassospathodus*. The S_3 of the former has an accessory anterior process branching from its anterior end, which is missing in the latter (Orchard 2005). Kozur (1989) suggested a derivation of *Paragondolella* from the Early Triassic genus *Pseudogondolella*, now viewed as a synonym of the genus *Borinella* (Orchard 2007). But *Borinella* went extinct in the late Smithian (Orchard 2007) and thus cannot be the direct forebear of *Paragondolella*. Chen (op. cit) favours Nicora's (1977) link of *Paragondolella bulgarica* (Budurov and Stefanov (1975)) with *Neogondolella* ex gr. *regale*, which are both close morphologically and overlap during the early Bithynian. The *Paragondolella* lineage leads thus from *P. regale* to *P. bulgarica, P. hanbulogi, P. praezsaboi, P. bystrickyi, P. excelsa, P. fueloepi* and *P. inclinata* (Budurov, in Muttoni et al. 2000).

Genera Incertae Sedis in Gondolellids

Rediversifications into independent peramorphic lineages took place for a number of Dienerian to Spathian morphotypes that do not harbour some of the neogondolellin features. These morphotypes were mostly assigned to the genus *Neogondolella*; however, their multi-elements may often be unavailable or differ from the neogondolellin apparatus. Several of these taxa did not evolve through Darwinian anagenesis but through atavistic reversal instead. Such rediversification from proteromorphs, as it repeated several times each rediversification, possibly issued from a distinct proteromorph, resulted in neogondolelli-morph, only similar in appearance yet not part of a same anagenetic lineage. Rediversifications into independent peramorphic lineages may be differentiated on the basis of platform shape and blade characteristics, as in Orchard (2007).

The genus *Siberigondolella* Kiliç and Hirsch comprises the Boreal lineage described as *Neogondolella* by Dagys (1984) and Klets and Yadrenkin (2001) that includes Griesbachian-Dienerian *S. griesbachensis* Orchard, *S. mongeri* Orchard, the Smithian *S. composita* Dagys, *S. altera* Klets, *S. sibirica* Dagys and *S. jakutensis* Dagys (Kiliç and Hirsch 2019).

Under the generic name of *Neogondolella* are the Spathian forms *N. jubata* Sweet, *N. dolpanae* Balini, Gavrilova and Nicora, *N. shevyrevi* Kozur and Mostler, *N. taimyrensis* Dagys, *N. elongata* Sweet, *N. paragondolellaeformis* Dagys, *N. amica* Klets, *N. captica* Klets and a large number of unnamed taxa (C, D, E, F, G, H and K), all derived by Orchard (2007) from Smithian *Borinella sweeti*. Among these, *N. elongata* Sweet, common in the USA (Orchard and Tozer 1997), was referred to the genus *Columbitella* (Orchard 2007), based on modifications to its apparatus (Orchard 2005). Having characteristics from pointed posterior platform and terminal cusp to reduced anterior platforms or a rounded posterior platform and brim surrounding the cusp, they produced successor species with a stronger cusp.

The Smithian genus ***Paullella*** (Orchard 2008) was established for the species "*Gladigondolella*" *meeki* (Paull (1983)) of the *Meekoceras* beds, Thaynes Formation, Idaho. The assemblage of *Paullella meeki, Neospathodus posterolongatus, N. pakistanensis* and *Guangxidella bransoni* appears as an outburst within a level equivalent to the lower part of the waageni zone (Golding 2014).

The Spathian genus ***Icriospathodus*** (Krahl et al. (1983)) was established for *Neospathodus collinsoni* (Solien 1979) (= *Spathoicriodus*, Budurov et al. 1987). Type locality: Thaynes Formation, near Salt Lake City, Utah, USA.

(D) Family Gladigondolellidae Ishida and Hirsch

After genus ***Gladigondolella*** (Mueller 1956), type species *Gladigondolella tethydis* (Huckriede), the probably monogeneric taxon has an octomembrate apparatus in which all elements fundamentally differ from their equivalents in Gondolellidae. For Kiliç et al. (2013), the two P_2 elements are sexual dimorphs. For Koike (1999), followed by Orchard (2005), one of these is considered as the P1 element of form-genus "*Cratognathus*". The basinal open marine habitat of the family was, as indicated by high fluorapatite $\delta^{18}O_{phos}$ values of ~21–21.5‰ in *Gladigondolella*, compared to 20‰ in coeval eurythermal *Paragondolella* (Rigo and Trotter 2014), deep cold oceanic. The family originated in the cool environment that first reappeared in the Late Spathian. For Orchard (2005), the denticulation in *Gladigondolella* and *Cratognathodus* resembles that of *Guangxidella robusta*, except for the significantly different M and S_3elements.

Wapitiodus might well be the Smithian reinitialisation of the Permian genus *Vjalovognathus* Kozur (1977), type species *V. shindyensis* Kozur and Mostler (1976), revised in Nicoll and Metcalfe (1998, p. 435–436). *Vjalovognathus* was restricted to the northern margin of eastern Gondwanaland, suggested to have been tolerant to cool temperate conditions, a characteristic shared by Gladigondolellidae (Rigo and Trotter 2014). Orchard (2005) named the genus *Wapitiodus* after Lake Wapiti, Sulphur Mountain Formation, British Columbia, Canada, and saw it as a diversification issued from *Kashmirella* (*Sweetognathodus*) *kummeli*, based on S_3 elements with an accessory anterolateral process branching from the cusp. The genus is included in subfamily Mullerinae (Orchard (2005)) that comprises the genera *Conservatella* and *Discretella* that Orchard (2005) established after the original descriptions of *Ctenognathodus conservativus* and *Ctenognathodus discretus* by Mueller (1956). *Neoprioniodus bransoni* (Mueller (1956)) belongs in genus *Guangxidella* (Zhang and Yang (1991)). However, the genera *Conservatella*, *Discretella* and *Guangxidella*, instead of related to *Kashmirella*, are rather the Smithian reinitialisation of Permian taxa like *Vjalovognathus*. *Guangxidella* is the origin of *Cratognathodus* and *Gladigondolella*.

5.3.3 Griesbachian Zonation

Brosse et al. (2016) have put in evidence six unitarian associations of up to 16 species in UAZ 4. The first Griesbachian level UAZ 3 counts *Clarkina planata* and *C. taylorae*; level UAZ 4 is marked by the coming in of the genus *Isarcicella* and level UAZ 5 is marked by *Isarcicella isarcica*, while UAZ 6 still yields *C. taylorae* (Figs. 5.5 and 5.6). The Griesbachian is thus marked by the *Clarkina taylorae* zone and the subdivisions of the *C. planata* subzone (UAZ 3–4) and *Isarcicella isarcica* subzone (UAZ 5), the latter being the most reliable marker. The Griesbachian is also marked by several *Hindeodus* specimens, depending on factors of depth and temperature, as discussed later. It is noteworthy that Zhao et al. (2007) in Chaohu, Anhui Province, China, have established the *Hindeodus typicalis* zone for the Griesbachian, instead of *Hindeodus parvus*, not found in Chaohu! *Hindeodus typicalis* (Sweet), found in the Elika Formation (Iran) by Hirsch and Suessli (1973), was recently illustrated in Maaleki-Moghadam et al. (2019; p. 377, fig. 13). These authors identified *Hindeodus typicalis* (Sweet), *H. praeparvus* Kozur and *H.* cf. *parvus*. Their complete *H.* cf. *parvus* specimen differs from *H. praeparvus* by a narrower and acute-angled main denticle, not high enough to be a typical *H. parvus*. While this sample indicates a stratigraphic position close to the Permian-Triassic

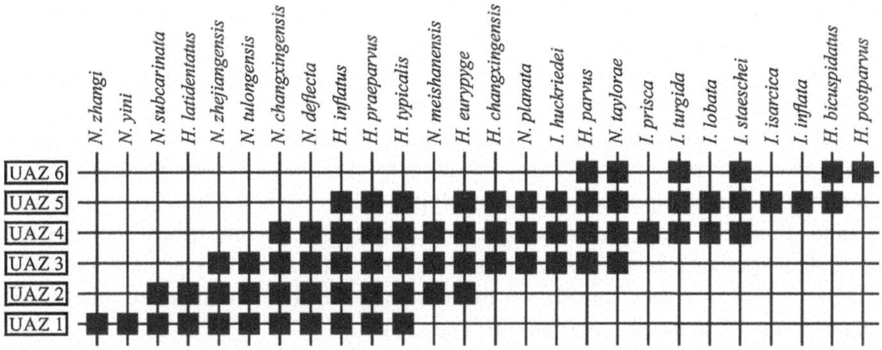

Fig. 5.5 The six unitarian association zones (UAZ 1–6) (after Brosse et al. 2016)

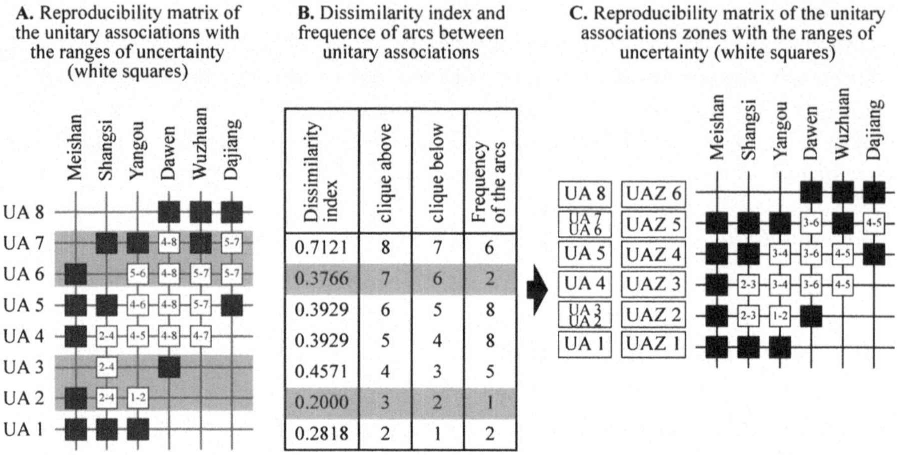

Dissimilarity index	clique above	clique below	Frequency of the arcs
0.7121	8	7	6
0.3766	7	6	2
0.3929	6	5	8
0.3929	5	4	8
0.4571	4	3	5
0.2000	3	2	1
0.2818	2	1	2

Fig. 5.6 The poor reproducibility of the UA2, UA3, UA4 and UA6: (**a**) the grey shades indicating the UAs that are merged into UAZs, (**b**) the lateral traceability of UAZ 1–6, (**c**) a sequence of eight UAs subsequently grouped into six UAZs of higher lateral reproducibility (Brosse et al. 2016)

boundary, it again wakes up doubts about the many alleged species of the highly variable genus *Hindeodus*.

5.3.4 *Conodont Morphogenesis*

The evolution of the genera in the subfamily Neogondolellinae consists of a number of reiterating trends: (1) displacement of the basal cavity from its posterior position towards the middle, accompanied by the modification of its shape from loop-like to amygdaloid; (2) reduction of the platform, that in most Early and Middle Triassic genera of the family borders the entire unit of adult specimens, by the formation of a free blade; and (3) splitting of the monolobate basal groove into a bilobate, forked platform. The evolutionary trends are paced by recurring retrogradations as protero-morphic neospathid morphs followed by accelerated rates of rediversified peramorphic speciations, such as the Smithian radiations of *Scythogondolella milleri, Borinella sibiriensis, B. nepalensis* and Spathian *Columbitella jubata*.

Evolutionary "simplifications" have been interpreted in terms of heterochrony such as paedomorphism, progenesis and neoteny (Gould 1977). Paedomorphosis is when the descendant species is underdeveloped relatively to the ancestor, smaller in size and simpler in shape, resembling juvenile ancestors; peramorphosis is when the descendant species transcends its ancestor in terms of size and shape; and a neotenous descendant is of the same size as the adult ancestor but is underdeveloped (simpler) in terms of shape (Lieberman 2011; p. 35).

In their cladogram of gondolellid taxa, Henderson and Mei (2007) consider the evolution of bifid S_3 elements as transitional between *Mesogondolella* and

Jinogondolella: the changes in ontogenetic developmental timing, including lack of a platform or its reduction to a narrow rib in juvenile specimens within the development of *Clarkina* (*Neoclarkina*). Newly, the sudden loss of platform in *Neospathodus* is attributed to proteromorphosis (Guex 2001) and the platform development in the succeeding peramorphic series to reinitialisation (Guex 2006). This is seen in *Borinella* and *Scythogondolella*, evolving from *Neospathodus*, consisting of peramorphic processes in which ontogeny was restored and additional stages are added in which the platform is developed in intermediate and adult forms.

The Early Triassic is characteristic not only for a high frequency of retrogradations that are followed by reinitialisation to peramorphic lineages, but also for the frequent appearance of totally new morphologies with previously unseen ornamentations, often of limited duration.

The Middle Triassic lineage of *Paragondolella* derived from *Kashmirella* and prevails in the more open marine scene.

A so far unidentified event during the Early Anisian of the central part of the Northern Tethys has precipitated the appearance of the forms *Kamuellerella-Ketinella-Gedikella.* These small-size ramiform units, found in the Turkish Istanbul zone (Gedik 1975; Kiliç et al. 2018), may suggest some extraordinary local warming event.

During the Early Carnian Pluvial interval the main evolutionary trends are (a) reduction of the platform with development of a free blade and (b) the trend of splitting the basal groove (Budurov 1977).

5.3.5 Palaeogeography (After Brosse et al. 2016)

The structurally distinctive oceanic patterns of the gradually closing Palaeotethys (N. and S. China) and opening Mesotethys (Pakistan, India) are essentially tropical. The Mesotethys opens into pan-latitudinal Panthalassa that also caps the Earth with an immense Boreal (Arctic) realm (Fig. 5.7).

Tethys The Early Triassic tropical Tethyan realms encompass the following zones: Griesbachian *typicalis*, *isarcica* and *carinata*; Dienerian *kummeli*, *dieneri* and *cristagalli*; Smithian *pakistanensis*, *waageni*, *conservativus* and *milleri*; and Spathian *jubata*, *triangularis*, *homeri* and *timorensis*.

The Northern Palaeotethyan area of Mangyshlak (West Kazakhstan) is characterised by the Homeri zone that, next to *Neospathodus homeri*, also yields *N. abruptus, N. symmetricus* and "*Neogondolella*" *dolnapae*. For Balini et al. (2000) this species has affinities to "*N.*" *jubata* (Sweet 1970).

Boreal The Canadian Arctic, British Columbia and Western USA during the Smithian comprise next to *Neospathodus waageni, Borinella buurensis, Scythogondolella milleri, S. mosheri, Guangxidella bransoni*, *Conservatella* spp.,

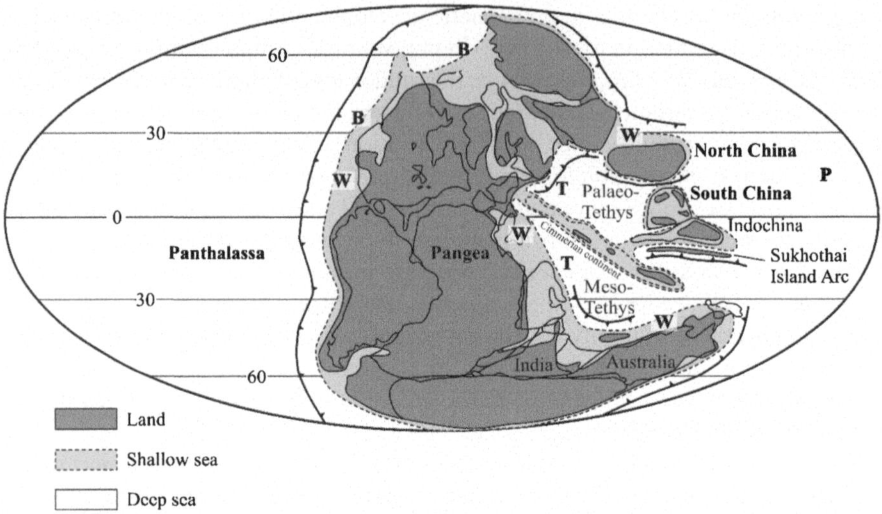

Fig. 5.7 Early Triassic paleogeography (Brosse et al. 2016). *T* Tethys, *B* Boreal, *W* Werfen, *P* Pacific

Discretella spp., *Paullella meeki*, *Wapitiodus robustus* and *N. posterolongatus*, some of them providing a rather Boreal identity.

In Spitsbergen, Nakrem et al. (2008) list next to *Clarkina meishanensis C. taylorae, C. orchardi, C. hauschkei, C. carinata, Neospathodus dieneri, Ns. pakistanensis, Ns. cristagalli, Ns. waageni, Scythogondolella milleri, Sc. mosheri, Sc. sweeti, Borinella buurensis, Bo. nepalensis, Eurygnathodus, Columbitella paragondolellaeformis, Neospathodus svalbardensis* and *Merillina peculiaris*. The Boreal belt of Siberia (Klets and Kopylova 2004; Fig. 2, Table 1) includes the cosmopolitan *Neospathodus* with *Scythogondolella* and the endemic species *Borinella buurensis* (Dagys), "*Neogondolella*" *composita* Dagys and *N. jakutensis* Dagys (Dagys 1984; Kuzmin and Klets 1990; Klets and Yadrenkin 2001; Klets and Kopylova 2004). For Klets and Kopylova (2007), *Neospathodus* replaced *Clarkina* and generated the above-mentioned segminiplanate "gondolellids", next to "*N.*" *taimyrensis* and "*N.*" *sibirica* that are widely distributed in the Smithian of the northern latitudes.

"Werfen"-Like Ellisoniids seem rather restricted to shallower shelf-conditions (Carr et al. 1984) with *Furnishius triserratus* in the Western USA and Western South Pacific (including Japan), East Asia and Eastern Europe. Similarly, the Alpine-Mediterranean Werfen province yields the Ellisoniids *Pachycladina* and *Hadrodontina* (Staesche 1964; Hirsch 1975; Aljinovic et al. 2006) that also extend to the "Cimmerian" Elikah Formation of the Alborz Mountains (BadriKolalo et al. 2015).

In the Far East (Vietnam), Shigeta et al. (2014) report the Lower Smithian *Neospathodus waageni* zone extending from the Owenites koeneni ammonoid zone to the Lower Middle Smithian *Flemingites rursiradiatus* and *Urdiceras tulongensis* ammonoids. The zone yields *Conservatella conservativa, Discretella discreta, D. robusta, Guangxidella bransoni, Neospathodus spitiensis* and *N. waageni.* In its lower part appear *Eurygnathodus costatus, Neospathodus cristagalli, N. dieneri, N. posterolongatus* and *Kashmirella novaehollandiae.* In the middle part appear *Smithodus longiusculus* and *Hadrodontina.*

The Upper Smithian Neospathodus pingdingshanensis zone represents the Xenoceltites variocostatus beds (= *Anasibirites novelise* zone to *Tirolites cassianus* zone) is coeval to the Werfen Formation of the Southern Alps and represents the Smithian-Spathian transition, yielding *Neospathodus pingdingshanensis, N. waageni, N. triangularis, Icriospathodus collinsoni, I. crassatus* and *I. zaksi.* The Lower Spathian *Triassospathodus symmetricus* zone is marked by ammonoids of the *Tirolites* zone and yields *Neospathodus triangularis, Triassospathodus symmetricus* and *T. homeri. Triassospathodus homeri* evolved from *T. symmetricus* in the Early Spathian, extending from Chios and N. Dobrogea to Oman, Salt Range, Kashmir, China and Japan.

Pacific Terranes The Abrek-Lazurnaya Bay area of Primorye, Russian Far East, yields the Griesbachian *Clarkina carinata*; Dienerian *Neospathodus* cf. *cristagalli, N. dieneri, Borinella* cf. *nepalensis* and *Eurygnathodus costatus*; Dienerian-Smithian *N. pakistanensis*; and Smithian *Ellisonia* cf. *peculiaris, Neospathodus* ex. gr. *waageni, N. novaehollandiae, N.* aff. *posterolongatus, N. concavus, N. spitiensis* and *Foliella gardenae* (Igo 2009).

In Japan, Koike (1988, 2004, 2016), Nakazawa et al. (1994) and Agematsu et al. (2014) described Ellisoniidae: *Ellisonia triassica, Hadrodontina aequabilis, H. agordina, H. anceps, Cornudina breviramulis, Pachycladina obliqua, P. eromera, Furnishius triserratus* and *Staeschegnathus perrii*; Anchignathodontidae: *Hindeodus typicalis* and *H. parvus;* Gondolellidae: *Neospathodus dieneri, N. conservativus, N. waageni, N.*cf. *robustus, N. homeri, Icriospathodus collinsoni, Platyvillosus costatus, P. hamadai, Kashmirella timorensis* and *Paragondolella regale*; Gladigondolellidae: form element *Cratognathodus kochi*, followed higher up by *Gladigondolella tethydis.* Nakazawa et al. (1994) see two different faunas in Japan, one resembling the Primorye fauna, while the other is closely related to the Tethyan fauna.

5.4 Conclusions

Early Triassic conodont faunas developed under the impact of succeeding ecological crises of their marine ecosystems. Early Triassic environmental changes (e.g. temperature) and opening of new niches were accompanied by the appearance of new genera.

From the three surviving families—Anchignathodontidae, Ellisoniidae and Gondolellidae—only the late Griesbachian, possibly new Spathognathid genus *Isarcicella* became an important short-lived event.

Among Gondolellidae, the genus *Clarkina* of the latter's subfamily Neogondolellinae had ten species surviving the Permian-Griesbachian boundary, producing the two additional short-living Griesbachian species *C. krystyni* and *C. discreta.* From the latter, the latest Griesbachian genera *Neospathodus* and *Kashmirella* emerged by retrogradation becoming the dominant Dienerian taxa that gave origin to the reinitialisation of all short- or long-term lineages with a "Neogondolella-like" appearance. Thus *Neospathodus* and *Kashmirella* were at the origin of the wealth of genera that radiated until the end of the Spathian. Among these are the segminiplanate Dienerian-Smithian genera *Borinella* and *Eurygnathodus*, the Smithian subfamily Scythogondolellinae and genus *Paullella* (not illustrated). Among uncertain segminate Spathognathodontid genera are Smithian *Guangxidella, Conservatella, Discretella* and *Wapitiodus* and the Spathian Gladigondolellid form-genus *Cratognathodus.* The uncertain subfamily Cornudininae and genus *Icriospathodus* remain unrepresented in our scheme.

Boreal reinitiated segminiplanate genera encompass the lineages of the Late Griesbachian-Smithian genus *Siberigondolella* Kiliç and Hirsch and of the Spathian genus C*olumbitella* Orchard.

The third surviving Paleozoic family Ellisoniidae is well represented in the Smithian-Spathian Thaynes/Werfenian environments by the genera *Furnishius, Pachycladina* and *Hadrodontina,* their huge size and massive build that during the Smithian may concur with an apatite peak.

A great wealth of new forms appeared and went extinct during the time span of less than 5 million years between the Permian-Triassic boundary mass extinction and the Spathian-Anisian environmental bottleneck. Of the lineages that appeared during the Early Triassic, none persisted beyond the Spathian, with the exception of genera *Paragondolella* and *Gladigondolella.* The former survived by repeated retrogradations until the Rhaetian final conodont extinction and the latter, *Gladigondolella*, which ascended in the Late Spathian with the reappearance of cooler pelagic conditions, was mostly found in equatorial environments until the end of the Julian.

Environmental instabilities are a key to morphogenesis in biology and palaeontology.

During the Early Triassic, each event was marked by short-lived entirely new morphologies, while the long-lasting "Gondolellid" segminiplanate retrograded into segminate proteromorphs. These occur as two parallel 4 Ma long lineages, each characterised by its own evolution of successive morphs, interpreted by some as genera. In this chapter, we analyse this succession in an attempt to define their taxonomy (Fig. 5.8).

The wealth of new morphs that appeared during the Early Triassic evokes the Cambrian explosion that Stephen Gould compared with the many models of the first automobiles (Gould 1996). In the animal world, a similar comparison can be made with Darwin's finches, well known for their phenotypic variability and evolution in

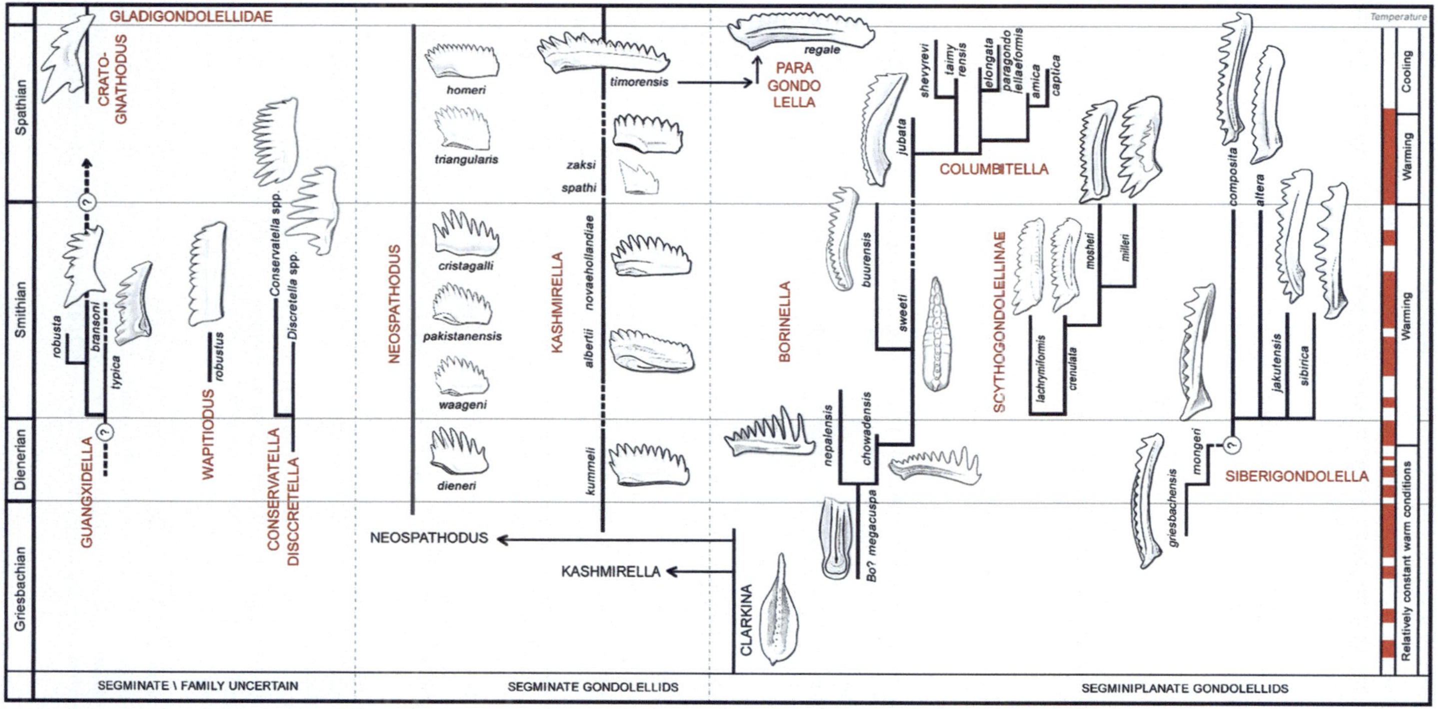

Fig. 5.8 Main Gondolellid segminiplanate, segminate and uncertain segminate lineages. The latest Griesbachian Gondolellid genus *Clarkina*, of subfamily Neogondolellinae, initialised the genera *Neospathodus* and *Kashmirella* by retrogradation. Becoming dominant Dienerian-Spathian taxa, *Neospathodus* gave origin to the reinitialisation of all lineages of "*Neogondolella*"-like appearance that radiated until the end of the Spathian. Among these are the segminiplanate Dienerian-Smithian genera Borinella and Eurygnathodus and the Smithian genus *Paullella* and subfamily Scythogondolellinae. Spathian *Kashmirella* was at the origin of the genus *Paragondolella*. Among uncertain segminate genera, Smithian *Guangxidella*, *Conservatella*, *Discretella* and *Wapitiodus* and Spathian *Cratognathodus* apparently initiated the family Gladigondolellidae. Boreal-reinitiated segminiplanate genera encompass the lineages of the Dienerian-Smithian genus *Siberigondolella* Kiliç and Hirsch and of the Spathian genus *Columbitella* Orchard

response to changing environmental conditions. Recently, McNew et al. (2017) documented epigenetic variations within each of the two species of Darwin's finches. This suggests that evidence of phenomena observed within extinct taxa in the fossil state may be compared with phenomena for which living organisms can provide the evidence in the laboratory.

Conodonts, as a biotic group, next to ammonoids, pollen and spores, crustaceans and vertebrates, provide proxy clues for environment and age assessments of the rocks in which they occur.

Acknowledgements We are indebted to Prof. H. Bucher (University of Zurich, Switzerland), Dr. E.S. Carter (University of Portland, USA), for reading the first draft of the manuscript, and to our reviewers. They all permitted us to improve the text.

References

Agematsu S, Sano H, Sashida K (2014) Natural assemblages of *Hindeodus* conodonts from the Permian-Triassic boundary sequence, Japan. Palaeontology 57(6):1277–1289

Aljinović D, Kolar-Jurkovsek T, Jurkovsek B (2006) The lower Triassic marine succession in Gorse Kotari Region (External Dinarides, Croatia): Lithofacies and conodont dating. Rivista italiana di Paleontologia e Stratigrafia 112:35–53

Angiolini L, Nicora A, Bucher H, Vachard D, Pillevuit A, Platel JP, Roger J, Baud A, Broutin J, Hashmi HA, Marcoux J (1998) Evidence of a Guadalupian age for the Khuff Formation of southeastern Oman: preliminary report. Rivista italiana di Paleontologia e Stratigrafia 104:329–340

BadriKolalo N, Bahaeddin H, Seyed Hamid V, Seyed AA (2015) Biostratigraphic correlation of Elikah formation in Zal section (Northwestern Iran) with Ruteh and type sections in Alborz Mountains based on conodonts. Iranian J Earth Sci 7:78–88

Balini M, Gavrilova VA, Nicora A (2000) Biostratigraphical revision of the classic Lower Triassic Dolnapa section (Mangyshlak, west Kazakhstan): Zentralblatt für Geologie und Paläontologie 1998, v. 1, p. 1441–1462

Barash MS (2006) Development of marine biota in the Paleozoic in response to abiotic factors. Oceanology 6:848–858. ISSN 0001-4370 (Original Russian Text, in Okeanologiya 46: 899–910)

Baud A, Magaritz M, Holser WT (1989) Permian-Triassic of the Tethys: carbon isotope studies. Geol Rundschau 78(2):649–677

Beauchamp B, Baud A (2002) Growth and demise of Permian biogenic chert along northwest Pangaea: evidence for end-Permian collapse of thermohaline circulation. Palaeogeogr Palaeoclimatol Palaeoecol 184:37–63

Bender H (1970) Zur gliederung der Mediterranen Trias II. Die Conodontenchronologie der Mediterranen Trias. Annales Geologiques des Pays Helleniques 19:465–540

Beyers JM, Orchard MJ (1991) Upper Permian and Triassic conodont faunas from the type areas of the Caches Creek complex, south-central British Columbia, Canada. Geol Surv Can Bull 417:269–297

Brayard A, Krumenacker LJ, Botting JP, Jenks JF, Bylund KG, Fara E, Vennin E, Olivier N, Goudemand N, Saucède T, Charbonnier S, Romano C, Doguzhaeva L, Thuy B, Hautmann M, Stephen DA, Thomazo C, Escarguel G (2017) Unexpected Early Triassic marine ecosystem and the rise of the Modern evolutionary fauna. Sci Adv 3(2):e1602159. https://doi.org/10.1126/sciadv.1602159

Brosse M, Bucher H, Bagherpour B, Baud A, Frisk ÅM, Guodun K, Goudemand N (2015) Conodonts from the Early Triassic microbialite of Guangxi (South China): implications for the definition of the base of the Triassic System. Palaeontology 58(3):563–584

Brosse M, Bucher H, Goudemand N (2016) Quantitative biochronology of the Permian-Triassic boundary in South China based on conodont unitary associations. Earth Sci Rev 155:153–171

Budurov KJ (1977) Revision of the late Triassic platform conodonts. Geologica Balcanica 7(3):31–48

Budurov KJ, Pantić S (1973) Conodonten aus den Campiler Schichten von Brassica (Westserbien). Bulgarian Acad Sci Bull Inst Seri Paleontol 22:49–64

Budurov KJ, Stefanov S (1972) Platform-conodonten und ihre Zonen in der Mittleren Trias Bulgariens. Mitt Ges Geol Bergbaustud 21:829–852

Budurov KJ, Stefanov S (1975) Middle Triassic conodonts from drillings near the town of Knezha. Palaeontol Stratigr Lithol 3:11–18. Sofia

Budurov KJ, Sudar M (1993) *Kozurella* gen. n. (Conodonta) from the Olenekian (Early Triassic). Geologica Balcanica 23:24

Budurov KJ, Sudar M (1994) *Borinella* Budurov and Sudar, nomen novum for the Triassic conodont genus *Kozurella* Budurov and Sudar, 1993. Geologica Balcanica 24(3):30

Budurov KJ, Sudar M, Gupta VJ (1987) *Spathoicriodus*, a new Early Triassic conodont genera. Bull Indian Geol Assoc 20:175–176

Budurov KJ, Buryi GI, Sudar MN (1988) *Smithodus* n. gen. (Conodonta) from the Smithian stage of the Lower Triassic. Mitteilungen der Gesellschaft der Geologie- und Bergbaustudenten, Osterreich 34/35:295–299

Burgess SD, Muirhead JD, Bowring SA (2017) Initial pulse of Siberian Traps sills as the trigger of the end-Permian mass extinction. Nat Commun 8:164

Buryi G (1979) Lower Triassic conodonts of southern Primorye. Institut Geologii i Geofiziki, Sibirskoye Otdeleniye, Akademiya Nauk SSSR, Moskva [In Russian]

Carpenter SJ, Lohmann KC (1995) $\delta^{18}O$ and $\delta^{13}C$ values of modern brachiopod shells. Geochim Cosmochim Acta 59:3749–3764

Carr TR, Paull RK, Clark DL (1984) Conodont paleoecology and biofacies analysis of the Lower Triassic Thaynes Formation in the Cordilleran Miogeocline. Geol Soc Am Spec Pap 196:283–294

Chen Y (2015) A multi-method study on Triassic conodonts. Dissertation zur Erlangung des Doktorgrades der Naturwissenschaften an der Karl-Franzens-Universität Graz Institut für Erdwissenschaften, 219 pp

Chen J, Beatty TW, Henderson C, Rowe C (2009) Conodont biostratigraphy across the Permian–Triassic boundary at the Dawen section, Great Bank of Guizhou, Guizhou Province, South China: implications for the Late Permian extinction and correlation with Meishan. J Asian Earth Sci 36:442–458

Chen B, Joachimski MM, Shen SZ, Lambert LL, Lai XL, Wang XD, Chen J, Yuan DX (2013a) Permian ice volume and palaeoclimate history: oxygen isotope proxies revisited. Gondwana Res 24:77–89

Chen Y, Twitchett RJ, Jiang H, Richoz S, Lai X, Yan C, Sun Y, Liu X, Wang L (2013b) Size variation of conodonts during the Smithian–Spathian (Early Triassic) global warming event. Geology 41(8):823–826

Chen Y, Jiang H, Lai X, Yan C, Richoz S, Liu X, Wang L (2015) Early Triassic conodonts of Jiarong, Nanpanjiang Basin, southern Guizhou Province, South China. J Asian Earth Sci 105:104–121

Clark DL, Sincavage JP, Stone DD (1964) New conodont from the lower Triassic of Nevada. J Paleontol 38:375–377

Clark DL, Sweet WC, Bergstrom SM, Klapper G, Austin RL, Rhodes FHT, Muller KJ, Ziegler W, Lindstrom M, Miller JF, Harris AG (1981) Conodonta. Part W, Miscellanea, Supplement 2 of R. A. Robison, ed. Treatise on invertebrate paleontology. Geological Society of America, Boulder, Colo., and University of Kansas, Lawrence

Clutson M, Brown D, Tanner L (2018) Distal processes and effects of multiple late Triassic terrestrial bolide impacts: insights from the Norian Manicouagan Event, Northeastern Quebec, Canada. In: Tanner L (ed) The late Triassic world. Topics in geobiology, vol 46. Springer, Cham

Cohen KM, Harper DAT, Gibbard PL, Fan JX (2018) The ICS international chronostratigraphic chart. Updated from Cohen KM, Finney SC, Gibbard PL, Fan JX (2013) Episodes 36:199–204, 2013, (c) International Commission on Stratigraphy, August 2018

Dagys AA (1984) Early Triassic conodonts of northern Middle Siberia (in Russian). Trans Inst Geol Geophys Acad Sci USSR Siberian Branch 554:1–69

De Wever P, O'Dogherty L, Gorican S (2007) The plankton turnover at the Permo-Triassic boundary, emphasis on radiolarians. Eclogae Geol Helv 99:S49–S62

Dorritie D (2004, 2005, 2006, 2007) Killer in Our Midst. www.killerinourmidst.com

Friesenbichler E, Richoz S, Baud A, Krystyn L, Sahakyan L, Vardanyan S, Peckmann J, Reitner J, Heindel K (2018) Sponge-microbial build-ups from the lowermost Triassic Chanakhchi section in southern Armenia: microfacies and stable carbon isotopes. Palaeogeogr Palaeoclimatol Palaeoecol 490:653–672

Galfetti T, Hochuli P, Brayard A, Vigran J (2007) Smithian-Spathian boundary event: evidence for global climatic change in the wake of the end-Permian biotic crisis. Geology 35(4):291–294

Garbelli C, Angiolini L, Brand U, Shen S-Z, Jadoul F, Posenato R, Azmy K, Cao C-Q (2015) Neotethys seawater chemistry and temperature at the dawn of the end Permian mass extinction. Gondwana Res 35:272–285

Gedik I (1975) Die Conodonten der Trias auf der Kocaeli-Halbinsel (Turkei). Palaeontogr Abt A 150:99–160

Golding M (2014) Biostratigraphy and sedimentology of Triassic hydrocarbon-bearing rocks in Northeastern British Columbia. PhD Thesis, The University of British Columbia, Vancouver, 388 pp

Golding M (2018) Reconstruction of the multielement apparatus of Neogondolella ex gr. regalis Mosher, 1970 (Conodonta) from the Anisian (Middle Triassic) in British Columbia, Canada. J Micropalaeontol 37:2124

Gould JS (1996) Full house: the spread of excellence from Plato to Darwin. Harmony, New York

Gould JS (1977) Ontogeny and Phylogeny. Belknap, Harvard Univ. Press, Cambridge, Mass. 490p

Gradstein FM, Ogg JG, Smith AG (2005) A geological time scale. Cambridge University Press, Cambridge, 589 pp

Guex J (2001) Environmental stress and atavism in ammonoid evolution. Eclogae Geol Helv 94:321–328

Guex J (2006) Reinitialization of evolutionary clocks during sublethal environmental stress in some invertebrates. Earth Planet Sci Lett 242:240–253

Guex J (2016) Retrograde evolution during major extinction crises. Springer briefs in evolutionary biology. Springer, Cham 77 pp

Haas J, Deméeny A, Hips K, Zajzon N, Weiszburg T, Sudar M, Palfy J (2007) Biotic and environmental changes in the Permian–Triassic boundary interval recorded on a western Tethyan ramp in the Bükk Mountains, Hungary. Global Planet Change 55:136–154

Hammer Ø, Jones M, Schneebeli-Hermann E, Hansen B, Bucher H (2019) Are Early Triassic extinction events associated with mercury anomalies? A reassessment of the Smithian/Spathian boundary extinction. Earth Sci Rev. https://doi.org/10.1016/j.earscirev.2019.04.016

Harland WB, Smith AG, Wilcock B (1964) Phanerozoic time-scale. Geological Society of London, 466 pp

Heindel K, Foster WJ, Richoz S, Birgel D, Roden VJ, Baud A, Brandner R, Krystyn L, Mohtat T, Kosun E, Twitchett RJ, Reitner J, Peckmann J (2018) The formation of microbial-metazoan bioherms and biostromes following the latest Permian mass extinction. Gondwana Res 61:187–202

Henderson CM (2006) Multielement definition of *Clarkina* Kozur. Permophiles 48:23–24

Henderson C, Mei S (2007) Geographical clines in Permian and lower Triassic gondolellids and its role in taxonomy. Palaeoworld 16:190–201

Hermann E, Hochuli PA, Mehay S, Bucher H, Bruehwiler T, Hautmann M, Ware D, Roohi G, Rehman K, Yaseen A (2011) Organic matter and palaeoenvironmental signals during the Early Triassic biotic recovery: the Salt Range and Surghar Range records. Sediment Geol 234:19–41

Heydari E, Hassanzadeh J, Wade WJ (2000) Geochemistry of central Tethyan Upper Permian and Lower Triassic strata, Abadeh region, Iran. Sediment Geol 137:85–99

Hirsch F (1975) Lower Triassic conodonts from Israel. Bull Geol Surv Isr 66:39–48

Hirsch F (1994a) Triassic conodont multielements versus eustatic cycles. Lausanne, Mémoires de Géologie 22:35–52

Hirsch F (1994b) Triassic conodonts as ecological and eustatic sensors. In: Embry AF (ed) Pangea: global environments and resources. Memoir, vol 17. Canadian Society of Petroleum Geologists, Calgary, pp 949–959

Hirsch F, Suessli P (1973) Lower Triassic conodonts from the Lower Elikah formation, Central Alborz, North Iran. Eclogae Geol Helv 66(3):525–531

Hochuli PA, Vigran JO, Hermann E, Bucher H (2010) Multiple climatic changes around the Permian–Triassic boundary event revealed by an expanded palynological record from mid-Norway. Geol Soc Am Bull 122:884–896

Holser WT, Schoenlaub HP, Attrep M, Boeckelmann K, Klein P, Magaritz M, Orth CJ, Fenninger A, Jenny C, Kralik M, Mauritsch H, Pak E, Schramm JM, Stattegger K, Schmoller R (1989) A unique geochemical record at the Permian/Triassic boundary. Nature 337:39–44

Horacek M, Koike T, Richoz S (2009) Lower Triassic $\delta^{13}C$ isotope curve from shallow-marine carbonates in Japan, Panthalassa realm: confirmation of the Tethys $\delta^{13}C$ curve. J Asian Earth Sci 36:481–490

Igo H (2009) Conodonts. In: Shigeta Y, Zakharov Y, Maeda H, Popov AM (eds) The Lower Triassic system in the Abrek Bay area, South Primoriye, Russia. National Museum of Nature and Science Monographs 38, Tokyo, pp 44–40

Joachimski MM, Lai X, Shen S, Jiang H, Luo G, Chen B, Chen J, Sun Y (2012) Climate warming in the latest Permian and the Permian-Triassic mass extinction. Geology 40:195–198

Kiliç AM, Hirsch F (2019) *Siberigondolella* n. gen., a Boreal Early Triassic Lanceolate Conodont. Turk J Zool 43(5):536–539

Kiliç AM, Plasencia P, Ishida K, Hirsch F, Yumun ZU (2013) Reflections on the apparatus of Triassic Gondolellacea (Conodonta) and the question of sexually dimorphs. In Albanesi GL, Ortega G (eds) Conodonts from the Andes: Proceedings of the 3rd international conodont symposium Asociacion Paleontologica Argentina, publicacion especial no. 13, pp 63–67

Kiliç AM, Plasencia P, Ishida K, Guex J, Hirsch F (2016) Proteromorphosis of *Neospathodus* (conodonta) during the Permian-Triassic crisis. Rev Micropaleontol 59:33–39

Kiliç AM, Plasencia P, Onder F (2018) Debate on skeletal elements of the Triassic conodont *Cornudina* Hirschmann. Acta Geologica Polonica 68(2):147–159

Kiparisova LD, Popov YD. (1956) Subdivision of the lower series of the Triassic system into stages. Doklady AN SSSR 109:842–845 (in Russian)

Kiparisova LD, Popov YD (1964) The project of the subdivision of the Lower Triassic into stages. In: XXII International Geological Congress, Reports of Soviet Geologists, pp 91–99 (in Russian)

Klets TV, Kopylova AV (2004) New findings of Triassic Conodontophoridae in northeastern Asia. News of paleontology and stratigraphy. Supplement to Geologiya i Geofizika (Russia)

Klets T, Kopylova S (2007) The problem of Triassic Gondolellid conodont systematics Conodontophorida, Conodonta). In: Lucas SG, Spielmann JA (eds) The Global Triassic. New Mexico Museum of Natural History and Science Bulletin 41, Albuquerque, NM, pp 131–133

Klets TV, Yadrenkin AV (2001) Lower Triassic conodonts from Kotelny island (taxonomic composition, correlation): news of paleontology and stratigraphy. Suppl J "Geologiya I Geofizika" 42(4):14–21. (in Russian)

Knell SJ (2012) The great fossil enigma, the search of the conodont animal. Indiana University Press Science, Bloomington, IN, 440 pp

Koike T (1982) Triassic conodont biostratigraphy in Kedah, West Malaysia. Geol Palaeontol Southeast Asia 23:9–51

Koike T (1988) Lower Triassic Conodont *Platyvillosus* from the Taho Limestone in Japan. Sci Rep Yokohama Natl Univ II 35:61–79

Koike T (1999) Apparatus of a Triassic conodont species *Cratognathodus multihamata* (Huckriede). Paleontol Res 3:234–248

Koike T (2004) Early Triassic Neospathodus (Conodonta) apparatuses from the Taho Formation, southwest Japan. Paleontol Res 8:129–140

Koike T (2016) Multielement conodont apparatuses of the Ellisoniidae from Japan. Paleontol Res 20:161–175

Korngreen D, Bialik OM (2015) The characteristics of carbonate system recovery during a relatively dry event in a mixed carbonate/siliciclastic environment in the Pelsonian (Middle Triassic) proximal marginal marine basins: a case study from the tropical Tethyan northwest Gondwana margins. Palaeogeogr Palaeoclim Palaeoecol 440:793–812

Korngreen D, Zilberman T (2017) The role of land-marine teleconnections in the tropical proximal Permian-Triassic Marine Zone, Levant Basin, Israel: insights from stable isotope pairing. Global Planet Change 154:44–60

Korte C, Kozur HW, Veizer J (2005) $\delta^{13}C$ and $\delta^{18}O$ values of Triassic brachiopods and carbonate rocks as proxies for coeval seawater and palaeotemperature. Palaeogeogr Palaeoclimatol Palaeoecol 226:287–306

Kozur H (1977) Beitrage zur Stratigraphie des Perms, Teil I: Probleme der Abgrenzung und Gliederung des Perms. Freiberger Forschungsheft C319:79–121

Kozur H (1988) Division of the gondolellid platform conodonts. In: Abstract. Proceedings of the fifth European conodont symposium, contributions 1. Courier Forchungsinstitut Senckenberg, vol 102, pp. 244–245

Kozur H (1989) The taxonomy of the gondolellid conodonts in the Permian and Triassic. Courier Forschungsinstitut Senckenberg 117:409–469

Kozur H (1994) *Chengyuania*, a new name for *Pseudogondolella* Kozur 1988 (Conodonta) [non *Pseudogondolella,* Yang, 1984 (hybodont fish teeth)]. Palaontologische Zeitschrift 68:529–530

Kozur H (1996) The conodonts *Hindeodus*, *Isarcicella* and *Sweetohindeodus* in the uppermost Permian and lowermost Triassic. Geol Croat 49(1):81–115

Kozur H (2004) Pelagic uppermost Permian and the Permian–Triassic boundary conodonts of Iran: part I: taxonomy. Hall Jb Geowiss 18:39–68

Kozur H (2006) Remarks to the base of Olenekian. Albertiana 34:66–70

Kozur H, Mostler H (1976) Neue Conodonten aus dem Jungpaläozoikum und der Trias. Geologische und Paläontologische Mitteilungen. Innsbruck 6:1–33

Kozur H, Mostler H, Krainer K (1998) *Sweetospathodus* n. gen. and *Triassospathodus* n. gen., two important lower Triassic conodont genera. Geol Croat 51:1–5

Krahl J, Kauffmann G, Kozur H, Richter D, Foerster O, Heinritzi R (1983) Neue Daten zur Biostratigraphie und zur tektonischen Lagerung der Phyllit-Gruppe und der Trypali-Gruppe auf der Insel Kreta (Greichenland). Geologische Rundschau 72:1147–1166

Krause AJ, Mills BJW, Zhang S, Planavsky NJ, Lenton TM, Poulton SW (2018) Stepwise oxygenation of the Paleozoic atmosphere. Nat Commun 9:4081. https://doi.org/10.1038/s41467-018-06383-y

Krystyn L, Bhargava ON, Richoz S (2007) A candidate GSSP for the Olenekian Stage: Mud at Pin Valley; district Rahul and Spiti, Himachal Pradesh (Western Himalaya), India. Albertiana 35:5–29

Kuzmin SP, Klets TV (1990) On biostratigraphy of Lower Triassic deposits of the Shevli River basin. In: Abstracts of the fourth FE RMSS, Khabarovsk, Russia, pp 146–147

Leu M, Bucher H, Goudemand N (2018) Clade-dependent size response of conodonts to environmental changes during the late Smithian extinction. Earth Sci Rev. https://doi.org/10.1016/j.earscirev.2018.11.003. 49 pp

Lieberman DE (2011) The evolution of the human head. The Belknap Press of Harvard University Press, Cambridge, MA. ISBN: 978-0674046368, ISBN-10. xii, 756 pp
Lucas S (2018) The GSSP method of chronostratigraphy: a critical review. Front Earth Sci. https://doi.org/10.3389/feart.2018.00191
Maaleki-Moghadam M, Rafiei B, Richoz S, Woods AD, Krystyn L (2019) Anachronistic facies and carbon isotopes during the end-Permian biocrisis: evidence from the mid-Tethys (Kisejin, Iran). Palaeogeogr Palaeoclimatol Palaeoecol 516:364–383
Maekawa T, Komatsu T (2014) Conodont succession. In Shigeta Y, Komatsu T, Maekawa T, Tran HD (eds) Olenekian (Early Triassic) stratigraphy and fossil assemblages in Northeastern Vietnam. Nat Mus Nat Sci Monogr 45:51–54. Tokyo, Japan
McNew SM, Beck D, Sadler-Riggleman I, Knutie S, Koop J, Clayton D, Skinner M (2017) Epigenetic variation between urban and rural populations of Darwin's finches. BMC Evol Biol 17:183. https://doi.org/10.1186/s12862-017-1025-9
McTavish RA (1973) Triassic conodont faunas from Western Australia. Neus Jahrbuch fur Geologie und Palaontologie Abhandlungen 143:275–303
Metcalfe I (2012) Changsinghian (Late Permian) conodonts from Son La, northwest Vietnam and their stratigraphic and tectonic implications. J Asian Earth Sci 50:141–149
Mosher LC (1968) Evolution of Triassic platform conodonts. J Paleontol 42(4):895–946
Mueller KJ (1956) Triassic conodonts from Nevada. J Paleontol 44:818–830
Murchey BL, Jones LD (1992) A mid-Permian chert event: widespread deposition of biogenic siliceous sediments in coastal, island arc and oceanic basins. Palaeogeogr Palaeoclimatol Palaeoecol 96(1):161–174
Muttoni G, Gaetani M, Budurov K, Zagorchev I, Trifonova E, Ivanova D, Petrounova L, Lowrie W (2000) Middle Triassic paleomagnetic data from northern Bulgaria: constraints on Tethyan magnetostratigraphy and paleogeography. Palaeogeogr Palaeoclimatol Palaeoecol 160:223–237
Nakazawa K, Ishibashi T, Kimura T, Koike T, Shimizu D, Yao A (1994) Triassic biostratigraphy of Japan based on various taxa. Lausanne, Mémoires de Géologie 22:83–101
Nakrem HA, Orchard MJ, Weitschat W, Hounslow M, Beatty TW, Mørk A (2008) Triassic conodonts from Svalbard and their Boreal correlations. Polar Res 27:523–539
Nicoll RS, Metcalfe I (1998) Early and Middle Permian Conodonts from the Canning and Southern Carnarvon Basins, Western Australia: their implications for regional biogeography and palaeoclimatology. Proc Roy Soc Victoria 110:419–461
Nicoll RS, Metcalfe I, Wang C (2002) New species of the conodont genus Hindeodus and the conodont biostratigraphy of the Permian–Triassic boundary interval. J Asia Earth Sci 20:609–631
Nicora A (1977) Lower Anisian platform-conodonts from the Tethys and Nevada; taxonomic and stratigraphic revision. Palaeontographica 157:88–107
Orchard MJ (1995) Taxonomy and correlation of Lower Triassic (Spathian) segminate conodonts from Oman and revision of some species of *Neospathodus*. J Paleontol 69:110–122
Orchard MJ (2005) Multielement conodont apparatuses of Triassic Gondolelloidea. Spec Pap Palaeontol 73:73–101
Orchard MJ (2007) Conodont diversity and evolution through the latest Permian and Early Triassic upheavals. Palaeogeogr Palaeoclimatol Palaeoecol 252:93–117
Orchard MJ (2008) Lower Triassic conodonts from the Canadian Arctic, their intercalibration with ammonoid-based stages and a comparison with other North American Olenekian faunas. Polar Res 27:393–412
Orchard MJ, Krystyn L (1998) Conodonts of the lowermost Triassic of Spiti, and new zonation based on *Neogondolella* succession. Rivista Italiana di Paleontologia e Stratigrafia 104:341–368
Orchard MJ, Tozer ET (1997) Triassic conodont biochronology, its calibration with the ammonoid standard, and a biostratigraphic summary for the Western Canada Sedimentary Basin. Bull Can Petrol Geol 45(4):675–692
Orchard MJ, Nassichuk WW, Lin R (1994) Conodonts from the Lower Griesbachian *Otoceras latilobatum* bed of Selong, Tibet and the position of the Permian-Triassic boundary. Memoir Can Soc Petrol Geol 17:823–843

Paull RK (1983) Definition and stratigraphic significance of the Lower Triassic (Smithian) conodont *Gladigondolella meeki* n. sp. in western United States. J Paleontol 57:188–192

Payne JL, Clapham ME (2012) End-Permian mass extinction in the oceans: an ancient analog for the twenty-first century? Annu Rev Earth Planet Sci 40:89–111

Payne J, Van de Schootbrugge B (2007) Life in Triassic oceans: links between plankton and benthic recovery and radiation, Chapter 9. In: Falkowski PG, Knol AH (eds) Evolution of primary producers in the sea. Academic Press, London, pp 165–189

Payne JL, Lehrmann DJ, Wei J, Orchard MJ, Schrag DP, Knoll AH (2004) Large perturbations of the carbon cycle during recovery from the end-Permian extinction. Science 305:506–509

Péron S, Bourquin S, Fluteau F, Guillocheau F (2005) Paleoenvironment reconstructions and climate simulations of the Early Triassic: impact of the water and sediment supply on the preservation of fluvial system. Geodinamica Acta 18(6):431–446

Perri MC, Farabegoli E (2003) Conodonts across the Permian boundary in the Southern Alps. Cour Forsch Inst Senckenberg 245:281–313

Petsios E, Thompson J, Pietsch C, Bottjer D (2017) Biotic impacts of temperature before, during, and after the end-Permian extinction: a multi-metric and multi-scale approach to modeling extinction and recovery dynamics. Palaeogeogr Palaeoclimatol Palaeoecol 513:86–99

Rampino MR, Eshet Y (2017) The fungal and acritarch events as time markers for the latest Permian mass extinction: an update. Geosci Front 9(1):147–154

Retallack GJ (2013) Permian and Triassic greenhouse crises. Gondwana Res 24:90–103

Retallack GJ, Veevers JJ, Morante R (1996) Global coal gap between Permian-Triassic extinction and Middle Triassic recovery of peat-forming plants. Geol Soc Am Bull 108:195–207

Rexroad CB, Furnish WM (1964) Conodonts from the Pella Formation (Mississippian), south-central Iowa. J Paleontol 38(4):667–676

Richoz S, Krystyn L, Baud A, Brandner R, Horacek M, Mohtat-Aghai P (2010) Permian-Triassic boundary interval in the Middle East (Iran and N. Oman): progressive environmental change from detailed carbonate carbon isotope marine curve and sedimentary evolution. J Asian Earth Sci 39:236–253

Rigo M, Trotter J (2014) Stenothermal habits of *Gladigondolella* and *Norigondolella*: constraints from $\delta^{18}O$ of conodont apatite. In: Abstract in 2014 GSA annual meeting, Vancouver, BC

Romano C, Goudemand N, Vennemann TW, Ware D, Schneebeli-Herman E, Hochuli PA, Brühwiler T, Brinkmann W, Bucher H (2013) Climatic and biotic upheavals following the end-Permian mass extinction. Nat Geosci 6:57–60

Rothman DH, Fournier GP, French KP, Alm EJ, Boyle EA, Cao C, Summons RE (2014) Methanogenic burst in the end-Permian carbon cycle. PNAS 111(15):5462–5467

Schaal E (2014) Permian-Triassic Global Change: the Strontium cycle and body size evolution in marine clades. Dissertation, Dept. Geol. & Environs. Sciences, Stanford University, 266 pp

Schneebeli-Hermann E, Kuerschner WM, Hochuli PA, Ware D, Weissert H, Bernasconi SM, Roohi G, Rehman K, Goudemand N, Bucher H (2013) Evidence for atmospheric carbon injection during the end-Permian extinction. Geology 41(5):579–582

Sellwood B, Valdes P (2006) Mesozoic climates: general circulation models and the rock record. Sediment Geol 190:269–287

Shevyrev AA (2005) Triassic biochronology: state of the art and main problems. Stratigr Geol Correlation 14:629–641

Shigeta Y, Zakharov Y, Maeda H, Popov A (2009) The Lower Triassic system in the Abrek Bay Area, South Primorye, Russia. Natl Mus Nat Sci Monogr 38:24–27

Shigeta Y, Komatsu T, Maekawa T, Tran HD (2014) Olenekian (Early Triassic) stratigraphy and fossil assemblages in Northeastern Vietnam. Natl Mus Nat Sci Monogr 45, 190–271

Silva-Tamayo JC, Lau KV, Jost AB, Payne JL, Wignall PB, Newton RJ, Eisenhauer A, Depaolo DJ, Brown S, Maher K, Lehrmann DJ, Altiner D, Yu M, Richoz S, Paytan A (2018) Global perturbation of the marine calcium cycle during the Permian-Triassic transition. GSA Bull 130(7-8):1323–1338

Solien MA (1979) Conodont biostratigraphy of the Lower Triassic Thaynes Formation, Utah. J Paleontol 53:276–306

Staesche U (1964) Conodonten aus dem Skyth von Suedtirol. Neues Jahrbuch für Geologie und Paläontologie 119:247–306

Stefani M, Furin S, Gianolla P (2010) The changing climate framework and depositional dynamics of Triassic carbonate platforms from the Dolomites. Palaeogeogr Palaeoclimatol Palaeoecol 290:43–57

Sun T, Joachimski MM, Wignall PB, Yan C, Chen Y, Jiang H, Wang L, Lai X (2012) Lethally hot temperatures during the Early Triassic greenhouse. Science 388:366–370

Sweet WC (1970) Permian and Triassic conodonts from a section at Guryul Ravine, Vihi District, Kashmir. Univ Kansas Palaeontol Contrib 49:1–10

Sweet WC (1992) A conodont-based high-resolution biostratigraphy for the PermoTriassic boundary interval. In: Sweet WC, Zunyi Y, Dickins JM, Hongfu Y (eds) Permo-Triassic events in the Eastern Tethys. Cambridge University Press, Cambridge, pp 120–133

Sweet WC, Mosher LC, Clark DL (1971) Conodont biostratigraphy of the Triassic. Geol Soc Am 127:441–465

Thomazo C, Vennin A, Brayard I, Bour I, Mathieu O, Elmeknassi S, Olivier N, Esquarguel G, Bylund KG, Jenks J, Stephen DA, Fara E (2016) A diagenetic control on the Early Triassic Smithian–Spathian carbon isotopic excursions recorded in the marine settings of the Thaynes Group (Utah, USA). Geobiology 14(3):220–236

Tian C, Dai J, Tian S (1983) Triassic conodonts. In: Chengdu Institute of Geology and Mineral Resources (ed) Paleontological atlas of Southwest China. Microfossils. Geological Publishing House, Beijing, pp 345–398

Tozer ET (1967) A standard for Triassic time. Geol Surv Can Bull 156:332

Tozer ET (1994) Canadian Triassic ammonoid faunas. Geol Surv Can Bull 467:1–663

Trotter JA, Williams IS, Nicora A, Mazza M, Rigo M (2015) Long-term cycles of Triassic climate change: a new $\delta^{18}O$ record from conodont apatite. Earth Planet Sci Lett 415:165–174

Veevers JJ (1994) Pangaea: evolution of a supercontinent and its consequences for Earth's paleoclimate and sedimentary environments. In: Klein GD (ed) Pangaea: paleoclimate, tectonics and sedimentation during accretion, zenith, and breakup of a supercontinent. Geological Society of America Special Paper 288, Boulder, CO, pp 12–24

Von Mojsisovics E, Waagen W, Diener C (1895) Entwurf einer Gliederung der pelagischen Sedimente des Trias-Systems. Sitzungsber Akad Wiss Wien 104:1271–1302

Wang ZG, Cao YY (1981) Early Triassic conodonts from Lichuan, Western Hubei. Acta Palaeontologica Sinica 20(4):363–373. (in Chinese with English abstract)

Wang YB, Tong JN, Wang JS, Zhou XG (2005) Calcimicrobialite after end-Permian mass extinction in South China and its palaeo-environmental significance. Chin Sci Bull 50(7):665–671

Ward PD (2018) Lamarck's revenge: heritable epigenetic consequences of stress and environmental change, from deep time to near future. The Stuart Agency. 240 pp

Wignall P, Benton M (1999) Discussion on Lazarus taxa and fossil abundance at times of biotic crisis—reply. J Geol Soc 156(3):453–456

Yang SR, Wang X, Hao WC (1986) Early and Middle Triassic conodonts sequence in Western Guangxi. Acta Scientiarum Naturalium Universitatis Pekinensis 4:90–106

Zakharov Y (1974) Importance of palaeobiogeographical data for the solution of the problem on the Lower Triassic Division. In: Zapfe H (ed) The stratigraphy of the Alpine-Mediterranean Triassic, vol 2. Oesterr. Akad. Wiss. Schiftenreihe Erdwissensch. Komm., pp 237–243

Zakharov Y, Biakov D, Horacek M (2014) Global correlation of basal Triassic layers in the light of the first carbon isotope data on the Permian–Triassic boundary in Northeast Asia. Russ J Pac Geol 8(1):1–17

Zapfe H (1983) Das Forschungsprojekt "Triassic of the Tethys Realm" (IGCP Projekt 4) Abschlussbericht. Oesterreichische Akad. Wissenschaften, Schriftenreihe der Erdwissenschaftliche Kommissionen 5:7–16

Zhang SX (1990) On the lower Triassic conodont sequence of western Guangxi. J Grad Sch China Univ Geosci 4(2):1–15

Zhang S, Yang Z (1991) On multielement taxonomy of the Early Triassic conodonts. Stratigr Paleontol China 1:17–47

Zhao L, Orchard MJ, Tong J, Zhang S, Sun Z (2007) Lower Triassic conodont sequence in Chaohu, Anhui Province, China and its global correlation. Palaeogeogr Palaeoclimatol Palaeoecol 252:24–38

Zhao L, Tong J, Sun Z, Orchard M (2008) A detailed Lower Triassic conodont biostratigraphy and its implications for the GSSP candidate of the Induan-Olenekian boundary in Chaohu, Anhui Province. Prog Nat Sci 18:79–90

Zhao L, Chen Z, Algeo T, Chen J, Chen Y, Tong J, Gao S, Zhou L, Hu Z, Liu Y (2013) Rare-earth element patterns in conodont albid crowns: evidence for massive inputs of volcanic ash during the latest Permian bio-crisis? Global Planet Change 105:135–151

Chapter 6
Developmental Plasticity Induced by Either External or Internal Environment Co-opts Ancient Regulatory Networks

Juan Nicolas Malagon, Sam Scanga, Ernest Ho, Armen Manoukian, and Ellen Larsen

Abstract We describe two examples of developmental plasticity during fruit fly development. One pertains to a nutritionally dependent eye color mutation and the other to morphogenesis of a secondary sex characteristic depending on local tissue conditions. In both cases, we suggest that plasticity is based on harnessing ancient pathways, metabolic pathways in the case of the eye color mutation, and cellular pathways in the case of tissue morphogenesis.

Keywords Insulin · Tissue mechanics · Nutrition · Development

6.1 Introduction

Understanding the mechanisms and dynamics of phenotypic adaptation to changing environments is often best viewed through the lens of developmental biology. West-Eberhard (2003) has made a comprehensive and compelling case that transitions between adapted phenotypes may occur through developmental plasticity. Developmental plasticity is the phenomenon in which the adult phenotype is influenced by an environment encountered during some stage in development. The transition from a plastic developmental condition to a single phenotypic state could occur if the environment stabilized and one of the phenotypic variations was subject to positive selection.

In this chapter we explore two examples of developmental plasticity. The first is associated with a particular spontaneously occurring genetic mutation where the

J. N. Malagon
Canadian Mennonite University, Winnipeg, MB, Canada

S. Scanga · E. Ho · A. Manoukian · E. Larsen (✉)
Cell & Systems Biology, University of Toronto, Toronto, ON, Canada

J. Guex et al. (eds.), *Morphogenesis, Environmental Stress and Reverse Evolution*, https://doi.org/10.1007/978-3-030-47279-5_6

genetic and developmental basis of external environmental influences is explored, and the second shows how internal tissue plasticity in a model organism casts light on factors associated with a variety of phenotypes in different species.

6.2 Investigating a Single Mutation with Nutritionally Induced Developmental Plasticity

A striking example of externally induced developmental plasticity is the difference between fertile queen honey bees and infertile worker bees, which depends on the special nutrition provided during the larval stage of the future queen. This, of course, is a species-wide plasticity. However, in order to understand the genetic basis underlying the evolution of such plasticity it may be helpful to examine de novo plasticity due to a single mutation, which we explore below.

In our laboratory we noticed that in some mutant stocks the frequency and degree of mutant expression were reduced in old, nutrient-reduced, fly bottles. One of these, *obake*, was confirmed in detail (Atallah et al. 2004). We wondered whether nutritional plasticity might buffer or canalize mutants to produce a wild-type phenotype during nutritional stress.

From a genetic perspective, phenotypic change in response to the environment might be mediated by a mutation in a protein-coding gene(s), changes in noncoding regions such as *cis*-regulatory elements, or intervention of transposable elements, the so-called jumping genes (Kidwell and Lisch 1997). These mechanisms are not mutually exclusive, providing a framework for considering possible genetic influences.

We have investigated possible genetic/metabolic mechanisms underlying nutritional plasticity of *vermilion* (v) mutants in the fruit fly (*Drosophila melanogaster*) eye color pathway. The red eyes of normal flies contain both an orange pigment and a brown pigment. When there is a loss of function mutation in the orange pigment pathway the eye color is brown and conversely such a mutation in the brown eye pathway leads to a bright orange phenotype. The mutations in the *v* gene have a bright orange eye phenotype, owing to a compromised enzyme, tryptophan 2,3-dioxygenase (TDO), in the first step of the pathway converting the amino acid tryptophan to kynurenine in the sequence leading to brown eye pigment (Walker et al. 1986).

The first mutation discovered in the *vermilion gene* (v^{1}) was a spontaneous allele described by Morgan and Bridges (1916). Subsequent work revealed a surprising phenotypic response under nutrient deprivation: the v^{1} mutant phenotype reverted to normal (Beadle et al. 1939). Another allele of *vermilion*, v^{36f}, does not respond to low nutrition by increasing brown eye pigment sufficiently to be visualized. v^{1} lends itself to further study since it was established that kynurenine is normally accumulated in cells of the anterior portion of the fat body of fruit fly larvae before its transfer to the developing eyes during pupation, where brown pigment is then formed (Rizki and Rizki 1963, 1968). Since kynurenine is autofluorescent, its concentration in fat body cells can be visualized (Rizki 1963; Rizki and Rizki 1963). Observations with fluorescence microscopy showed that v^1 fly larvae produced very little autofluorescent kynurenine in fat body cells under normal food conditions. The molecular nature of several vermilion mutations has been determined, and both v^1 and v^{36f} contain transposable elements in different regions of the RNA transcript of the gene for TDO (Walker et al. 1986; Searles et al. 1990).

Our experimental approach was to test a potential regulator of kynurenine ex vivo in three different genotypes, a wild-type DGRP [28] inbred line (Mackay et al. 2012), which was crossed into both the v^1*and* v^{36f} strains to reduce the possible influence of different background genes. Fat bodies from yeast-fed larvae at the late third instar (crawling stage) were dissected out and transferred to 1.5 mL microfuge tubes containing 1 mL of Schneider's media. The fed group and the group to be starved were placed into 1 mL 20% sucrose for 6 h. Half of the tubes in the fed group and half in the deprived group were also exposed to a potential regulator of kynurenine accumulation. This experimental design produced 12 different gene X treatment combinations: three genotypes each with two nutritional conditions, fed or starved, and two treatment conditions, control or added potential regulator.

The most likely regulators of nutrition are members of the insulin-TOR nutritional sensing pathways. Insulin has been implicated in diverse processes hinging on nutritional state, not only caste determination in honey bees, but also insect ornaments and weapons (Emlen et al. 2012), and the evolution of eusociality in ants (Chandra et al. 2018). Here we report the effect of insulin on kynurenine accumulation in fat bodies. Our hypothesis was that insulin would rescue kynurenine production in v^1 fed flies, but not in v^{36f} fat bodies.

Figure 6.1 shows representative photos of fat body cells with green autofluorescent kynurenine accumulation. As predicted from eye color in fed and starved adult eyes, the v^1 fed fat body (e) had low fluorescence, while starved fat body cells had abundant kynurenine (g). As hypothesized, substantial kynurenine rescue in the v^1 fed + insulin fat body image is seen (f), whereas comparably treated v^{36f} (j) are not rescued. In Fig. 6.2 Tukey's test to determine those treatments not significantly different from each other at $p < 0.05$ is represented by a continuous black line beneath the treatments. Notably, v^1 starved + insulin has the highest value of all, and is significantly higher than the v^1 starved control. As predicted, v^{36f} had low fluorescence values regardless of treatment, although its central distribution of fluorescence requires more investigation. The low accumulation of the DGRP wild-type line when starved (Fig. 6.1c, d) was not predicted, although it was found that relative to other DGRP strains, DGRP [28] was below the mean for starvation resistance (Mackay et al. 2012 supplement).

The rescue of kynurenine accumulation by exogenous insulin in fed v^1 suggests that TDO enzyme function associated with this mutation was restored compared with the v^{36f} mutation. Putting together data from larval physiology, the DNA basis of *v* mutations and molecular action of a recessive allele at an independent gene locus, which suppresses the mutant phenotype of v^1, a mechanism for nutritional plasticity can be proposed.

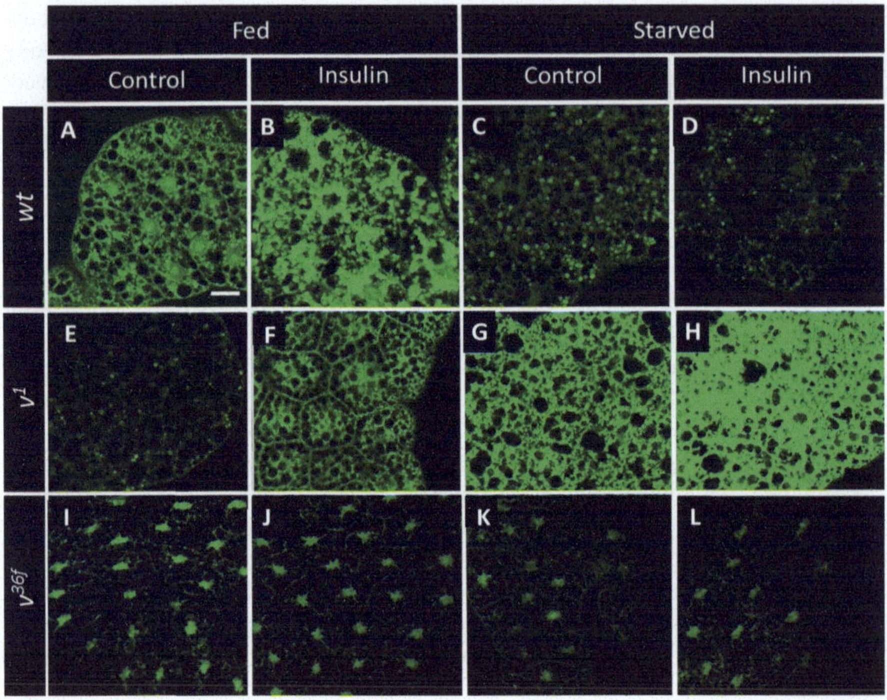

Fig. 6.1 Confocal images displaying kynurenine autofluorescence distribution in fruit fly fat bodies. Anterior portions of fat bodies of late third instar larvae/white pupae from DGRP [28](wt), *vermilion*[1] (v^1), *and vermilion*[36f] (v^{36f}) fly lines containing the DGRP [28] genetic background. Staged larvae were grown on standard yeast-supplemented fly food at 25 °C up to late third instar (96 h after egg-laying/crawling stage). Late third instar larvae were then removed from the food, washed in PBS, and transferred to slides containing BSS buffer. Fat body tissues were dissected from the anterior half of the larvae and transferred to 1.5 mL microfuge tubes containing 1 mL of Schneider's insect media (fed) or 1 mL of 20% sucrose solution (starved). Insulin was then added to half the tubes for each nutrient condition and allowed to incubate for another 6 h at 25 °C
Fat bodies were fixed and mounted for fluorescence microscopy and representative images from each treatment are presented. In the control groups lacking insulin, robust, generalized fluorescence is only found in fed wt (**a**) and starved v^1 (**g**). Insulin treatment rescued fluorescence in v^1 fed (**f**) and appears to have dramatically increased kynurenine fluorescence in v^1 starved (**h**). v^{36f} appears to respond similarly to all treatments (**i–l**) with centralized fluorescence somewhat resembling that found in wt controls (**a**) with low intensities also seen in *wt* starved (**c–d**) and v^1 fed (**e**). Scale bar: 30 μm

It is known that the fly fat body produces a *Drosophila* insulin-like factor (dilp6) in quantity during the post-feeding period before pupation, and also in response to starvation (Kannan and Fridell 2013). This suggests that insulin is increased naturally in starved v^1 fat bodies, but not in fed fat bodies, thus producing an unsuspected paracrine effect of boosting TDO function, and consequently kynurenine accumulation. The problem remains explaining the mechanism of increased TDO

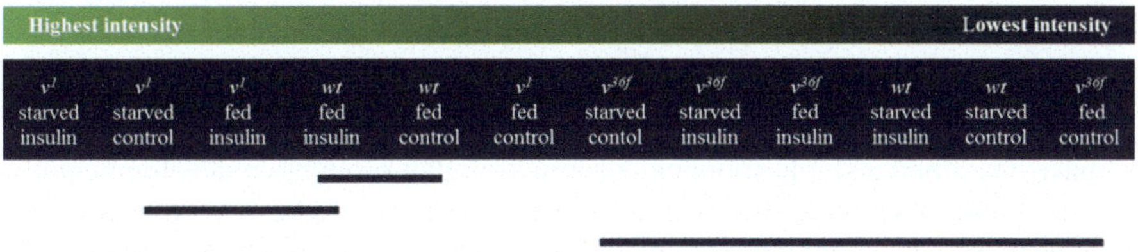

Fig. 6.2 Distribution of kynurenine between different treatments studied. To quantify the amount of kynurenine fluorescence, 10 cells from each of the 5 fat bodies in the 12 different treatments were measured using *ImageJ* 1.48v software (*NIH*). These data were analyzed using Analysis of Variance Software in R version 3.3.2, car library. The highly significant differences for all treatments were further analyzed using Tukey's Honest significant difference test (*Sigma Plot*). The treatment combinations appear left to right from the highest to lowest fluorescence means. Where two treatment values are underlined, no significant difference was noted at $P \leq 0.05$. These results support the observation that starved v^1 supplied with insulin had the highest level of kynurenine. v^1 starved control and fed insulin conditions had fluorescence intensities greater than wt fed control. v^1 fed control had higher intensities than any of the v^{36f} genotype treatments and the starved *wt* whether in the control- or insulin-treated group

function in the v^1 mutant during starvation, and insulin's putative role. Clues may come from the possibility that the insulin-TOR nutrition-sensing network may utilize strategies similar to those of the suppressor sable mutation (su^s). su^s is thought to act to stabilize the RNA transcribed from the DNA of the TDO gene (Kuan et al. 2009). In order to make a protein, the RNA must be spliced into pieces to make a transcript which can then be translated into a protein. During this complex period, RNA molecules deemed to be imperfect are eliminated. The transposable element (TE) in v^1 is not part of the final translated transcript, but does have an internal splice site which can be used by splicing machinery to remove the TE from the transcript (Fridell et al. 1990).

RNA stability is apparently compromised during RNA splicing when there is good nutrition, but where su^s is present, tests have shown that transcripts containing parts of the TE and those where the TE is removed both increase quantitatively in v^1 flies. The increase in TDO transcripts under this condition suggests that su^s stabilizes RNA during the splicing process (Fridell and Searles 1994; Geyer et al. 1991; Kuan et al. 2009). The TE in v^{36f}, on the other hand, is within the part of the transcript that is translated into protein (Walker et al. 1986; Searles et al. 1990), and this appears to be a reason it produces less functional enzyme (Tartof 1969). We speculate that insulin either directly or indirectly influences the stability of RNA during splicing, thus increasing the quantity of translatable transcript, although there may be other mechanisms involved. It would be interesting to determine whether mutations other than v^1 that show reductions in mutant expression during starvation are rescued with exogenous insulin under normal nutritional conditions. Others have suggested that insulin sensitivity may be crucial in modulating nutritional plasticity through the effects of the forkhead gene FOXO (Tang et al. 2011).

6.2.1 The Evolutionary Implications of a Single-Gene Mutation That Produces Developmental Plasticity

We have described a single spontaneous mutation that produces a developmentally plastic, visible phenotype that is nutritionally dependent and can be traced to the metabolism of a discrete group of cells. We provide evidence that this plasticity is a result of co-opting the insulin pathway, which is an ancient and nutritionally sensitive pathway. In species-wide developmental plasticity, the genetic basis is almost certainly polygenic rather than a mutation at a single locus, yet, even in these cases, we suggest that ancient pathways will likely be involved, and that even if the initial plasticity were associated with a single mutation, polygenic modifiers may be selected over time, concealing this history. For example, vitamin D signaling appears to be involved in diapause of a vertebrate such as the annual killifish (Romney et al. 2018), dauer formation in *Caenorhabditis rhabditis* (Fielenbach and Antebi 2008), and possibly can induce diapause-like metabolism in taxa whose life histories do not include a dormant state (Romney et al. 2018). It is likely that in evolution, the genetics around vitamin D signaling has been modified in such vastly different organisms.

Developmental plasticity has been heralded as a stepping stone to fixation of new phenotypes in evolution (West-Eberhard 2003). If phenotypic developmental plasticity can be initiated in individual genes with a reasonable frequency, this may reduce the time needed for adaptation compared to that required if several polygenic loci must be mutated for the same effect. Given the relatively high rate of TE transposition in particular organisms, some of it associated with stress (García Guerreiro 2012), this is one route for fairly rapid adaptation to changing environments.

Developmental plasticity is also a potential route for what may be termed reverse evolution, in which phenotypes of species within a clade may seesaw back and forth over time, as is true of sex comb positioning (Kopp 2011), a process in some developing fly species, to be discussed below (West-Eberhard 2003).

6.3 Internal Cellular Environment: Cell Shape, Size, and Movement Are Crucial Factors in Morphogenesis and Its Evolution

Sex combs are a male-specific group of leg bristles, with spectacular morphological diversity during evolution (Atallah 2008; True 2008; Kopp 2011) (Fig. 6.3a). Some of the common morphological variations observed in *Drosophila* sex combs include tooth number (Ahuja and Singh 2008), tooth pigmentation, and comb orientation (Tanaka et al. 2009) (Fig. 6.3a, b). Sex combs are used in different courtship behaviors among *Drosophila* species, leading to an increase in the amount of successful mating (Ng and Kopp 2008; Kopp 2011). Careful developmental studies show that the cellular environment surrounding the comb can play an important role during evolution (Atallah et al. 2009a, b; Tanaka et al. 2009; Malagón et al. 2014). In the

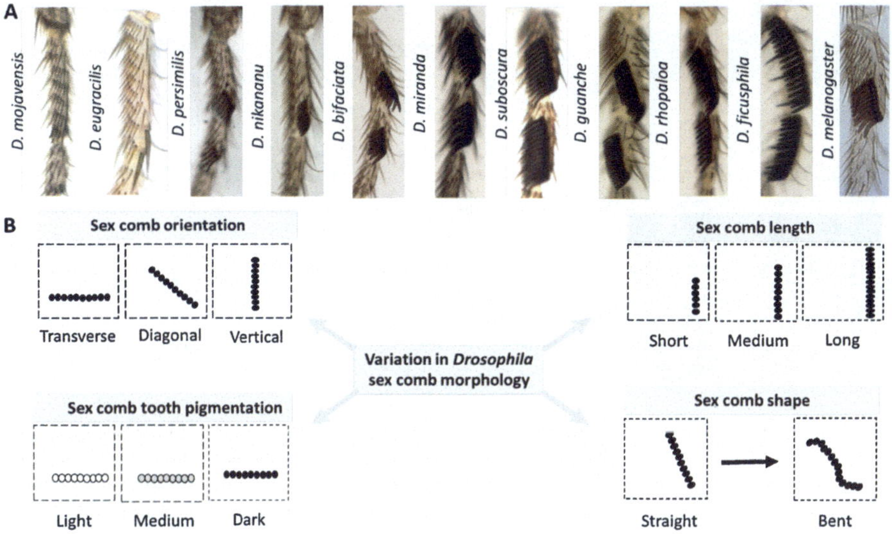

Fig. 6.3 Morphological variation observed in sex combs among *Drosophila species*. (**a**) Adult legs of different *Drosophila* species. (**b**) Schematics displaying examples of morphological modifications in each of the four sex comb characteristics. By using either selection or genetic mutations in *Drosophila melanogaster*, we reproduced the type variation observed in other *Drosophila* species (modified from Ho et al. 2018)

next section, we describe changes in the internal cellular environment of a sex comb during development, and then explain its role in evolution.

Despite the incredible morphological variation observed among *Drosophila* sex combs, this bristle row usually displays three similar features in its cellular environment. First, sex combs are always surrounded by an epithelial tissue, which has few individual bristles embedded within it (Atallah et al. 2012; Liang et al. 2013; Malagon 2013) (Fig. 6.4). The second feature is the presence of bristle rows proximal to the comb (Malagón et al. 2014), which are used as cleaning devices for the eyes and wings (Kopp 2011). The third and final feature is the presence of the joint in the distal region of the comb. Due to an abundance of genetic tools, *D. melanogaster* has been used as a system to study how these three features of the cellular environment bias comb evolution.

6.3.1 *The Cellular Environment Plays a Major Role in Determining Comb Orientation in Single Rotating Combs*

In *D. melanogaster*, the adult sex comb is a static, vertical straight row of bristles located in the first tarsal segment of the foreleg. In contrast, the developing comb is a dynamic structure that moves and bends during morphogenesis. The *D. melanogaster* comb rotates during development from a transverse to a vertical position

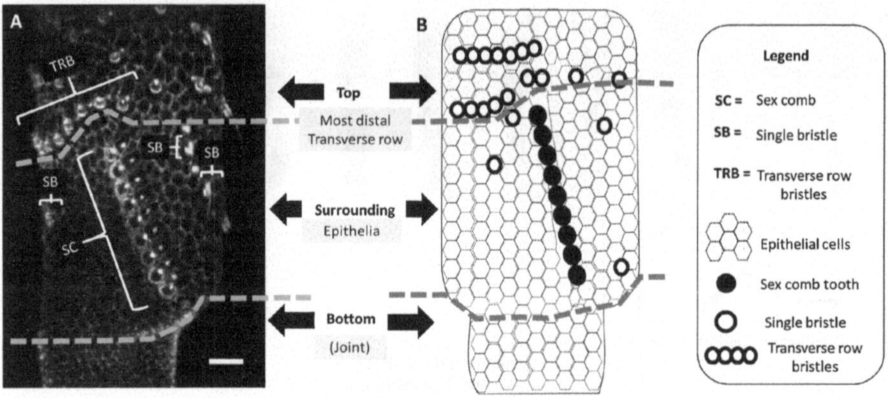

Fig. 6.4 The internal environment of the *Drosophila* sex combs. (**a**) Confocal image and (**b**) schematic of a developing *D. melanogaster* comb and its internal environment. The internal environment is composed of a combination of epithelial tissue and different types of bristles. Three regions are discerned: top (most distal transverse row), surrounding epithelia, and bottom (joint). Scale bar: 10 μm

during the pupal stage (Fig. 6.5a). Developmental studies have shown that the motive force for the rotation is the cellular environment rather than the comb itself (Ho et al. 2018). In vivo and in silico approaches suggest a simple mechanical mechanism: cell growth underneath the comb pushes this row of bristles, passively moving the comb and leading to a dramatic reorganization of the epithelial tissue above the comb (Liang et al. 2013; Malagon 2013; Ho et al. 2018). Malagon (2013) suggested that small modifications to this simple cellular "push" mechanism can explain the rapid evolution in *Drosophila* comb orientation (Ho et al. 2018; Liang et al. 2013; Malagon 2013).

The orientation of *Drosophila* sex combs can be divided into three main categories: transverse (0°–30°), diagonal (30°–60°), and vertical (60°–90°) (Liang et al. 2013; Malagon 2013; Ho et al. 2018) (Fig. 6.1a). Interestingly, phylogenetic analyses showed that changes in comb orientation display a rapid and frequent variation during evolution (Atallah et al. 2009a, b, 2012; Tanaka et al. 2009). In *D. melanogaster*, many mutations affecting comb development phenocopy the diversity of comb orientations, suggesting that the high evolvability in comb orientation is due to subtle modifications in the push mechanism (Liang et al. 2013; Malagon 2013; Ho et al. 2018) (Fig. 6.3a). Analysis of the cellular processes modified by those mutations showed that a reduction in cell growth also causes a reduction in comb movement. In contrast, genetic perturbations affecting the comb itself have little or no effect on the rotation (Tanaka et al. 2011; Atallah et al. 2014). As a result, only the cellular environment surrounding the *D. melanogaster* comb plays a major role in determining comb orientation and its evolution (Fig. 6.5d).

Although we have only concentrated on the epithelial tissue surrounding the sex comb, the other two components of the cellular environment, most distal

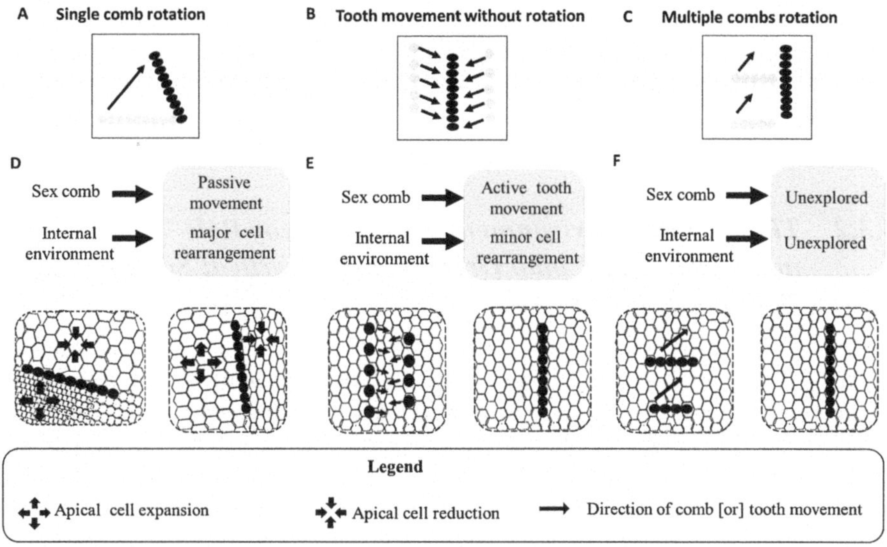

Fig. 6.5 Developmental mechanisms involved in comb orientation and role of internal environment. (**a**, **b**) Schematics of developmental mechanisms responsible for comb orientation. Single comb rotation (SCR). (**b**) Sex comb without rotation (SCWR). Teeth are born in a vertical position and move toward each other. (**c**, **d**) Schematics displaying cellular changes observed during morphogenesis of comb orientation. (**e**, **f**) Schematics illustrating the formation of sex combs and the types of cell changes surrounding them responsible for final positioning in different developmental scenarios

transverse row and joint, also play an important role in shaping comb evolution (Fig. 6.4). For example, the most distal transverse row sets the limit for comb length in *Drosophila melanogaster*. Artificial selection experiments in *D. melanogaster* showed that comb length can dramatically reduce or increase in bristle number in a short time (24 generations) (Ahuja and Singh 2008). Despite this incredible flexibility of *D. melanogaster* comb length, the cellular environment surrounding the comb determines a limit in the maximum comb length. Malagón et al. (2014) demonstrated that *D. melanogaster* long combs are unable to rotate due to a length threshold imposed by the closest transverse row, which can block comb rotation. Morphological studies of other *Drosophila* species support this idea, showing that long combs are accompanied by various leg modifications to make room for rotation, including increase in tarsal segment length and removal of blocking transverse rows (Malagon and Larsen 2015). Finally, a similar space limitation has been observed with the joint between the first and second tarsal segments: Reduced space blocks the rotation of the bottom section of the comb. Preliminary studies found similar atypical bent shapes close to the joint in long combs in both *D. melanogaster* lines and *Drosophila* species (Lee et al. 2011; Malagón et al. 2014).

Although *D. melanogaster* demonstrates that the cellular environment is important for sex comb evolution, the study of other *Drosophila* species complements and enriches this discussion, showing variations on the theme of achieving a vertical sex comb.

6.3.2 The Cellular Environment Plays a Minor Role in Determining Comb Orientation in Vertical Sex Combs Without Rotation

Studying other *Drosophila* sex combs shows the existence of different developmental mechanisms to achieve a vertical comb orientation (Atallah et al. 2009a, b, 2012; Tanaka et al. 2009) (Fig. 6.3). For example, a vertical comb can be formed without rotation (VCWR). The sex comb teeth appear already in a vertical stripe, although scattered and separated. Then they move individually and attach to one another forming a row (Fig. 6.5b). The mechanical challenge of moving a bristle row compared to moving individual bristles involves a completely different role of the cellular environment. Contrary to the case of the rotating comb in *D. melanogaster*, in VCWR, the bristle cells themselves are responsible for the movement, while the epithelium surrounding the individual teeth passively allows this movement.

6.3.3 Evolutionary Implications

Drosophila sex combs provide an insight into some of the ways the cellular environment may influence morphological evolution. For example, *D. melanogaster* demonstrates that the cellular environment of a phenotypic trait can be fundamental to understanding its ontogeny and evolution (Tanaka et al. 2009; Liang et al. 2013; Malagón et al. 2014; Malagon and Khan 2016). As previously shown, the epithelial tissue surrounding the comb and other bristle rows highly influences comb orientation, length, and shape in this species (Tanaka et al. 2009; Liang et al. 2013; Malagón et al. 2014; Malagon and Khan 2016). However, studying other *Drosophila* combs also shows that a completely different scenario can occur as shown by the VCWR. Interestingly, these opposite roles of the cellular environment seem to be easily interchangeable during sex comb evolution.

The flexibility of the cellular environment is evidently due to the co-option of ancient morphogenetic processes already operating in *Drosophila* legs. For example, cell expansion is a fundamental developmental process in most if not all developing organisms, and in *Drosophila* legs it is prominent during evagination (Fristrom and Chihara 1978). The SCR mechanism reuses this cell behavior to push a bristle row from a transverse to a vertical orientation. Similarly, the developmental mechanism to form horizontal bristle rows from initially separated cells is an entrenched feature in all *Drosophila* species in the form of transverse rows, and in the VCWR mechanism this strategy is reused with a slight modification: individual sex comb teeth attach along the vertical rather than the horizontal axis.

6.4 Conclusion

We have described some of the underpinnings of environmentally related developmentally plastic responses. In one case, the plasticity was due to a single mutation. In the other case, variation in a morphological trait was described for a clade associated with internal plastic potential. In both cases, the developmental mechanisms are rooted in ancient pathways, the former at a metabolic level and the latter at a cellular level. These findings are in concert with a growing appreciation that evolutionary novelties are often rooted in new uses of old pathways, referred to as exaptations (Gould and Vrba 1982).

Acknowledgements We thank Carol Schwartz for alerting EL to the nutritional peculiarities of the v^1 mutation, Ian Dworkin for providing the script for the analysis of variance in R, and Reuven Dukas for recommending and providing the DGRP [28] strain used. We are grateful for helpful comments from the following reviewers: Lewis Held, Rama Singh, Carol Schwartz, and Joel Atallah.

References

Ahuja A, Singh R (2008) Variation and evolution of male sex combs in Drosophila: nature of selection response and theories of genetic variation for sexual traits. Genetics 179:503–509

Atallah J (2008) The development and evolution of complex patterns: the Drosophila sex comb as a model system. University of Toronto

Atallah J, Dworkin I, Cheung U et al (2004) The environmental and genetic regulation of obake expressivity: morphogenetic fields as evolvable systems. Evol Dev 6:114–122

Atallah J, Liu N, Dennis P et al (2009a) Cell dynamics and developmental bias in the ontogeny of a complex sexually dimorphic trait in Drosophila melanogaster. Evol Dev 11:191–204

Atallah J, Liu N, Dennis P et al (2009b) Developmental constraints and convergent evolution in Drosophila sex comb formation. Evol Dev 11:205–218

Atallah J, Watabe H, Kopp A (2012) Many ways to make a novel structure: a new mode of sex comb development in Drosophilidae. Evol Dev 14(6):476–483

Atallah J, Vurens G, Mavong S et al (2014) Sex-specific repression of dachshund is required for Drosophila sex comb development. Dev Biol 386:440–4477

Beadle G, Tatum E, Clancy C (1939) Food level in relation to the rate of development and eye pigmentation in Drosophila melanogaster. Biol Bull 75:447–462

Chandra V, Fetter-Pruneda I, Oxley P et al (2018) Social regulation of insulin signaling and the evolution of eusociality in ants. Science 361:398–402

Emlen D, Warren I, Johns A et al (2012) A mechanism of extreme growth and reliable signalling in sexually selected ornaments and weapons. Science 337(6096):860–864

Fielenbach N, Antebi A (2008) C. elegans dauer formation and the molecular basis of plasticity. Genes Dev 22:2149–2165

Fridell R, Searles L (1994) Evidence for a role of the Drosophila melanogaster suppressor of sable gene in the pre-mRNA splicing pathway. Mol Cell Biol 14:859–867

Fridell RA, Pret A, Searles L et al (1990) A retrotransposon 412 insertion within an exon of the Drosophila melanogaster vermilion gene is spliced from the precursor RNA. Genes Dev 4:559–566
Fristrom D, Chihara C (1978) The mechanism of evagination of imaginal discs of Drosophila melanogaster. Dev Biol 570:564–570
García Guerreiro M (2012) What makes transposable elements move in the Drosophila genome? Heredity 108:461–468
Geyer P, Chien A, Corces V et al (1991) Mutations in the su(s) gene affect RNA processing in Drosophila melanogaster. Proc Natl Acad Sci U S A 83:404–408
Gould J, Vrba E (1982) Exaptation—a missing term in the science of form. Paleo Paleobiology 8:4–15
Ho E, Malagon J, Ahuja A et al (2018) Rotation of sex combs in Drosophila melanogaster requires precise and coordinated spatio-temporal dynamics from forces generated by epithelial cells. PLoS Comput Biol 14:e1006455
Kannan K, Fridell Y (2013) Functional implications of Drosophila insulin-like peptides in metabolism, aging, and dietary restriction. Front Physiol 4:1–8
Kidwell M, Lisch D (1997) Transposable elements as source of variation in animals and plants. Proc Natl Acad Sci U S A 94:7704–7711
Kopp A (2011) Drosophila sex combs as a model of evolutionary innovations. Evol Dev 13:504–522
Kuan Y, Brewer-Jensen P, Bai W et al (2009) Drosophila suppressor of sable protein [Su(s)] promotes degradation of aberrant and transposon-derived RNAs. Mol Cell Biol 29:5590–5603
Lee J, Malagon J, Larsen E (2011) The case of fly sex combs: using a model organism to infer mechanisms of morphological evolution. JULS 5:29–30
Liang Y, Larsen E, Malagon J (2013) Cell dynamics of sex comb morphogenesis in Drosophila melanogaster. JULS 7:26–35
Mackay T, Richards S, Stone E (2012) The Drosophila melanogaster genetic reference panel. Nature 482:173–178
Malagon J (2013) Sex combs in motion: cellular processes involved in sex comb rotation in Drosophila melanogaster. PhD Thesis. University of Toronto
Malagon J, Khan W (2016) Evolution of allometric changes in fruit fly legs: a developmentally entrenched story. Acta Biol Colombiana 21:509. https://doi.org/10.15446/abc.v21n3.53650
Malagon J, Larsen E (2015) Heredity and self-organization: partners in the generation and evolution of phenotypes. Int Rev Cell Mol Biol 315:153–181
Malagón J, Ahuja A, Sivapatham G et al (2014) Evolution of Drosophila sex comb length illustrates the inextricable interplay between selection and variation. Proc Natl Acad Sci U S A 111:E4103–E4109. https://doi.org/10.1073/pnas.1322342111
Morgan T, Bridges C (1916) Sex-linked inheritance in Drosophila. Carnegie Inst Wash 237:1–88
Ng C, Kopp A (2008) Sex combs are important for male mating success in Drosophila melanogaster. Behav Genet 38:195–201
Rizki T (1963) Genetic control of cytodifferentiation. J Cell Biol 16:513–520
Rizki T, Rizki R (1963) An inducible enzyme system in Drosophila melanogaster. J Cell Biol 15:87–92
Rizki T, Rizki R (1968) Allele specific patterns of suppression of the vermilion locus in Drosophila melanogaster. Genetics 59:477–485
Romney A, Davis E, Corona M et al (2018) Temperature-dependent vitamin D signaling regulates developmental trajectory associated with diapause in an annual killifish. Proc Natl Acad Sci U S A 115:12,763–12,768
Searles L, Ruth R, Pret A et al (1990) Structure and transcription of the Drosophila melanogaster vermilion gene and Several mutant alleles. Mol Cell Biol 10:1423–1431
Tanaka K, Barmina O, Kopp A (2009) Distinct developmental mechanisms underlie the evolutionary diversification of Drosophila sex combs. Proc Natl Acad Sci U S A 106:4764–4769
Tanaka K, Barmina O, Sanders L et al (2011) Evolution of sex-specific traits through changes in HOX-dependent doublesex expression. PLoS Biol 9:e1001131. https://doi.org/10.1371/journal.pbio.1001131

Tang H, Smith-Caldas S, Driscoll M et al (2011) FOXO regulates organ-specific phenotypic plasticity in Drosophila. PLoS Genet 7:1–12
Tartof K (1969) Interacting gene systems. I. The regulation of tryptophan pyrrolase by the vermilion-suppressor of vermilion system in Drosophila. Genetics 62:781–795
True J (2008) Combing evolution. Evol Dev 10:400–402
Walker A, Howells A, Tearle R (1986) Cloning and characterization of the vermilion gene of Drosophila melanogaster. Mol Gen Genet 202:102–107
West-Eberhard M (2003) Developmental plasticity and evolution. Oxford University Press, Oxford

Chapter 7
Extreme Morphological Plasticity Within *Orbulina*-“*Praeorbulina*-Like” Assemblages Related to Environmental Stress

Ahmed Belhadji, Annachiara Bartolini, Linda Rossignol, Lahcène Belkebir, and Jean Guex

Abstract Planktic foraminifera, unicellular microzooplankton with a calcitic shell, have produced an exceptional fossil record, revealing an invaluable archive of biodiversity, morphological and evolutionary changes.

The evolutionary lineage starting from *Trilobatus* Spezzaferri 2015 (= “*Globigerinoides*”) culminating in *Orbulina universa* d’Orbigny 1839 is a fascinating example of peramorphic spherisation lineage (increasing involution, coupled with increasing shell curvature).

This chapter focuses on the extreme morphological variability observed in the *Orbulina* group in some horizons from Chélif Basin in Algeria, just preceding the well-known Messinian (Late Miocene) salinity crisis in the Mediterranean basin. Surprisingly, in such horizons, spherical *Orbulina universa* lineage end-member specimens coexist with ancestor-like morphotypes, such as *Orbulina suturalis* Brönnimann 1951 and the supposed extinct *Praeorbulina* Olsson 1964, as well as with malformed specimens. Many authors considered in fact that *Praeorbulina* last occurred within the Langhian stage in the Middle Miocene.

A. Belhadji · L. Belkebir
Laboratoire de Paléontologie Stratigraphique et Paléoenvironnements, Université d’Oran 2 Mohamed Ben Ahmed, Oran, Algeria
e-mail: belhadji.ahmed@univ-oran2.dz; belkebir.lahcene@univ-oran2.dz

A. Bartolini (✉)
Centre de recherche en paléontologie—Paris, UMR 7207, CNRS-MNHN-SU, Muséum National d’Histoire Naturelle, CP38, Paris, France
e-mail: bartolini@mnhn.fr

L. Rossignol
Laboratoire Environnement et Paléoenvironnement Océaniques et Continentaux UMR CNRS 5805, Université de Bordeaux, Pessac, France
e-mail: linda.rossignol@u-bordeaux.fr

J. Guex
Geopolis—University, Lausanne, Switzerland
e-mail: jean.guex@unil.ch

J. Guex et al. (eds.), *Morphogenesis, Environmental Stress and Reverse Evolution*, https://doi.org/10.1007/978-3-030-47279-5_7

A similar recovery of individuals which show an intergradation between a typical *Orbulina* morphology and morphologies close to the ancestors *Orbulina suturalis* and *Praeorbulina* was also reported in Last Glacial Maximum sediments from the northern Arabian Sea. In this Late Pleistocene case, AMS ^{14}C data showed clearly unreworked character of this "*Praeorbulina*-like" populations.

We discuss the possible link between this extreme morphological plasticity of *Orbulina* group in specific time horizons and possible stress conditions of the water column.

Keywords Planktic foraminifera · Morphological plasticity · Reverse evolution · Environmental stress

7.1 The *Trilobatus–Praeorbulina–Orbulina* Evolutionary Lineage

Planktic foraminifera have produced an exceptional fossil record, revealing an invaluable archive of biodiversity, morphological and evolutionary changes. They are widely utilised for the biostratigraphy of Cretaceous and Cenozoic marine sediments, as well as proxies for paleoceanographic reconstructions (Kucera 2007). The"*Globigerinoides*"–*Praeorbulina–Orbulina* evolutionary lineage characterises the upper Burdigalian to Langhian interval (lower-middle Miocene) and has been employed in most of the planktic foraminifer biostratigraphic zonations both in open ocean regions and in the Mediterranean (e.g., Blow 1969; Bolli and Saunders 1985; Iaccarino 1985; Berggren et al. 1995; Wade et al. 2011; Turco et al. 2011).

Recently, Spezzaferri et al. (2015), combining fossil and molecular genetic data, distinguished within "*Globigerinoides*" group the existence of two distinct lineages (*Globigerinoides and Trilobatus*) that evolved independently and erected the new genus *Trilobatus. Trilobatus* is paraphyletic and gave rise to the *Praeorbulina–Orbulina* and *Sphaeroidinellopsis–Sphaeroidinella* lineages.

In the following we will focus on the line of evolution, which leads to *Orbulina universa* d'Orbigny 1839, via *Trilobatus sicanus* (de Stefani 1952), *Praeorbulina glomerosa curva* (Blow 1956), *P. glomerosa glomerosa* (Blow 1956), *P. glomerosa circularis* (Blow 1956) and *Orbulina suturalis* Brönnimann 1951 (Fig. 7.1). As the evolution proceeds, the last chamber tends to completely envelop all the earlier ones, while the apertures increase in number but decrease in size (Blow 1956). In later stage *Orbulina* the apertures migrate from the sutural position and they are distributed over the surface of the last chamber (Blow 1956). Moreover they become rounded large-pore "openings" (= areal apertures), well distinct from the smaller pore openings. The rounded large-pore openings act as apertures, allowing exchanges of food and other particles, while the small-pore openings are real pores

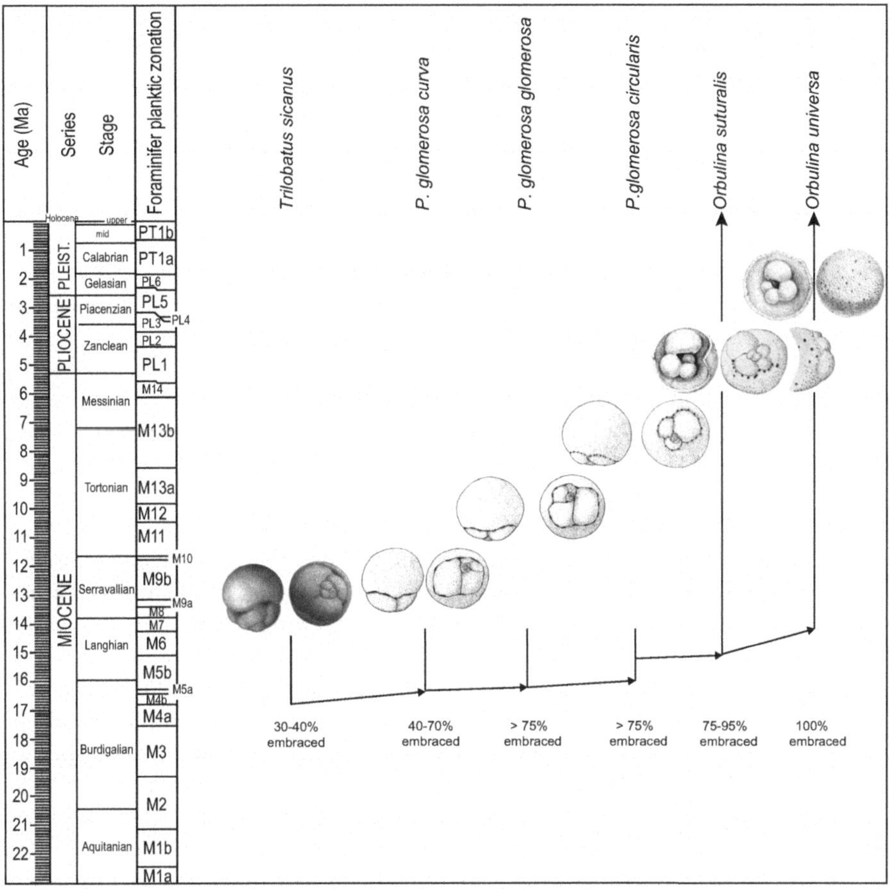

Fig. 7.1 *Trilobatus–Praeorbulina–Orbulina* evolutionary lineage (Blow 1956), modified after Boudagher-Fadel (2015). Stratigraphical ranges of *Trilobatus sicanus* (de Stefani 1952), *Praeorbulina glomerosa curva* (Blow 1956), *Praeorbulina glomerosa glomerosa* (Blow 1956), *Praeorbulina glomerosa circularis* (Blow 1956), *Orbulina suturalis* (Brönnimann 1951) and *Orbulina universa* (d'Orbigny 1839). The age of the First Occurrence datum of *Trilobatus sicanus* is according to Turco et al. (2011). The ages of the other datum events are according to the planktic foraminiferal biochronology and age calibration of Berggren et al. (1995), updated by Wade et al. (2011). The chronostratigraphic framework of Wade et al. (2011) has been modified for the Pliocene-Pleistocene interval in agreement with the more recent global chronostratigraphical correlation table for the last 2.7 million years (Cohen and Gibbard 2016). The percentage of last chamber embracement of the earlier portion of the test (=% embraced) is after Blow (1956). Images sources: *Trilobatus sicanus* holotype drawings from Blow (1969); *Praeorbulina glomerosa curva*, *Praeorbulina glomerosa glomerosa* and *Praeorbulina glomerosa circularis* holotype drawings from Blow (1956); *Orbulina suturalis* holotype drawings from Brönnimann (1951); *Orbulina suturalis* pre-adult stage and *Orbulina universa* drawings from Postuma (1971)

(Spero 1988; Schiebel and Hemleben 2017). Blow (1956) emended *Orbulina* d'Orbigny to include the presence of areal apertures.

The evolutionary lineage starting from a trochospiral *Trilobatus* (= "*Globigerinoides*") and culminating in spherical *Orbulina* had been suggested as early as 1940 by Cushman and Dorsey (Cushman and Dorsey 1940). On the basis of material collected from the Pozon Formation (Venezuela), Blow (1956) traced more in detail the evolution of the genus *Orbulina* from *Trilobatus trilobus* (Reuss 1850) and proposed two independent evolutionary bioseries, the first leading to *Orbulina universa* and the second to *Orbulina bilobata* d'Orbigny 1846. Two-chambered forms have been separated under the species *O. bilobata*, and even given generic rank (*Biorbulina*, Blow 1956). However such separation is controversial and according to several authors *Orbulina bilobata* is best regarded as variant (ecophenotype) of *O. universa* species (Schiebel and Hemleben 2017). High food availability might lead to the formation of a double sphere; *Biorbulina* tests were in fact observed to form when individuals are "overfed" in laboratory culture (Hemleben et al. 1989; Schiebel and Hemleben 2017).

Olsson (1964) erected the genus *Praeorbulina*, which includes the group of the species that are immediately ancestral to *Orbulina*. *Praeorbulina* is characterised by low trochospiral test with final chamber, globular to spherical, embracing the earlier portion of the test from 40% to over 75% (Fig. 7.1). It differs from *Orbulina* in the lack of areal apertures and from *Trilobatus* (= "*Globigerinoides*") in the lack in the adult stage of a distinct umbilicus and umbilical aperture (Olsson 1964). The first evolutionary appearance of the genus *Praeorbulina* (i.e. the *Praeorbulina* datum) is contentious in the literature, and it depends on the generic attribution of the species *sicanus* (De Stefani 1952) and its relationship with *Trilobatus bisphericus* (Todd 1954) and *Praeorbulina glomerosa curva* (Blow 1956) (see Turco et al. 2011 for more details). For example some authors considered *Trilobatus* (= *Globigerinoides*) *bisphericus* as junior synonym of *T. sicanus* and considered the first occurrence of *Praeorbulina glomerosa curva* as the first evolutionary appearance of the genus *Praeorbulina* (Blow 1956, 1969; Kennett and Srinivasan 1983), while other authors (Jenkins et al. 1981; Bolli and Saunders 1985; Berggren et al. 1995; Wade et al. 2011) considered *T. bisphericus* and *T. sicanus* as belonging to distinct genus referring *bisphericus* to the genus *Trilobatus* (= Globigerinoides) and *sicanus* to the genus *Praeorbulina*. In this last case, *Praeorbulina sicana* is believed to be an intermediate form in the lineage between *G. bisphericus* and *P. glomerosa curva* and the first occurrence of *Praeorbulina sicana* is considered as the *Praeorbulina* datum.

Turco et al. (2011), studying foraminiferal assemblages from several Mediterranean successions and comparing them with those from mid- and low-latitude Atlantic Ocean cores, concluded that the taxonomic concepts of Blow (1956) and Kennett and Srinivasan (1983), which take both the population variability and the gradual stratigraphical evolution of the biocharacters, can be better applied than those of Jenkins et al. (1981), which are mainly based on the characteristics of the holotypes without considering the variability of the species. Turco et al. (2011) showed that the evolution of *P. glomerosa curva* from *T. sicanus* was very gradational. The main evolutionary changes within *T. sicanus* populations leading to *Praeorbulina* are the appearance of specimens with three apertures and the gradual increase in the outline sphericity. Near-spherical individuals of *T. sicanus* with three apertures can be considered very close to *P. glomerosa curva*, which in turn is characterised by a spherical outline, at least four apertures with the primary aperture nearly undistinguishable from the supplementary ones and a close umbilicus.

Following Turco et al. (2011), in this chapter (Fig. 7.1) we consider the FO of *P. glomerosa curva* as the FO of *Praeorbulina* genus (= *Praeorbulina* datum). According to Wade et al. (2011), the first occurrence (FO) of *P. glomerosa curva* (*Praeorbulina* datum) is recorded near the top of the planktic foraminifer subzone M5a (16.3 Ma), in Burdigalian stage (Fig. 7.1). The genus *Praeorbulina* is considered a short-ranging biostratigraphical index (total occurrence of about 2.1 Myr), close to the early Miocene–middle Miocene transition (Kennett and Srinivasan 1983), within the planktic foraminifer biozones M5b-M6 (Wade et al. 2011) (Fig. 7.1). A bioseries within *Praeorbulina* is supposed to extend relatively rapidly from *Praeorbulina glomerosa curva* (FO 16.30 Ma), *Praeorbulina glomerosa glomerosa* (FO 16.20 Ma), *Praeorbulina glomerosa circularis* (FO 16.00 Ma) and in turn *Orbulina suturalis* (FO 15.10 Ma, *Orbulina* datum) (age calibrations from Wade et al. 2011).

In *Orbulina suturalis*, the last spherical chamber is not entirely enveloping the earlier test yet (Blow 1956). Tests show all transitions from forms in which the last chamber envelops about 75% of the earlier part to forms in which the earlier chambers are only just visible. *O. suturalis* is distinguished from *Praeorbulina glomerosa circularis* by the areal apertures over the surface of the last chamber. The initial evolutionary appearance of *O. suturalis* (*Orbulina* datum) has been recorded at the base of planktic foraminifer biozone M6, 15.1 Ma (Wade et al. 2011), within the Langhian stage (Fig. 7.1). Its stratigraphical distribution is up to recent (Kennett and Srinivasan 1983).

The first occurrence of *Orbulina universa* is reported imprecisely within the planktic foraminifer biozone N9 (Kennett and Srinivasan 1983). N9 planktic foraminifer biozone (Blow 1969) is equivalent to planktic foraminifer biozone M6 (Berggren et al. 1995), whose age is estimated to be 15.10–14.24 Ma (Lourens et al. 2004; Wade et al. 2011) (Fig. 7.1). *O. universa* is still extant. At the end of the evolutive series, it is characterised by a spherical test formed at the terminal ontogenic stage. The last spherical chamber is entirely enveloping the pre-adult trochospiral test, which is usually reduced. During its ontogeny, *O. universa* may attain different stages and test morphologies before the spherical adult terminal stage: a *Turborotalita*-like juvenile stage (first whorl of the test), a *Globigerina*-like pre-adult stage and sometimes, when small secondary apertures are visible on the spiral size, a *Globigerinoides*-like stage (Spero 1988; Schiebel and Hemleben 2017).

Three genotypes of *O. universa* are recognised to date, which differ in the size of pores and apertures, and they are probably related to certain water bodies and trophic conditions (De Vargas et al. 1999; Morard et al. 2009; André et al. 2014).

7.2 The Messinian Pre-salinity Crisis Case from the Mediterranean Basin

The Late Messinian is well known by a paleoenvironmental event of great magnitude in the Mediterranean area commonly known as the "Messinian salinity crisis" (MSC) (Ruggieri et al. 1967; Hsü et al. 1973), the beginning of which is estimated around 5.97 Ma (Manzi et al. 2013, 2018). However, in some marginal areas the first evidences of evaporitic conditions seem to have been occurred earlier around 6.14 Ma (Rouchy and Caruso 2006). The MSC is one of the giant evaporitic events of Earth's history, characterised by a huge volume of evaporites (at least 10^6 km^3) accumulated in the Mediterranean during a short time interval, i.e. 640 kyear (Krijgsman et al. 1999). One of the major causes evoked was a reduced water inflow from the Atlantic Ocean to the Mediterranean Sea, due to the interplay of tectonically driven constriction of the Atlantic gateways together with global-regional climatic and sea-level changes (Rouchy and Caruso 2006; Garcia-Castellanos and Villasenor 2011).

The Chelif Basin of Northwestern Algeria (Fig. 7.2), one of the largest Messinian marginal (peripheral) sub-basins (Rouchy et al. 2007), was located not so far from the Gibraltar arc seaway. So it was most likely highly sensitive to Atlantic water inflow fluctuations. It displays an almost complete sedimentary record of the Messinian salinity crisis and of the Miocene–Pliocene transition, characterised by the pre-evaporitic marly-diatomitic (Tripoli unit), evaporites and post-evaporitic "Lago–Mare" deposits (Rouchy et al. 2007). The evaporites (only calcium-sulphate deposits) reach local thickness of several hundreds of meters.

In the Kef el Biod section (Fig. 7.3), outcropping in the Chelif region (Fig. 7.2), well-preserved planktic foraminifer assemblages have been found throughout the pre-evaporitic marly-diatomitic, laminated-marly and limestone-marly units. High abundance peaks (up to 10–38%) of "*Praeorbulina*-like" individuals have been found in some punctual layers, coinciding with high abundance of abnormal tests found in total foraminiferal assemblages (Fig. 7.3, Plate 7.1 (10, 12)).

"*Praeorbulina*-like" morphotypes (Plate 7.1 (4–6, 9 and 11)) co-occur with *Orbulina suturalis* (Plate 7.1 (2, 3 and 8)) and *O. universa* (Plate 7.1 (1, 7)).

However, at a closer examination by secondary electronic microscope (SEM), it is notable that the specimens we named "*Praeorbulina*-like", according to Rossignol et al. (2011), present a continuum of transitional characters between *Praeorbulina* and *O. suturalis* (Plate 7.1).

The specimens illustrated in Plate 7.1 (4–6, 9 and 11) show close similarities with *Praeorbulina glomerosa*, according to the original description of Blow (1956). They present a relatively weak last chamber embracement (50–70%), and, consequently, the smaller penultimate and third chambers are well apparent and marked by sutures more or less deeply incised. Some individuals present numerous apertures lying in the suture separating the last chamber from the earlier chambers, as short narrow slits, typical of *Praeorbulina glomerosa glomerosa* (e.g. Plate 7.1 (9)) or as subcircularis-circularis openings, typical of *Praeorbulina glomerosa*

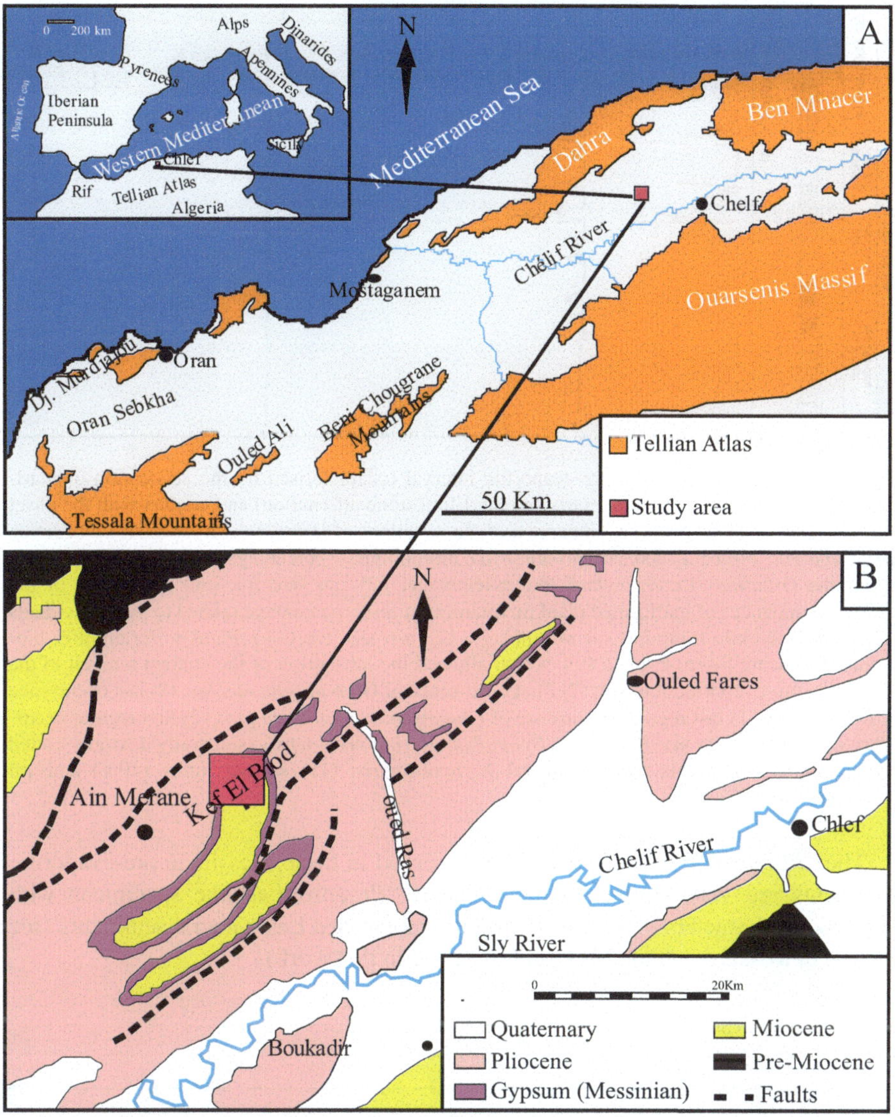

Fig. 7.2 (**a**) Location map of the Kef el Biod section, Chelif region, Northwestern Algeria (after Perrodon 1957). (**b**) Geological map of the eastern part of the Chelif basin (after Tauecchio and Marks 1973).

circularis (e.g. Plate 7.1 (11)). However, all the individuals observed under the SEM present also large rounded openings (= circular apertures) over the surface of the last chamber, which is considered as distinctive character of *Orbulina* genus, according to Blow (1956).

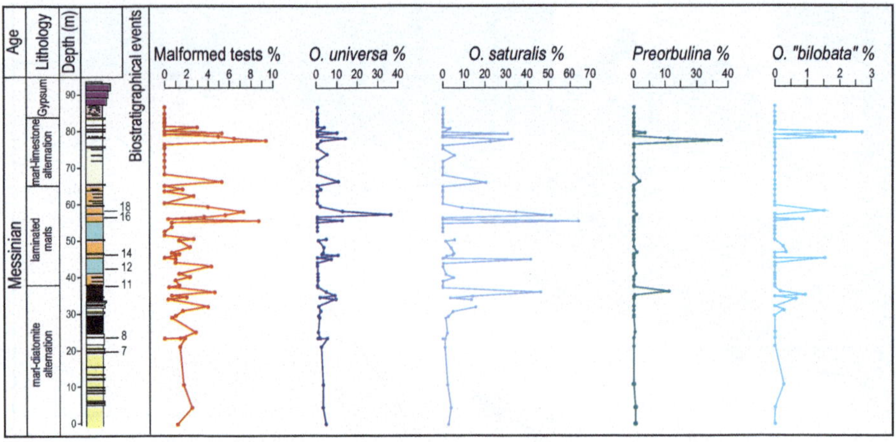

Fig. 7.3 Kef el Biod section, pre-evaporitic interval (characterised by the succession of marl-diatomite alternation, laminated marls and marl-limestone alternation) and contact with the onset of the evaporites (= Gypsum). Comparison of the variations of *Orbulina universa*, *Orbulina suturalis*, *Orbulina* "bilobata" and *Praeorbulina*-like morphotypes (percentage related to 100 individuals of the *Orbulina*-"*Praeorbulina*"-like assemblage, 125 μm size fraction) with those of the relative abundances of malformed planktic foraminifer tests (percentage related to 300 individuals of the total planktic foraminiferal assemblage, 125 μm size fraction). Planktic foraminifer bio-events (following Sierro et al. 2001), which allowed the attribution of the studied interval to the Messinian pre-evaporitic interval: (7) First occurrence of *Globorotalia nicolae*; (8) last occurrence of *G. nicolae*; (11) last regular occurrence of *Globorotalia miotumida* gr.; (12) first regular occurrence of *Turborotalita multiloba*; (14) *Neogloboquadrina acostaensis* (sinistrally/dextrally coiled change); (16) first influx (80%) sinistral *N. acostaensis*; (18) second influx (40%) sinistral *N. acostaensis*

The "*Praeorbulina*-like" specimens observed in these Messinian pre-evaporitic sediments are very similar to those called "*Praeorbulina*-like specimens with ambiguous characters" found in northern Arabian Sea Last Glacial sediments (see Plate 6 in Rossignol et al. 2011, and Plate 7.2 in this work).

Rossignol et al. (2011) as "*Praeorbulina*-like" specimens which ambiguous characters (see Plate 7.2).

Sample KB36 (55.3 m): (7) *Orbulina universa*. (8) *Orbulina suturalis*. (9) "*Praeorbulina*-like" specimen showing close similarities with *Praeorbulina glomerosa glomerosa* (presence of several crescentic slitlike apertures along the suture separating the last chamber from the earlier chambers), with the exception of the presence of areal circular apertures on the surface of the last chamber. (10) Abnormal *Orbulina* specimen. (11) "*Praeorbulina*-like" specimen similar to *Praeorbulina glomerosa circularis* (the last chamber embracing the earlier part of the test to about 70–75%, presence of several apertures circularis to subcircularis located along the suture separating the last chamber from the earlier chambers) with the exception of the presence of areal apertures on the surface of the last chamber. (12) Abnormal "*Praeorbulina*-like" specimen characterised by the presence of an additional chamber

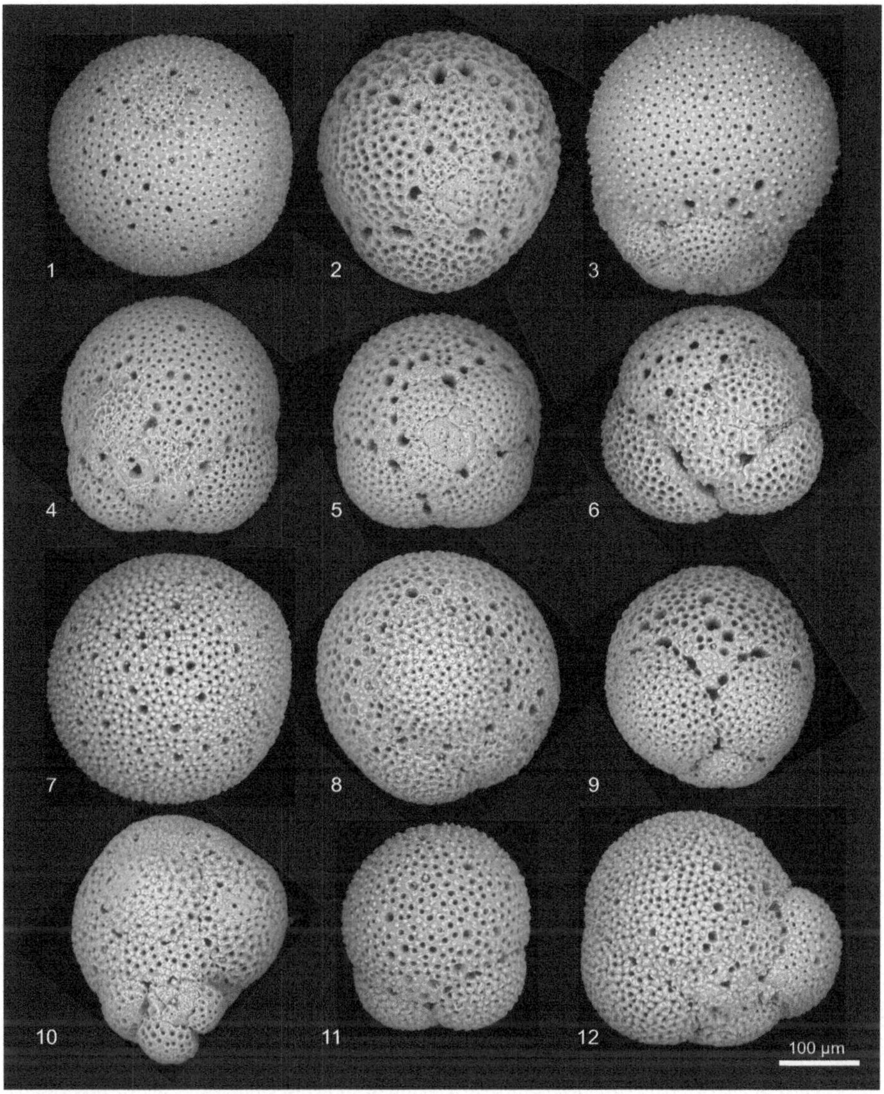

Plate 7.1 SEM microphotographs of some examples of *Orbulina*-"*Praeorbulina*-like" specimens picked from two samples (KB20 and KB36) coming from the Messinian pre-evaporitic interval of Kef el Biod section (Chelif Basin, Northwestern Algeria). In each of the two samples the individuals show a continuum of transitional characters between *O. suturalis* and "*Praeorbulina*-like" morphotypes.
Sample KB20 (40.5 m): (1) *Orbulina universa* d'Orbigny 1839. (2) *Orbulina suturalis* Brönnimann 1951. (3) *Orbulina suturalis.* (4–6) "*Praeorbulina*-like" specimens showing close similarities with *Praeorbulina glomerosa* (the last chamber embraces from 40% to 75% of earlier chambers and it presents sutural apertures, as short narrow slits or subcircularis openings lying in the suture separating the last chamber from the earlier chambers), with the exception of the presence of areal circular apertures over the surface of the last chamber (typical of *Orbulina* genus). These individuals are very similar to those found in the Last Glacial Maximum interval and described by

7.3 The Last Glacial Maximum Case from the Gulf of Oman, Northern Arabian Sea

The Gulf of Oman is a semi-confined marine basin, whose hydrological system is today highly influenced by advection of warm and highly saline near-surface waters, originating from the Persian Gulf, as well as by the Asian-Indian monsoon-climate system (Rossignol et al. 2011). Monsoonal winds influence the local upwelling regime intensity and consequently the primary productivity and the formation of an oxygen minimum zone, which extends today from 150 to 1250 m depth in the northern Arabian Sea (Reichart et al. 1998).

A study based on planktic foraminiferal census analyses of three Late Pleistocene sediment cores retrieved in the Gulf of Oman (KS01, MD04-2849 and MD04-2861) highlighted high abundance (up to 10–18%) of *Orbulina suturalis* together with numerous specimens morphologically close to *Praeorbulina*, co-occurring together with *Orbulina universa* within the Last Glacial Maximum interval, 19–23 Ka CAL--B.P (Rossignol et al. 2011). AMS ^{14}C dating was obtained on a sample of "*Praeorbulina*-like" specimens, giving an age of 19 CAL-Ka BP, which shows unreworked character of this so young *Praeorbulina* occurrence.

The observed "*Praeorbulina*-like" specimens show a large morphological variability intergrading from morphological characteristics close to those of the genus *Praeorbulina* to specimens resembling *Orbulina suturalis* (Plate 7.2). Detailed morphological SEM examinations of specimens picked from a same sample (e.g. highest abundance peak of KS01 at 20 ka CAL-B.P.) revealed the co-occurrence of *O. universa* (Plate 7.2 (1)) with morphotypes similar to *Trilobatus sicanus* (Plate 7.2 (12)), *Praeorbulina glomerosa curva* (Plate 7.2 (7–9)), *Praeorbulina glomerosa glomerosa* (Plate 7.2 (6)), *Praeorbulina glomerosa circularis* (Plate 7.2 (4, 5)) and *O. suturalis* (Plate 7.2 (2, 3)). Remarkably, all ancestor morphotypes of *O. universa* evolutionary lineage (Fig. 7.1) seem to be present in these Late Glacial

original description of *Praeorbulina glomerosa circularis* and this likely causes the uncovering of the umbilical opening (typical character of the genus *Trilobatus*). (5) "*Praeorbulina*-like" specimen showing the presence of several apertures circularis to subcircularis located along the basal suture of the last chamber, typical character of *Praeorbulina glomerosa circularis* (Plate 4, Fig. a, Rossignol et al. 2011). (6) "*Praeorbulina*-like" specimen showing similarities with *Praeorbulina glomerosa glomerosa* (presence of several crescentic slitlike apertures along the basal suture of the last chamber), but also with *Orbulina suturalis* (presence of circular areal apertures on the surface of the last chamber). (7) "*Praeorbulina*-like" specimen showing similarities with *Praeorbulina glomerosa curva* (last chamber subglobular, embracing between 40 and 70% of the earlier tests, few narrow slit apertures lying in the suture separating the last chamber from the earlier chambers). (8–10) "*Praeorbulina*-like" specimens presenting ambiguous characters (Plate 6, Figs. a, b, e, Rossignol et al. 2011). The specimens illustrated in (8 and 9) present similarities with *Praeorbulina glomerosa curva* (last chamber embracing between 40 and 70% of the earlier chambers, few narrow slit apertures lying in the suture separating the last chamber from the earlier chambers), with the exception of the presence of circular areal apertures on the surface of the last chamber (typical character of *Orbulina* genus). (11) *Praeorbulina*-like specimen presenting ambiguous characters. (12) Specimen presenting morphological features close to *Trilobatus sicanus* (Plate 1, Fig. c, Rossignol et al. 2011)

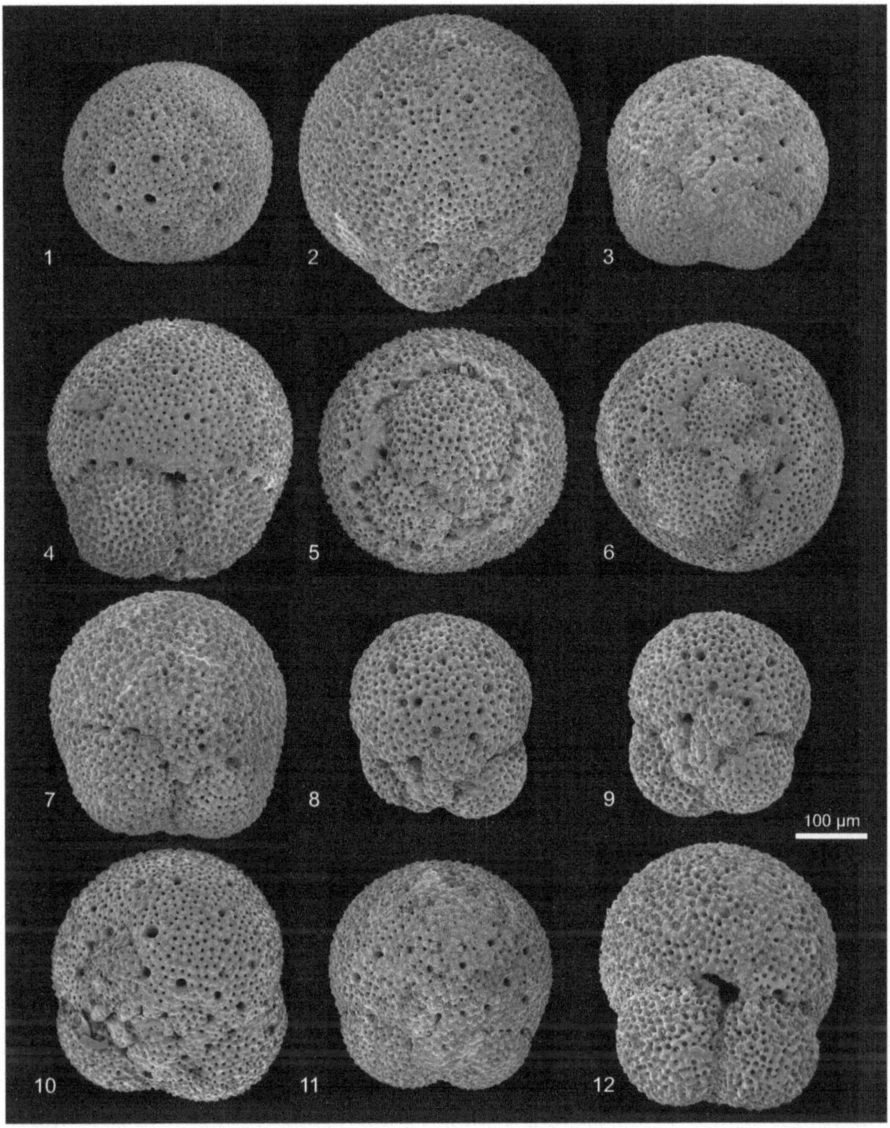

Plate 7.2 SEM microphotographs of some examples of *Orbulina*-*Praeorbulina*-like specimens picked in the highest abundance peak within the Last Glacial Maximum interval of KS01 core (754 cm depth, 20 ka CAL-B.P.), northern Arabian Sea (Rossignol et al. 2011). (1) *Orbulina universa* (Plate 5, Fig. h, Rossignol et al. 2011). (2) *Orbulina suturalis*. (3) *Orbulina suturalis* (Plate 5, Fig. a, Rossignol et al. 2011). (4) "*Praeorbulina*-like" specimens with ambiguous characters (Plate 6, Fig. d, Rossignol et al. 2011). This individual shows striking resemblances to *Praeorbulina glomerosa circularis* (presence of several apertures circularis to subcircularis located along the suture separating the last chamber from the earlier chambers) with the exception of the presence of areal apertures on the surface of the last chamber (typical of *Orbulina* genus). In addition, the last chamber embraces the previous part of the test to a lesser extent (about 50%) with respect to the

stratigraphical layers. However, as remarked for the Messinian, also in the case of Late Pleistocene, many "*Praeorbulina*-like" specimens present ambiguous characters by showing the presence of large areal pores (= apertures over the surface of the last chamber), in combination with a weak last chamber embracement (Plate 7.2 (4, 6, 8–11)) (Rossignol et al. 2011).

7.4 Ancestor-Looking "*Praeorbulina*-Like" Occurrences: Link to Environmental Stress?

The evolutionary lineage starting from trochospiral *Trilobatus* culminating in spherical *Orbulina universa* (Fig. 7.1) can be considered as a "textbook" example of peramorphic spherisation lineage, where the increasing involution is coupled with increasing shell curvature (Guex 2006). At the end of lineage, in the *O. universa* the increasing involution and spherisation are accentuated at adult stage ("peramorphosis"), when the last spherical chamber envelops entirely the pre-adult trochospiral test (Fig. 7.1). Time-oriented morphological transformations ("evolutionary trends") characterised by the increasing of involution (tightly coiled) starting from ancestral evolute (open coiling) shell morphologies have been well illustrated in several Mesozoic ammonoid lineages, in nautiloids and certain gastropods, as well as planktonic and benthic foraminifera (Guex 2001, 2006). Sphericity has evolved repeatedly in planktonic foraminifera (Pearson et al. 1997), as well as in other protists, as radiolarians (Guex et al. 2012, 2014).

In the studied cases previously illustrated in this chapter (Messinian pre-evaporitic and Last Glacial Maximum intervals), we show the sudden appearance in punctual stratigraphic layers of ancestor-looking forms ("*Praeorbulina*-like" morphotypes) together with the end-lineage spherical *Orbulina universa*. How to explain these astonish co-occurrences?

First of all, a trivial case of reworking of these "*Praeorbulina*-like" specimens coming from Last Glacial Maximum sediments can be excluded because of ^{14}C dating (Rossignol et al. 2011). In the case of Messinian, we remark on the SEM photos that a majority of specimens named "*Praeorbulina*-like", according to Rossignol et al. (2011), display transitional morphological characters that are intermediary between *Praeorbulina sensu stricto* and *Orbulina suturalis* (Plate 7.1). They present in fact an ancestral architectural morphology of the test (a weak last chamber embracement of the early chambers), typical of *Praeorbulina sensu stricto*, but they also display new characters, such as the presence of areal apertures (= large pores) over the surface of the last chamber, which are characteristic of the genus *Orbulina*. *Orbulina* genus FO datum was later of *Praeorbulina* genus FO datum (Fig. 7.1). Thus, the observed "*Praeorbulina*-like" specimens found in the Messinian samples do not appear to be reworked older Burdigalian-Langhian *Praeorbulina sensu stricto*. Many "*Praeorbulina*-like" individuals coming from Last Glacial Maximum sediments show also remarkable resemblances with *Praeorbulina sensu stricto*, except for the presence of areal apertures over the last chamber (Rossignol et al. 2011) (Plate 7.2).

The *Praeorbulina*-like occurrences seem to display a very discontinuous stratigraphic record in both the Messinian from the Chelif Basin (new data) and the Late Pleistocene from the northern Arabian Sea (Rossignol et al. 2011). They are generally absent or very rare in the foraminiferal assemblages and increase suddenly in abundance (>10%) in punctual and specific stratigraphic layers (Figs. 3 and 4 in Rossignol et al. 2011). Two possible hypotheses might explain their discontinuous stratigraphical behaviour.

A possible explanation is that the standing stock of their populations is generally low below the threshold that allows their occurrence in the fossil record. Their chance of fossilisation would be enhanced under certain ecological conditions favouring their sudden explosive developing (Rossignol et al. 2011).

As alternative, it would seem a case of reversal of evolutionary trends ("retrograde evolution") (Guex 2006; Cabej 2012) related to environmental stress. In this case, *Praeorbulina*-like morphotypes, presenting areal apertures, might derive from *Orbulina suturalis*, inverting the evolutive trend. Untightening the coil, the result is a less embarrassing of the last chamber over the early whorls. Thus *Praeorbulina*-like morphotypes might be homeomorphic of the Burdigalian-Langhian *Praeorbulina*, with the exception of the distribution of the apertures. The two characters, the embracement of the last chamber over the early trochospire (= tightening of the trochospiral coil of the shell) and the distribution of apertures, which would have guided the evolutionary lineage of *Orbulina universa* (Blow 1956) (Fig. 7.1) seem to be independent. The presence of areal apertures, once acquired, seems to be in fact a much more conservative character with respect to the tightening of the coil of the shell, which seems to be a more plastic character to environmental variations. Examples of morphological retrogradation in several marine organisms (ammonoids, protists), called "proteromorphosis", have been evidenced during mass extinction periods, suggesting that the evolutionary clock of these organisms can be reinitialised by environmental stress (Guex 2006, 2016).

An argument that could support the hypothesis that *Praeorbulina*-like could also be a case of evolutionary reversal related to stressful environmental conditions is its co-occurrence with high percentages of abnormal individuals in the foraminiferal assemblages (Fig. 7.3). In the Kef el Biod section high peaks of *Praeorbulina*-like specimens, up to 10–38%, coincide in fact with a remarkable high percentage, up to 5–10%, respectively, of malformed specimens in total foraminiferal assemblages (Fig. 7.3) (Belhadji et al. in prep.). Generally in the planktic foraminiferal assemblages abnormal specimens rarely exceeded 1.5% of the total assemblage (Mancin and Darling 2015). In the Kef el Biod samples the highest rate of morphological abnormalities was detected in the *Orbulina* populations (Plate 7.1 (10, 12)), up to 30% and with more than half of the samples exceeding the 10% (Belhadji et al. in prep.). There are many examples in the literature showing the influence of environmental factors, particularly stress, on the morphology of living benthic and planktonic foraminifera (Vénec-Peyré 1981; Hemleben et al. 1989; Schiebel and Hemleben 2017 opt. cit.; Vénec-Peyré et al. 2020 and references herein). Thus the increase in percentage of abnormal specimens in the foraminiferal assemblages in the pre-evaporitic sediments might be a symptom of stressful conditions of seawater column in the Chelif basin, just preceding the Messinian salinity crisis. The Chelif Basin was located not so far from the Gibraltar arc seaway, so it was most likely highly sensitive to Atlantic water inflow fluctuations linked to the interplay of ongoing tectonic constriction of the Atlantic gateways and sea-level variations. In the Kef el Biod section, maxima values of morphological abnormalities targeting indeed specific levels are characterised by transient sharp decreasing of "cold" species abundance and decreasing of $\delta^{18}O$ values (Belhadji et al. in prep.). These specific levels in the pre-evaporitic interval might be related to transient episodes of reduction of North Atlantic water inflow (increasing restriction) and/or of increase in freshwater input (Simon et al. 2017) in Chelif Basin, preluding the Messinian salinity crisis. Several other works have already highlighted that the onset of the Messinian salinity crisis was not an abrupt event, but it was preceded during the pre-evaporitic interval by several steps marked by more rapid changes in salinity and bottom water oxygenation (Rouchy and Caruso 2006). Thus rapid and extreme climate fluctuations, changes in salinity, severe stratification of seawater column and development of oxygen-depleted conditions in the water column might be a combined interplay of different stressor factors favouring the developing of ancestor "*Praeorbulina*-like morphotypes" and abnormal specimens in the *Orbulina* populations.

Although a parallel between the environmental conditions, which might have favoured the "*Praeorbulina*-like" development in the Chelif Basin during the pre-evaporitic Messinian interval and in the Gulf of Oman during the last maximum glacial interval, might be premature, some common factors might have played out either way, such as the increasing restriction (confinement) together with the decreasing water-column ventilation. Today the Gulf of Oman is a semi-confined marine basin, and the decrease of sea level during the Last Glacial Maximum interval might have increased its confinement. Furthermore, the dynamics of the oxygen minimum zone might have played a crucial role. In western Arabian Sea, the deepening of the OMZ has been showed to occur preferentially during cold intervals in concordance with a reduced advection of oxygen-rich deepwater masses of North Atlantic and Antarctic origin (Schmiedl and Leuschner 2005). As already suggested by Rossignol et al. (2011), such change in water-column ventilation might have forced the morphological plasticity within *Orbulina* group.

7.5 Conclusion Remarks

The presented case studies in this chapter highlight the interest of distinguished "*Praeorbulina*-like" morphotypes within the intra-species variability of *Orbulina suturalis*. Probably, their occurrence in the post-Langhian stratigraphical record has been neglected, or considered in the variability of the species *Orbulina suturalis* (given the presence of the areal openings over the last chamber) or considered as reworked. Our preliminary observations suggest the "*Praeorbulina*-like" acme abundances as a potential marker of peculiar and possible stressful environmental conditions.

However further investigations on quantitative morphometric and geochemical analyses need to better define the taxonomic rank and ecological preferences of "*Praeorbulina*-like" morphotypes with respect to *Praeorbulina sensu stricto* and *Orbulina*. Moreover it would need to complete a long gap record between the Messinian to the Late Pleistocene.

References

André A, Quillévéré F, Morard R, Ujiié Y, Escarguel G, de Vargas C, de Garidel-Thoron T, Douady CJ (2014) SSU rDNA divergence in planktonic foraminifera: molecular taxonomy and biogeographic implications. PLoS One 9(8):e104641. https://doi.org/10.1371/journal.pone.0104641

Berggren WA, Kent DV, Swisher CC, Aubry M-P (1995) A revised Cenozoic geochronology and chronostratigraphy. In: Berggren WA, et al. (eds) Geochronology, time scales and global stratigraphic correlation. Lamont-Doherty Earth Observatory Biology and Paleo Environment, Society of Sedimentary Geology (SEPM), Special Publication, 54:129–212

Blow WH (1956) Origin and evolution of the foraminiferal genus *Orbulina* d'Orbigny. Micropaleontology 2:57–70

Blow WH (1969) Late middle Eocene to recent planktonic foraminiferal biostratigraphy. Bull Am Paleontol 39:59–271

Bolli HM, Saunders JB (1985) Oligocene to Holocene low latitude planktonic foraminifera. In: Bolli HM, Saunders JB, Perch-Nielsen K (eds) Plankton stratigraphy. Cambridge University Press, Cambridge, pp 155–262

Boudagher-Fadel MK (2015) Biostratigraphic and geological significance of planktonic foraminifera. UCL Press, London, 306 pp

Brönnimann P (1951) The genus Orbulina d'Orbigny in the Oligo-Miocene of Trinidad. B W I Contributions from the Cushman Foundation for Foraminiferal Research 2(4):135

Cabej N (2012) Epigenetic principles of evolution. Elsevier Insights, London, p 804

Cohen KM, Gibbard PL (2016) Global chronostratigraphical correlation table for the last 2.7 million years, v. 2016a. International Commission on Stratigraphy, IUGS. www.stratigraphy.org

Cushman JA, Dorsey AL (1940) Some notes on the genus *Candorbulina*. Contrib Cushman Lab Foramin Res 16:40–42

De Vargas C, Norris R, Zaninetti L, Gibb SW, Pawlowski J (1999) Molecular evidence of cryptic speciation in planktonic foraminifers and their relation to oceanic provinces. Proc Natl Acad Sci U S A 96:2864–2868

De Stefani T (1952) Su alcune manifestazioni di idrocarburi in provincia di Palermo e descrizione di foraminiferi nuovi. Plinia 3(4):1–13

Garcia-Castellanos D, Villasenor A (2011) Messinian salinity crisis regulated by competing tectonics and erosion at Gibraltar arc. Nature 480:359–363

Guex J (2001) Environmental stress and atavism in ammonoid evolution. Eclogae Geol Helv 94:321–328

Guex J (2006) Reinitialization of evolutionary clocks during sublethal environmental stress in some invertebrates. Earth Planet Sci Lett 424:240–253

Guex J (2016) Retrograde evolution during major extinction crises. SpringerBriefs in Evolutionary Biology, Heidelberg, 75 p

Guex J, O'Dogherty L, Carter ES, Gorican S, Dumitrica P, Bartolini A (2012) Geometrical transformations of selected Mesozoic radiolarians. Geobios 45:541–554

Guex J, Caridroit M, Kuwahara K, O'Dogherty L (2014) Retrograde evolution of *Albaillella* during the Permian-Triassic crisis. Rev Micropaleontol 57(2):39–43

Hemleben C, Splinder M, Anderson OR (1989) Modern planktonic foraminifera. Springer-Verlag, New York, 363 p

Hsü KJ, Ryan WBF, Cita MB (1973) Late Miocene desiccation of the Mediterranean. Nature 242:240–244

Iaccarino S (1985) Mediterranean Miocene and Pliocene planktonic foraminifera. In: Bolli HM, Saunders JB, Perch-Nielsen K (eds) Plankton stratigraphy. Cambridge University Press, Cambridge, pp 283–314

Jenkins DG, Saunders JB, Cifelli R (1981) The relationship of *Globigerinoides bisphericus* Todd 1954 to *Praeorbulina sicana* (De Stefani) 1952. J Foraminiferal Res 11:262–267

Kennett JP, Srinivasan MS (1983) Neogene planktonic foraminifera, A phylogenetic atlas. Hutchinson Ross, Stroudsburg, PA, 265 p

Krijgsman W, Hilgen FJ, Raffi I, Sierro FJ, Wilson DS (1999) Chronology, causes and progression of the Messinian salinity crisis. Nature 400:652–655

Kucera M (2007) Planktonic foraminifera as tracers of past oceanic environments. In: Hillaire-Marcel C, de Vernal A (eds) Proxies in late Cenozoic paleoceanography. Elsevier, Amsterdam, pp 213–254

Lourens LJ, Hilgen FJ, Shackleton NJ, Laskar J, Wilson D (2004) The Neogene period. In: Gradstein FM, Ogg JG, Smith AG (eds) Geological time scale 2004. Cambridge University Press, Cambridge, pp 409–440

Mancin N, Darling K (2015) Morphological abnormalities of planktonic foraminiferal tests in the SW Pacific Ocean over the last 550ky. Mar Micropaleontol 120:1–19. https://doi.org/10.1016/j.marmicro.2015.08.003

Manzi V, Gennari R, Hilgen F, Krijgsman W, Lugli S, Roveri M, Sierro FJ (2013) Age refinement of the Messinian salinity crisis onset in the Mediterranean. Terra Nova 25:315–322

Manzi V, Gennari R, Lugli S, Persico D, Reghizzi M, Roveri M, Schreiber BC, Calvo R, Gavrieli I, Gvirtzman Z (2018) The onset of the Messinian salinity crisis in the deep Eastern Mediterranean basin. Terra Nova 30:189–198

Morard R, Quillévéré F, Escarguel G, Ujiie Y, de Garidel-Thoron T, Norris RD, de Vargas C (2009) Morphological recognition of cryptic species in the planktonic foraminifer *Orbulina universa*. Mar Micropaleontol 71:148–165

Olsson RK (1964) *Praeorbulina* Olsson, a new foraminiferal genus. J Paleo 38:770–771

Pearson PN, Shackleton NJ, Hall MA (1997) Stable isotopic evidence for the sympatric divergence of *Globigerinoides trilobus* and *Orbulina universa* (planktonic foraminifera). J Geol Soc London 154:295–302

Perrodon A (1957) Etude géologique des bassins néogènes sub-littoraux de l'Algérie occidentale. Publ Serv Carte Géol Algérie n s Alger Bull 12:382

Postuma JA (1971) Manual of planktonic foraminifera. Elsevier for Shell Group, The Hague, pp 1–406

Reichart GJ, Lourens LJ, Zachariasse WJ (1998) Temporal variability in the northern Arabian Sea Oxygen Minimum Zone (OMZ) during the last 225,000 years. Paleoceanography 13:607–621

Rossignol L, Eynaud F, Bourget J, Zaragosi S, Fontanier C, Nadine E-Z, Lanfumey V (2011) High occurrence of *Orbulina suturalis* and "*Praeorbulina*-like specimens" in sediments of the northern Arabian Sea during the Last Glacial Maximum. Mar Micropaleontol 79:100–113

Rouchy JM, Caruso A (2006) The Messinian salinity crisis in the Mediterranean Basin: a reassessment of the data and an integrated scenario. Sediment Geol 188(/189):35–67

Rouchy JM, Caruso A, Pierre C, Blanc-Valleron M-M, Bassetti MA (2007) The end of the Messinian salinity crisis: evidences from the Chelif Basin (Algeria). Palaeogeogr Palaeoclimatol Palaeoecol 254:386–417

Ruggieri G, Adams CJ, Ager DV (1967) The Miocene and latter evolution of the Mediterranean Sea. Aspects of Tethyan Biogeography. Systematic Association Publication, London, England, p 283

Reuss AE (1850) Neue Foraminiferen aus den Schichten des österreichischen Tertiärbeckens. Denkschriften der Kaiserlichen Akademie der Wissenschaften 1:365–390

Schiebel R, Hemleben C (2017) Planktic foraminifers in the modern ocean. Springer-Verlag, Berlin Heidelberg, p 358

Sierro FJ, Hilgen FJ, Krijgsman W, Flores JA (2001) The Abad composite (SE Spain): a Messinian reference section for the Mediterranean and the ATPS. Palaeogeogr Palaeoclimatol Palaeoecol 168:141–169

Simon D, Marzocchi A, Flecker R, Lunt DJ, Hilgen FJ, Meijer PT (2017) Quantifying the Mediterranean freshwater budget throughout the late Miocene: new implications for sapropel formation and the Messinian salinity crisis. Earth Planet Sci Lett 472:25–37

Spero HJ (1988) Ultrastructural examination of chamber morphogenesis and biomineralization in the planktonic foraminifer *Orbulina universa*. Mar Biol 99:9–20

Spezzaferri S, Kucera M, Pearson PN, Wade BS, Rappo S, Poole CR, Morard R, Stalder C (2015) Fossil and genetic evidence for the polyphyletic nature of the planktonic foraminifera "*Globigerinoides*", and description of the new genus *Trilobatus*. PLoS One 10(5):e0128108. https://doi.org/10.1371/journal.pone.0128108

Schmiedl G, Leuschner DC (2005) Oxygenation changes in the deep western Arabian Sea during the last 190,000 years: Productivity versus deepwater circulation. Paleoceanography 20(2):PA2008. https://doi.org/10.1029/2004PA001044

Tauecchio P, Marks P (1973) The Messinian deposits of the Chelif Basin near El Asnam, Northern Algeria. In: Drooger CW (ed) Messinian events in the Mediterranean. North-Holland, Amsterdam, pp 188–191

Turco E, Iaccarino SM, Foresi LM, Salvatorini G, Riforgiato F, Verducci M (2011) Revisiting the taxonomy of the intermediate stages in the *Globigerinoides-Praeorbulina* lineage. Stratigraphy 8(2-3):163–187

Todd R (1954) Probable occurrence of Oligocene on Saipan. Am J Sci 252 (11):673–682

Vénec-Peyré MT (1981) Les Foraminifères et la pollution: étude de la microfaune de la Cale du Dourduff (Embouchure de la Rivière de Morlaix). Cahiers de Biologie Marine 22:25–33

Vénec-Peyré M-T, Bartolini A, Weber M, Lipps JH (2020) Morphological deformation of foraminiferal tests caused by intertidal oil spills (black tides). In: Guex J et al. (eds): Morphogenesis, Environmental Stress and Reverse Evolution. Springer Nature Switzerland AG 978-3-030-47278-8

Wade BS, Pearson PN, Berggren WA, Pälike H (2011) Review and revision of Cenozoic tropical planktonic foraminiferal biostratigraphy and calibration to the geomagnetic polarity and astronomical time scale. Earth Sci Rev 104:111–142

Chapter 8
Steps of Morphogenesis and Iterative Evolution of Imperforate Larger Foraminifera in Shallow Carbonate Shelves During Mesozoic Times: Possible Relations to Symbiotic and Abiotic Factors

Michel Septfontaine

Abstract

1. The microgranular/agglutinated imperforate larger foraminifera (ILF, chiefly "lituolids") of Mesozoic shallow marine carbonate shelves are a polyphyletic group of K-strategists, ecologically homogeneous inhabitants of the photic zone (nutrient poor) and hosting symbionts as their larger porcelaneous recent equivalents. They contrast with deeper water "lituolid" taxa and other deeper marine dwellers (hyaline perforate and planktonics r-strategists). They are narrowly linked to the carbonate platform history and evolution, birth and demise (correlated to paleotectonics or sea-level variations), and global climatic or volcanic events, OAE, etc.
2. General principles used in evolutionary sciences can be applied to decipher some morphological pathways among foraminifera through time to reconstruct phylogenetical arborescences, e.g., embryology (heterochronies of development), recapitulation law (Haeckel), size increase (Depéret), parallel evolution, and biostratigraphy.
3. Iterative evolution is the main mode of morphological variation through time and biostratigraphy helps to reconstruct the time repartition of diverging advanced homeomorphic taxa and their distinction. The new characters are issued from a constant, more tolerant stock of smaller and simpler primitive atavic forms by the transformation of previous structures or the addition of new ones. Advanced homeomorphic taxa are temporarily preserved and their morphological innovations linked probably to (epi)genetic constraints and symbiotic history increasing their energy efficiency in a rather stable (but cyclical) environment.
4. Morphological innovations appear slowly during the Jurassic and more rapidly (and frequently) during the Cretaceous following a probable saltative and allo-

M. Septfontaine (✉)
Independent researcher (retired), Froideville, Switzerland
e-mail: septfontaine.m@bluewin.ch

J. Guex et al. (eds.), *Morphogenesis, Environmental Stress and Reverse Evolution*, https://doi.org/10.1007/978-3-030-47279-5_8

patric mode of evolution during favorable periods after main environmental crisis, anoxia, and sea-level changes. The Cretaceous is a time of great diversification (adaptive radiation of microgranular lituolids and larger porcelaneous tests).

5. Morphological changes at a scale of one or two stages in one evolutive bioseries (general trend: size and internal complications increase) cannot be linked to global events or other abiotic factors (e.g., the orbitopsellids in the Pliensbachian) but are rather driven by internal epigenetic factors related to algal symbiosis and coevolution. The constancy of the model (text Fig. 8.2) through time in variable clades could be explained that way. The structural incompatibility between pseudokeriotheca/alveolae within the wall and marginal microstructures (radial partitions or hypodermic network) could be explained by a biological incompatibility of bacterial symbionts to live aside with larger microalgae as dinoflagellates. Local discontinuous events like cyclic (metric) parasequences are common on carbonate platforms but have no apparent influence on morphogenesis or population composition.
6. Conversely, global environmental crises affect deeply the larger foraminifera (chiefly advanced larger tests) and population compositions at a higher taxonomic rank (families, etc.), replacing in the aftermath the larger solid tests by smaller forms with a fragile (thin) test during anoxic crisis (T/J—Late Pliensbachian—Late Cenomanian, etc.). Larger forms are too specialized and complicated in their architecture and are particularly sensitive to environmental changes (Hallock, Paleobiology 11(2):195–208, 1985). They cannot adapt their test morphology ("retrograde evolution," Guex, Retrograde Evolution During Major Extinction Crisis, 2016) back to some kind of more tolerant simpler form because their low reproduction rates impede a rapid reaction to environmental stress, even for minor adjustments. Major sea-level and facies changes (deeper water conditions or large-scale emersions) cause the brutal disappearing of this very sensitive group of foraminifera. However, some large discoid taxa and other smaller siphovalvulinids-valvulinids cross the K/T environmental crisis, interpreted as survivors in the Paleocene.
7. Moreover, gigantism is probably a disadvantage against physical perturbations like tidal currents and storms in an otherwise stable carbonate environment. In a first evolutionary step, increasing of test size and flattening of chambers are selective advantages in terms of energy (increasing of chemical exchanges with more symbionts) and stability on flat surfaces; but near the end of the evolutive bioseries, the same traits become a drawback concerning test anchorage and risk of displacement in unfavorable microenvironments.

Keywords Mesozoic · Larger imperforate foraminifera · Iterative evolution · Carbonate shelves · Symbiosis · Extinction

8.1 Introduction

Recent larger foraminifera (fusiform alveolinids and discoidal porcelaneous forms as *Sorites*, *Amphisorus*, and *Marginopora*) in shallow warm waters of low latitudes were extensively studied during the twentieth century. They can serve as references or living models for extinct microgranular-agglutinant imperforate larger foraminifera (ILF) with a similar general morphology on carbonate platforms submitted to a sub- to supratidal cyclical (parasequences) type of environment considered here as "stable". Recent equivalent morphotypes are epiphytic as the discoid porcelaneous genus *Sorites* (Richardson 2009) or live in coarse calcareous sands associated with reef buildups and other carbonate depositional environments (ramps or restricted lagoons in rare cases).

Most genera host symbiotic microorganisms, e.g., nitrifying bacteria, non-photosynthetic, for the agglutinating *Spiculidendron* in shaded fore-reef habitats (Richardson and Rützler 1999), algae, dinoflagellates, and diatoms in shallower waters for porcelaneous or hyaline taxa. The symbionts are free living in the endoplasmic cell, or stocked, densely packed, in chamberlets near the surface of the translucid calcareous test (Hallock 1982, 1985; Richardson 2006). These authors give a general review of previous studies on this vast subject, illustrating that algal symbiosis is energetically highly advantageous under nutrient-deficient (oligotrophic) conditions, as is the case in hermatypic coral reefs (or bivalve mounds) environments. Thus, larger shells, hosting numerous symbionts, are advantageous in stable environments with restricted food resources. Hallock (1985) notes that, already in the 1970s, Ross (1974) and Lee et al. (1979) "*have postulated that evolution in larger Foraminifera is related to adaptation to algal symbiosis*." These conclusions based on the modern (porcelaneous) equivalents of fossil imperforate larger microgranular-agglutinated foraminifera (ILF) can serve as a guide for the understanding of the complex (iterative) morphogenetical evolution of this group of microorganisms during Jurassic and Cretaceous times.

However, general palaeogeographical and environmental conditions, even on carbonate shallow platforms in the past, were notably different from the recent, particularly in terms of size (width and length) along the tropical-equatorial belt (e.g., about 8000 km from Morocco to Mexico to the west or Oman and Iran to the east). In spite of the huge extension of Mesozoic carbonate platforms along the southern margin of the Tethys, most assemblages (and isolated genera) can be considered as synchronous on each location of this domain, except local migrations in some restricted areas, like the protected eutrophic lagoon of the "Mytilus beds" in the Briançonnais domain till Sardinia (Septfontaine 1984a).

Moreover, Jurassic and Cretaceous "lituolids" have developed complex original inner microstructures (from the Liassic onwards) that are not known in the Tertiary and in recent taxa. Thus, maximum caution must be taken when comparing and trying to decipher the function of such microstructures (Hottinger 2006) and their role in the individual life of the foraminifera, and more generally the relation (epigenetic or genetic trend in a stable environment) with the past biotopes in an evolutionary perspective, deduced only from morphological characters. On the other hand, as stated above, symbiosis is an important mechanism of evolution in the diversification of eukaryotic clades, and this form of coevolution is widespread in all taxa of the foraminifera (Brasier 1988; Richardson and Rützler 1999).

Delayed maturation and growth to large size (K-strategist bearing symbiotic microorganisms or "greenhouse" management) are only advantageous under stable conditions where resources are limited (Hallock 1985). But a rapid change of environmental factors (anoxia, temperature, acidity, water depth during sea-level drop or rise, etc.) will doom the giant cell with its symbionts to extinction. Moreover, some morphological limitation of size and complexity related to physical environmental conditions (hydrological instabilities due to tidal currents, storms) may play an important role regarding the fixation of the large tests on different substrates.

8.2 Summary of General Principles

Some basic principles in evolutionary sciences can help and guide the reconstruction of phyletic arborescences. They were already discussed by Septfontaine (1988), *as a working hypothesis*, with some additions in the field of geology, stratigraphy, and sedimentology. These principles are classical and were already used for Metazoa. An application for eukaryotic Protozoa seems reasonable, keeping in mind possible exceptions; this approach completes the classical cladistic-phylogenetic (morphological) method in biology (Hennig 1966; Lecointre and Le Guyader 2001). Obviously, the advantage of defining principles is that they can easily be refuted or at least modified!

1st Principle: The phyletic relations are to be found exclusively among the stock of paleoecologically specialized dwellers of shallow carbonate platforms (in the photic zone), from the protected lagoonal environment to the outer platform and fore-reef talus. The ancestors of each group (if polyphyletic) are to be found in these depositional environments (excluding the open marine basinal environments) and are necessarily small with a simple morphology; they can be sometimes recognized in the first ontogenetic steps of the test (nepionic stage) of more evoluted individuals.

2nd Principle: The stratigraphic and paleogeographic repartition of genera during Mesozoic times (Septfontaine 1981, 1988; Bassoullet 1997) allows identification of the homeomorphic taxa (in one individual clade). Each morphological novelty is rapidly widespread all along the tropical belt in N and S margins of the Thetysian domain (cosmopolitan genera), and no true endemism is known. Local absence is due to facies changes. In some cases (e.g., the Briançonnais domain, N margin with restricted eutrophic conditions) steps of migrations were observed in the lagoon (Septfontaine 1984a).

3rd Principle: Genetic proximity is deduced from the morphological closeness of characters through space and time (cladistic approach), basically the only possible method in paleontology with the risk to gather parallel isomorphic clades of different origins. The "weight" of characters should be estimated: presupposed

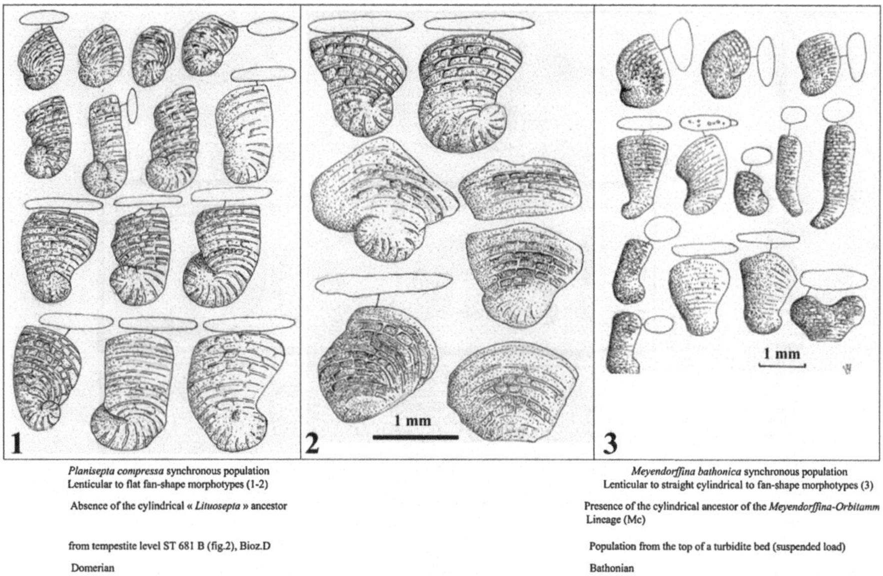

Fig. 8.1 Synchronous populations. 1–2: in a sample with *Planisepta compressa* from a tempestite level (Middle Liassic of Morocco); 3: in a sample with *Meyendorffina bathonica* (Bathonian of Swiss Prealps). The shape diversity (lenticular to flabelliform-reniform, with flattening of tests) is similar to a recent porcelaneous population of *Peneroplis pertusus* (Hottinger 2006). These contemporaneous morphotypes give a spectrum of possible variations of shape through time, as depicted in Fig. 8.2 for the genus *Orbitopsella*

minor or secondary characters may be of prime importance. The morphological variations (shape of chambers and contour of the test) in a given synchronous and abundant population (displaced tests recognized in tempestites or turbidites, an instant mechanism, Fig. 8.1) give a picture of the possible variations of that species through time as resumed horizontally and vertically in the general virtual model text; see Fig. 8.2. The model is shared by different clades of microgranular - imperforate foraminifera, and probably by isomorphic porcelaneous or hyaline - perforate larger discoid taxa (convergences and parallel evolution, MacGillavry 1978) in a similar environment.

4th Principle: Ontogenesis and heterochrony in individual developmental steps, combined with delayed maturation (Gould 1977; Chaline 1987), are considerable and simple means of morphogenetic innovations from existing structures to new ones via minor genetic mutations (neoteny, progenesis, or acceleration). The test of foraminifera, built by incrementation of successive chambers, can retain (or not) the atavic morphology in early steps of ontogenesis in some advanced taxa (recapitulation law). This particularity permits to decipher the phylogenetic history of some advanced clades.

5th Principle: The role of sedimentary sequential evolution in carbonate platforms (from parasequences to third-order sequences) regarding morphogenesis or assem-

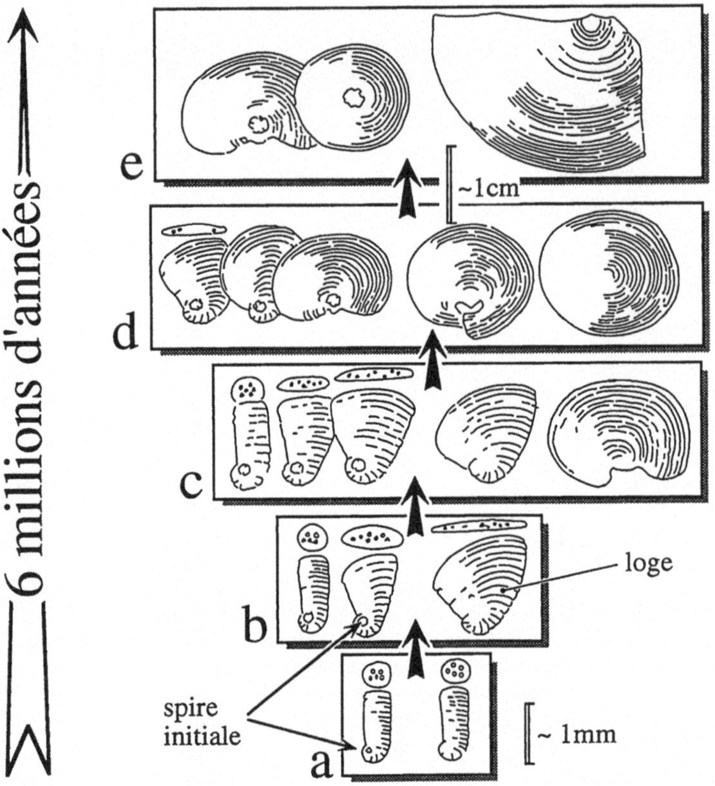

Fig. 8.2 Excerpt from Septfontaine (1988). *Virtual phylogenetical model* of Mesozoic larger discoid foraminifera (microgranular-agglutinated and porcelaneous) based on observations of evolutionary trends in the *mesoendothyrids* and *orbitopsellids s.l.* (*Mesoendothyra-Lituosepta-Orbitopsella*) of Morocco (from Late Sinemurian to Carixian; Septfontaine 1984b, 1988) and Mallorca. The last morphological steps (e) of the K-strategist giant cell disappear in a constant environment far below the beginning of the T-OAE crisis starting in Late Domerian (see also Table 8.6). Some internal factors, as a deleterious relation with symbiotic algae or a mechanical incompatibility of the large discoid shell regarding stability and attachment to the substrate (pseudopods through limited openings), may explain the disappearance of the population. The stratigraphical and sedimentological general setting should allow distinguishing between a time control (vertical) and a facies control (horizontal, "epigenetic," a–e) of the morphogenesis. Population biodiversity (and trend) is marked by chamber flattening and enlargement
The model has been observed in several Jurassic clades like the bioseries: *Amijiella-Alzonella*; *Haurania-Timidonella*; *Meyendorffina-Orbitammina*, etc. or Cretaceous bioseries: *Cuneolina-Dicyclina* and some others (Table 8.2)
Macrospherical individuals are to the left (rounded large embryo); microspherical reniform to discoidal tests to the right. Note that in very advanced populations (*Orbitopsella dubari* or *Cyclorbitopsella*) the chambers of macrospheric individuals are completely cyclical

blages of foraminifera can be estimated as well as their importance (or not) in disappearance or changes of populations (turnovers); sedimentological studies allow evidence for population composition or concentration due to sorting, reworking, or displacement of empty tests by currents or hurricanes (Septfontaine 1985; Davaud and Septfontaine 1995), all bias influencing paleoecological conclusions.

6th Principle: Morphogenesis of microstructures in the wall (pseudokeriotheca and alveolae) or in the marginal zone of the chambers (hypodermic network or radial partitions defining chamberlets) can be tentatively related to symbiosis with microorganisms and their coevolution with the foraminiferal cell: filamentous cyanobacteria (keriotheca of fusulinids in Vachard et al. 2004), diatoms, dinoflagellates, or other microalgae.

8.3 Incompleteness of the Fossil Record on Carbonate Platforms Impedes Evolutionary Studies: A Situation Linked to the Unstable Environmental Context (Hiatuses, Emersions, and Drownings)

The imperforate larger foraminifera on shallow carbonate platforms represent a particularly difficult group of microfossils to deal with for evolutionary and morphogenetical studies and interpretation, unlike the open marine planktonic foraminifera or radiolarians with a larger and more continuous record, as discussed by Guex (2016). The author introduces a new model of "retrograde evolution" in the Cretaceous rotaliporiids, well documented during the Cenomanian anoxic crisis. The main practical reasons (rarely invoked since Darwin 1859) that impede the comprehension of the mode of evolution (reverse, progressive versus saltative, etc.) and morphogenetical history of ILF are listed below and must be kept in mind:

1. It is not possible to consider a rather sufficient large population (and its morphological variations, principle 5 above) at a given level (generally from a hard rock sample) because of a rather poor abundance related to their mode of reproduction (K-strategists with a low reproduction rate compared to the r-strategist planktonics in one bed). Thus the number of morphotypes yielded for estimation of morphological variation and plasticity is commonly poor. In the Mesozoic interval, larger foraminifera are studied in thin sections, or scarcely as loose specimens extracted from shaly levels.
2. Their vertical (stratigraphical) occurrence in outcrops is generally episodical and the populations available are rather poor in individuals except for some intervals (top of parasequences in inter- to supratidal environments with an accumulation of dead transported tests by the tide, hurricanes, etc.) (Fig. 8.3, and Septfontaine 1985). This situation is classical in the Mesozoic carbonates along the southern and northern margins of the alpine and eastern Tethys. Obviously, the lack of continuity in sedimentation and major environmental changes related to paleotectonics and orbital cyclicity (emersions and hiatuses, Bosence et al. 2009;

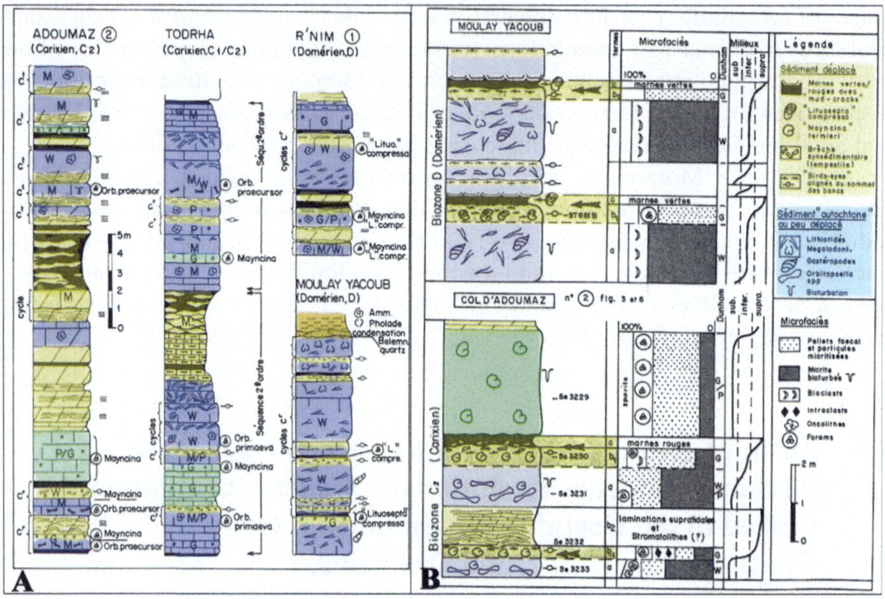

Fig. 8.3 (**A**) The Aganane formation (Pliensbachian) in different locations of the High Atlas and Rif foreland (Moulay Yacoub) in Morocco. (**B**) Detail of metric "shallowing upward" parasequences and storm beds (in green). This cyclical type of deposit, in the photic zone, is widespread in space and time along the southern margin of the Mesozoic Neotethys carbonate shelves from Morocco to Oman (via Italy and the Dinarids) and on the northern margin in Mallorca or Sardinia. These depositional conditions represent the general sedimentological setting where imperforate larger foraminifera (ILF) live and evolve, a rather unstable environment characterized by rapid and multiple (metric to decametric) sea-level changes, emersions, etc. Paradoxically, this environmental instability does not influence the morphogenetic variability, obviously depending on other factors like symbiotic coevolution. Important abiotic factors driving to microfaunal extinction or turnover are anoxia (Table 8.6) and global drowning
Figures excerpt from Septfontaine (1985)

Strasser (2016), versus drowning and deepwater conditions) explain the episodical and scarcity of microfossil-rich intervals compared to planktonics. It impedes the study of morphological pathways and diversity in space and time (either of genetic or of ecophenotypic, epigenetic origin) and finally to argue for the evidence of evolutionary trends. However, exceptions are in the liassic carbonates of Morocco (Hottinger 1967; Septfontaine 1985) and will be discussed with the orbitopsellid "plexus" (Fig. 8.2 and Table 8.6).

3. The stratigraphical calibration (e.g., with ammonite biozones) is difficult in particular in the eastern Tethys area, due to the almost absence of any valuable pelagic time marker. Moreover, bioevents are not well marked (rarity of macrofossils) and sequential stratigraphy is useless over long distances because of local paleotectonics in most studied cases (Jura-Briançonnais-Sardinia, Italy, Dinarids, Morocco, Oman, etc.) related to a general situation of the Tethyan

opening and rifting (and later closing) of the northern and southern margins during the Mesozoic.

8.4 Taxonomic Considerations: Weight and Hierarchy of Changing Characters—Origin from an Atavic Stock of Simple Forms

8.4.1 Review of Characters

This short review concerns chiefly the group of shallow water foraminifera with an imperforate microgranular-agglutinated wall (with or without pseudokeriotheca) classically designed as "lituolids"[1], most of them housing symbionts (Brasier 1988) in the wall and the marginal structures (Plate 8.1) of the chambers (as recent miliolids) related to the static protoplasm (Hottinger 1967). The presence of these hollow cavities offers relative protection to the symbionts against cytoplasmic movements or disorders. It is an argument for possible symbiotic coevolution forcing morphogenetic changes in a general trend similar to the one resumed in Fig. 8.2, regarding discoid shell shape and apparition of variable microstructures in the marginal zone of the chambers through time. Moreover, the size/volume/surface increase (Dollo, see Guex 2003) plays an important role in bioseries, advanced larger forms hosting more symbionts like recent *Cycloclypeus* living in deeper waters (Song et al. 1994). Conversely size decreases (and taxa turnover) aftermath anoxic events is a response to unfavorable conditions (Table 8.6). Equivalences of terminology for morphological descriptive terms of internal microstructures by various authors are given by Septfontaine (1981).

1. **The wall** is of a rather constant composition (secreted Mg-calcite with some agglutinated material) and presents a microstructure of diversely oriented microgranular calcitic rhombs (electron micrographs, Septfontaine 1971; Banner et al. 1991). In rare cases (*Meyendorffina, Pseudocyclammina, Pseudoeggerella*, etc.), the wall is covered by a fine surficial layer of agglutinated quartz grains. There exists clearly a relationship between the arrangement of non-oriented (isotropic) microstructure and the absence of ornamentation at the surface of the smooth imperforate test covered by a tectum (compared to the hyaline radial calcitic perforate tests of the Rotaliina exhibiting many ornamentations along the sutures, ridges, knobs, etc.). These ornamentations (or their absence) can also be a mea-

[1] **Lituolacea** De Blainville, 1825 in Loeblich and Tappan (1964, 1988) [part], incompletely defined in 1988 ("*planispirally enrolled in early stage, later may uncoil*") as the authors do not include trochospirally enrolled forms throughout the test in their definition. Many genera are grouped in the order **Loftusiida** by Kaminsky 2014. In the present study, the term "Lituolids" includes only inhabitants of shallow carbonate platforms.

sure of environmental changes related to water depth or periods of anoxic crises (Guex 2016).

In thin sections, the wall shows sometimes the presence of a *pseudokeriotheca*, a term incompletely defined as "*uniform parallel, radial elements covered by some kind of tectum*" by Hottinger (2006). This microstructure is indeed morphologically very close to the analogous keriotheca of fusulinids and endothyrids (Rigaud et al., 2014), but arranged on one row only (see also a review of terminology already discussed in Septfontaine (1981)); the interpretation as tiny tubules (5 μ) covered by a tectum (Banner et al. 1991) seems obvious (Plate 8.1 (1–6)).

According to Vachard et al. (2004), this microstructure of tiny alveolae served to house symbiotic algae or cyanobacteria in some fusulinids. It is common in many lituolid genera *but incompatible with radial partitions or hypodermic network*. This may be due to an incompatibility between symbiotic inhabitants, filamentous bacterial versus larger dinoflagellates. The size of these diverse chamberlets may be related to, and is consistent with, the size of photosynthetic symbionts (20–250 μ for microalgae).

Some genera develop coarser cylindrical alveolae within the wall (covered by a fragile tectum (6 μ)) which can be removed by mechanical abrasion in some mesoendothyrids, erroneously attributed to a "perforate" (an artifact) family "Lituoliporidae" by Gusic and Velic (1978). Transitional microstructures between pseudokeriotheca and alveolae in the wall do exist from the Triassic (Rigaud et al. 2014) to the Jurassic, and the Cretaceous onward (*Wernlina*, *Paleomayncina*, *Everticyclammina*, etc.) (Plate 8.1 (7–10)).

2. **The mode of coiling** is of prime importance in the definition and separation of Mesozoic clades of ILF and probably of polyphyletic origin (Septfontaine et al. 1991; BouDagher 2008; Mikhalevitch 2013; Kaminsky 2014; see Tables 8.1, 8.2, 8.3, 8.4, and 8.5 for details). Two main classical morphotypes or lineages are recognized from the Early Liassic to the Late Cretaceous:

    (a) The **planispiral** to uniserial (discoid) type or mesoendothyrids and hauraniids (Tables 8.2 and 8.3)
    (b) The **trochospiral** to uniserial (cylindrical-conical) type or valvulinids and pfenderinids/kurnubiids (Tables 8.4 and 8.5)

3. **The hypodermic network** and theM vertical radial partitions (see below) are Jurassic innovations dividing the margin of the chambers, already present in the Late Sinemurian. They are shared by different clades and probably diverge in form and size during the Mesozoic depending on the type (and size) of hosted symbionts. The hypodermic network can be irregular, with wavy subvertical and subhorizontal/oblique lamella (Bassoullet et al. 1976). The subvertical (with trapezoidal contour) lamella are detached from the floor of the chambers and attached to the roof ("stalactitic microstructure" in the genus *Alzonella*, etc.) by its subtriangular base (in vertical section) near the center of the chambers (Plate 8.1 (11–15)). The network can be more regular in some Late Jurassic-Cretaceous taxa like *Choffatella*, a deeper water genus.

4. **The vertical radial partitions** (a Late Sinemurian novelty) are attached to the external wall, contrary to the subdivisions of some Pfenderinids like *Satorina* issued from the central mass, not reaching the exterior wall. They divide the marginal zone of the chambers into chamberlets as in recent porcelaneous discoid equivalents. Horizontal subdivisions are absent in both cases (Plate 8.1 (16–20)). The partitions are plain or trapezoidal in shape or contour. The thickness at the base increases gradually toward the center of the chambers, so that the partitions show a characteristic triangular shape in vertical sections, with a "bifid" appearance in basal section. Near the central zone, the partition system generally does not coalesce at the base with the central infill, defining an "annular free zone" connected with the "crosswise oblique stolon system" sensu (Hottinger 1967). This later architectural arrangement is a Bathonian innovation appearing firstly with the genus *Meyendorffina* and later in the Early Cretaceous Orbitolinids (by homology) and *Eclusia* (Platc 8.1 (19–20)) described in Septfontaine (1971), Cherchi and Schroeder (1975).
5. **The pillars** (cylindrical or half-conical) are directly linked to the multiple apertures in the center of the chambers arranged in different stolon systems as in valvulinids and orbitolinids. The pillars are triangular in vertical sections and can be confused with the triangular sections of the radial partitions near the central zone of the chambers close to the central calcitic filling of some genera like *Meyendorffina*.
6. **The apertures** are associated with the central stolon system, straight or crosswise-oblique (Hottinger 1967) in discoid morphotypes, in relation to the dynamic protoplasm flux. Trochospiral valvulinids and pfenderinids show single (interiomarginal) or multiple apertures (pores) covered by a septal flap sometimes "trematophore-like," and modified in the pfenderinids in a central vertical half-tube (half columella) or "tooth plate" sensu Hofker in Hottinger (1967). The associated characteristic triangular-vertical partitions defining the *Eclusia-Ilerdorbis* morphotype appear later during Bathonian times in valvulinids and pfenderinids like *Meyendorffina*; see Tables 8.4 and 8.5.

8.4.2 Discussion

It is possible to draw some general trends concerning the relation between coiling and some internal microstructures as follows:

- The planispiral mesoendothyrids and trochospiral valvulinids/pfenderinids[2] are devoid of any subdivision in the marginal zone of the chambers. Instead, a pseudokeriotheca is always present in the (thick) wall covered by a tectum; pillars can

[2]The special case of the valvulinids *Pseudodictyopsella-Dictyopsella* and the "Kurnubiids" with a hypodermic network (without pseudokeriotheca) is discussed in Tables 8.4 and 8.5.

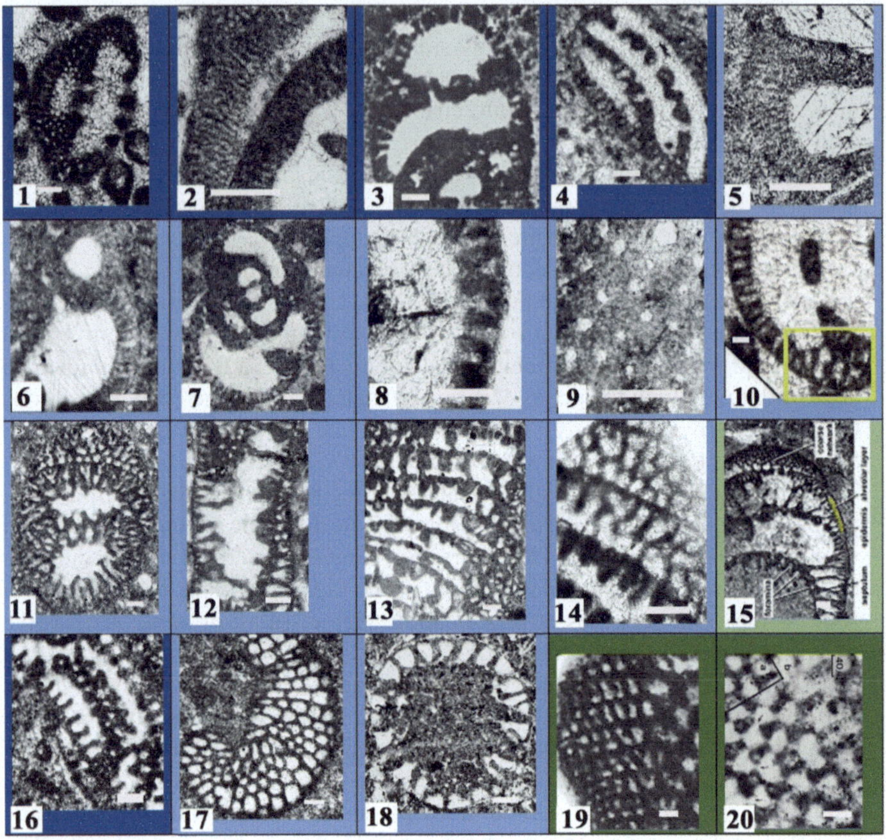

Plate 8.1 Wall microstructures: **pseudokeriotheca** and **alveolae** (1–10) with transitions (diam. 5–15 μ) and divisions developed in the marginal zone of the chambers: **hypodermic network** (11–15) (length 40–200 μ) or **radial vertical partitions** (16–20) (length 100 μ, without transitions). The walls (3) (*Paleomayncina* in Loeblich and Tappan 1988) and (10) (*Dubrovnikella* in Schlagintweit and Velic 2011, thickness 70 μ) show clearly the transition, respectively, between **pseudokeriotheca** and **alveolae** and **alveolae** to the **hypodermic network** (in *Dubrovnikella*, yellow rectangle)

Colors: dark blue = Middle Liassic; middle blue = Dogger; dark green = Early Cretaceous; light green = Late Cretaceous

The sizes (diam.) of alveolae and chamberlets vary from 5 to 20 μ (pseudokeriotheca-alveolae) to 40 μ or more (hypodermic network-radial partitions). This variation is possibly related to the size (or number) of symbionts, e.g., bacterial filaments (Vachard et al. 2004), dinoflagellates, or other algae, in a perspective of coevolution

List of genera (scale white bar ca 50 μ):

1 *Paleomayncina* (pseudokeriotheca and multiple openings)

2 *Biokovina* (pseudokeriotheca, excerpt from Gusic (1977))

3 *Paleomayncina* (transition pseudokeriotheca-alveolae, excerpt from Loeblich and Tappan (1988))

4 *Paleomayncina* (advanced flabelliform to reniform type)

5 *Chablaisia* (pseudokeriotheca)

6 *Paleopfenderina* (pseudokeriotheca)

(continued)

7 *Mesoendothyra* (transition pseudokeriotheca-alveolae)
8 *Everticyclammina* (alveolae)
9 *Everticyclammina* (detail of alveolae)
10 *Dubrovnikella* (transition alveolae to lamellae of the hypodermic network (yellow frame), excerpt from Schlagintweit and Velic 2011))
11 *Pseudocyclammina* (hypodermic network, long vertical lamellae below)
12 *Alzonella* (hypodermic network with long irregular vertical lamellae as *Persiacyclammina* (15))
13 *Alzonella* (idem, peloids form irregular vertical lamellae)
14 *Anchispirocyclina* (hypodermic network, irregular lamellae)
15 *Persiacyclammina* (fine hypodermic network with important development of main vertical lamellae toward the center of the chambers as in *Alzonella* (12)—excerpt from Schlagintweit and Rashidi 2017a)
16 *Orbitopsella* (vertical regular radial partitions, left)
17 *Meyendorffina* (vertical regular radial partitions, tangential/axial section)
18 *Meyendorffina* (vertical radial partitions becoming triangular in section near the central calcitic filling; basal section)
19 Orbitolinid (hypodermic network (left) and triangular main vertical partitions or lamellae, right)
20 *Eclusia* (triangular sections of radial partitions near the center of the chambers in a "crosswise oblique" stolons or "marginoporiform" system) (Hottinger and Caus 1982)

be present in the central zone of the chambers, associated with the cribrate openings and the dynamic protoplasm.

- Conversely, the planispiral-to-discoidal hauraniids and biserial dicyclinids never develop a pseudokeriotheca, but instead develop variable marginal microstructures (hypodermic network or radial partitions, see above) associated to the tectum.
- Central calcitic filling (plus pillars) with secondary ways formed by resorption/dissolution (tunnel, groove, spiral canal, or solid columella) develops only in trochospiral tests.

The hypodermic network and vertical radial partitions are specific Jurassic constructions unknown during Paleozoic and Triassic times. These two basically different architectures are both built from the exterior wall (exoskeleton auct.), and probably of polyphyletic origin during Early Liassic times, without morphological transitions between both. However, vertical regular "lamellae" of the hypodermic network (Plate 8.1 (19)) of orbitolinids can be assimilated morphologically to vertical radial partitions of other Jurassic taxa. It must be noted that the regular hypodermic network of orbitolinids is not known in the Jurassic, and can be interpreted as a novelty appearing during Early Cretaceous times.

These microstructures (Plate 8.1) appear iteratively through time in different clades distinguished by their mode of coiling (Tables 8.1, 8.2, 8.3, 8.4, and 8.5), and are considered of secondary importance in taxonomy. Nevertheless, an alternative proposition can be envisaged and the reverse (wall and marginal microstructures first, and coiling as a secondary character) can also be a relevant solution but not retained here. Likewise, the pillars and other structures associated with the cribrate aperture can be considered as secondary (but advanced) characters appearing as isomorphic (analogous) traits in different clades during their evolution.

Table 8.1 General test shapes and innovations of imperforate larger foraminifera

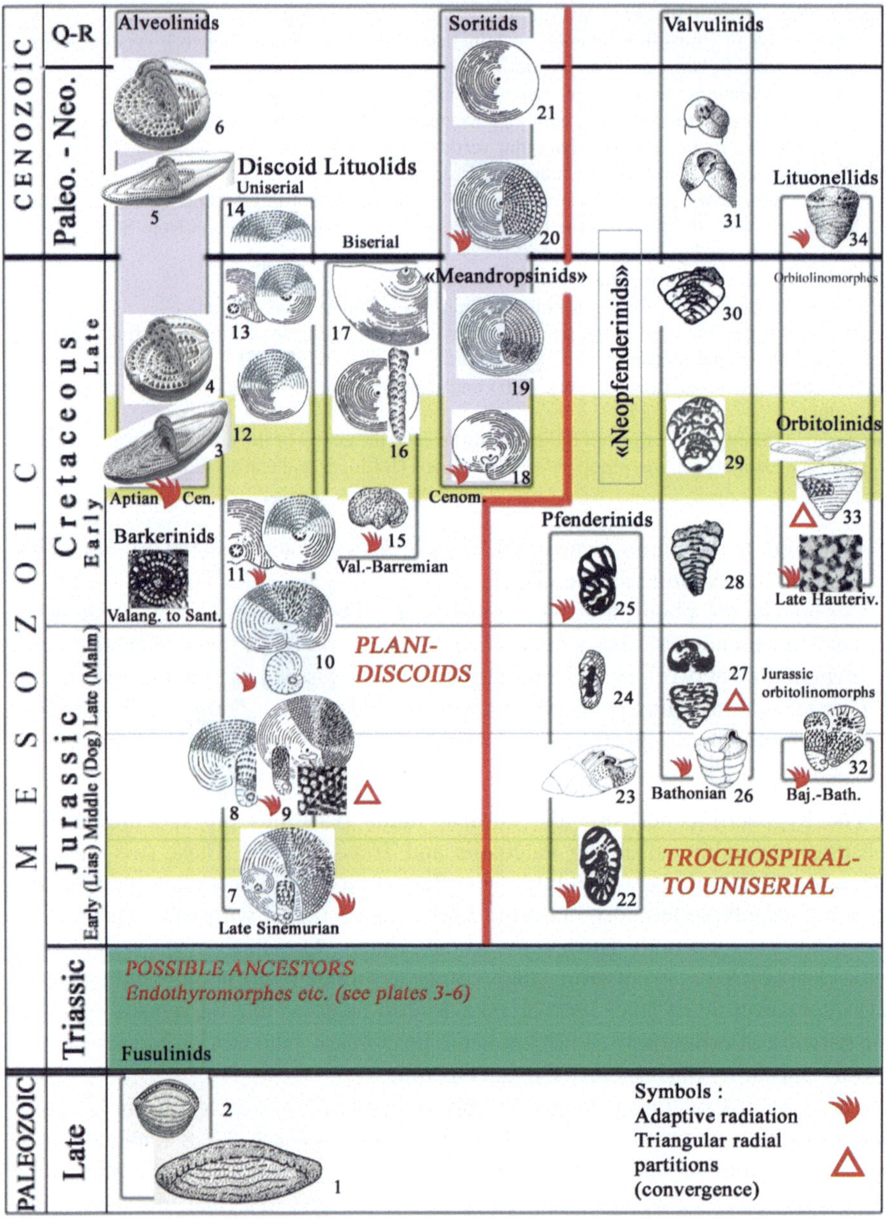

Morphological novelties and iterative evolution among ILF (and microgranular isomorphic porcelaneous alveolinids and soritids, in light grey) during Phanerozoic times. General coiling and shape **planispiral-discoid** (7–21) or **fusiform-globular** (1–2; 3–6; and Barkerinids) left of the red line versus **elongate-trochospiral** coiling of the test on the right. Note the first original appearance (7 and 22) of planispiral-discoid tests on the one hand, and trochospiral (pfenderinids)

(continued)

Table 8.1 (continued)

morphotypes on the other (in the Late Sinemurian) after the main OAE crisis of the Late Triassic. Some innovations of microstructures (hosting symbionts) in the marginal zone of the chambers are characteristic of the Mesozoic (hypodermic network; radial partitions with triangular axial section, see Plate 8.1), while others like the pseudokeriothecal wall represent a Paleozoic heritage. These microstructures are distributed iteratively in variable major clades: (1) strepto to planispiral or (2) trochospiral to uniserial/biserial. Note that the **biseriality** in discoid ILF foraminifera (15–17) is an Early Cretaceous novelty as well as the appearance of globular tests in Barkerinids

Yellow bands are periods of successive anoxic crisis (volcanism), e.g., OAE in the Toarcian-Aalenian and in the mid-Cretaceous (Albian-Cenomanian-Turonian, OAE 1b-2). For detail of anoxic periods see the caption of Table 8.2. Green band emphasizes the Triassic period with its specific microfauna (involutinids, endothyrids, see Plate 8.3, and a general dwarf microgranular microfauna). Aragonitic wall of involutinids (not shown) persists during the Mesozoic

Note that later anoxic periods during the Mesozoic have no major influence on appearance or demise of main morphotypes and microstructures, which are more related to coevolutive genetical processes between the host cell and probable symbiotic bacteria or algae

Fusulinid icons from Cushman (1933); *alveolinid icons from Reichel in Loeblich and Tappan* (1964)

(1) ***Triticites*** (icon excerpt from Cushman 1933); (2) ***Schwagerina*** (excerpt from Cusman 1933); (3) ***Praealveolina***; (4) ***Ovalveolina***; (5, 6) ***Borelis***; (7) ***Orbitopsella*** (icon excerpt from Septfontaine 1988); (8) ***Alzonella***; (9) ***Meyendorffina-Orbitammina***; (10) ***Timidonella***; (11) ***Balkhania***; (12) ***Pseudochoffatella***; (13) ***Neobalkhania***; (14) ***Vania***; (15) ***Cuneolina***; (16) ***Dicyclina***; (17) ***Suraqalatia*** (Giant Dicyclinid, diam. 7 cm); (18) ***Pastrikella (Broeckina)***; (19) ***Meandropsina*** (redrawn from Loeblich and Tappan 1964); (20) ***Orbitolites***; (21) ***Sorites-Amphisorus-Marginopora***; (22) ***Pseudopfenderina***; (23) ***Paleopfenderina***; (24) ***Kurnubia***; **(25)** ***Pfenderina***; **(26)** ***"Praevalvulina" = Valvulina (Redmondoides)***; **(27)** ***Parurgonina***; **(28)** ***Praechrysalidina***; **(29)** ***Chrysalidina***; **(30)** ***Accordiella***; **(31)** ***Valvulina***; **(32)** ***Gutnicella***; **(33)** ***Cribellopsis-Eclusia and Orbitolina***; **(34)** ***Lituonella***

However, the estimation of the hierarchical weight of characters is most often a subjective construction in a pure cladistic approach. An "a priori" decision can be refuted by further studies and new data. Then by experience within a particular group, it is questionable to classify morphological characters into specific, generic, or even higher taxonomic categories, as morphological elements (traits) do not have intrinsic significance or weight, as discussed by Fleury (2018) for larger miliolids. A character susceptible to distinguishing between species could be found again, but at a generic level and conversely a character permitting to distinguish genera will be found in the separation of diverse species. It may be the case for some lituolids, e.g., the test shape and contour of pfenderinids/*Nezzazatinella* or *Planisepta* and others discussed in Tables 8.2, 8.3, 8.4, and 8.5. Such complexity is due to the fact that the diverse morphological traits evolve independently, the ones from the others and at a different speed ("mosaic evolution," see Dollo in Gould 1977). Associations of characters should be considered to identify the evolutionary trends in their complexity and to distinguish them in a more general perspective (through space and time). This method needs a good global understanding of the considered group of foraminifera and the results must be shaded (or completed) with the addition of stratigraphical and paleoecological data.

Table 8.2 Polyphyletic **planispiral** morphogroups of flabelliform–reniform to discoid adult-shape larger foraminifera, **uniserial** (A to E) or **biserial** (F) with (or without) various "exoskeletal" marginal microstructures hosting endosymbionts, with (or without) central pillar "endoskeleton"

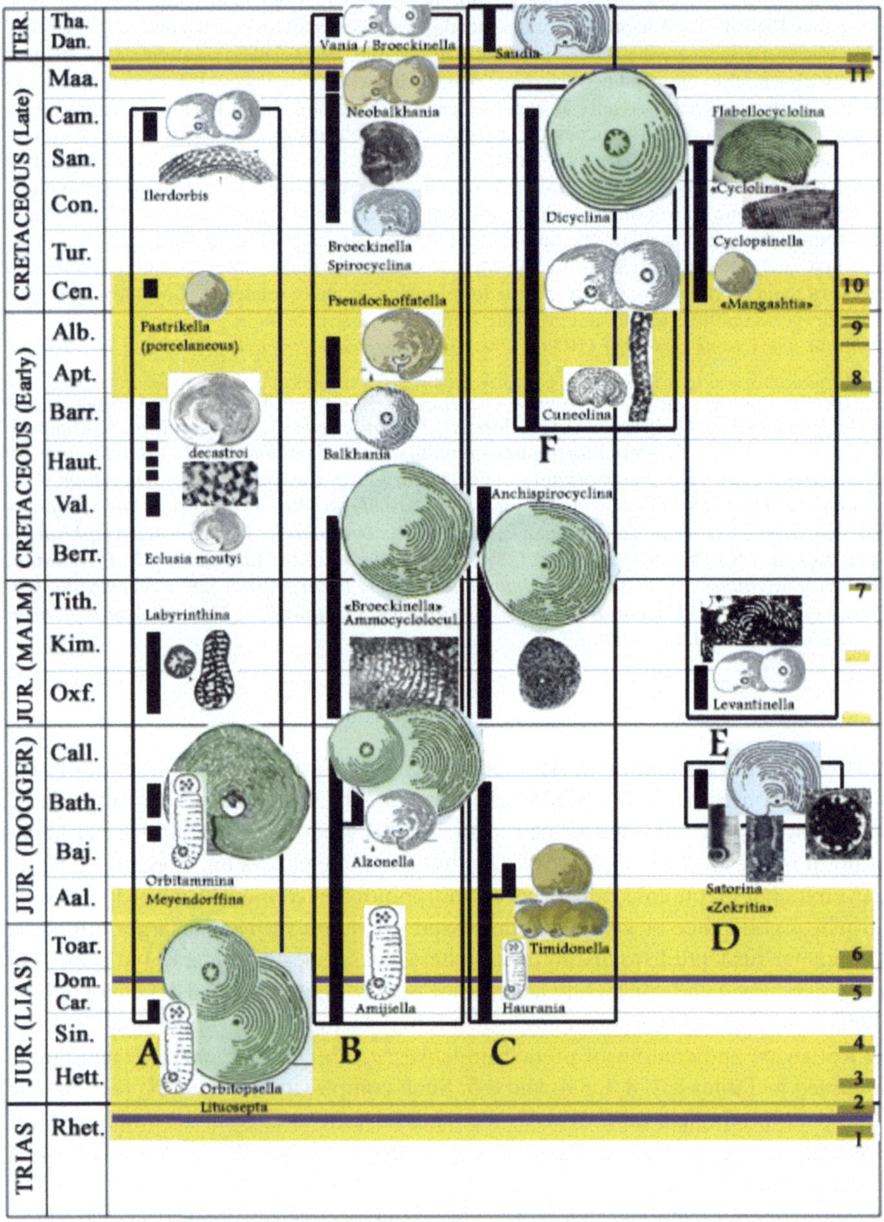

(continued)

Table 8.2 (continued)

Aperture multiple, cribrate, or pores aligned

Left to right: (A) with marginal radial partitions (orbitopsellids) becoming triangular in section near the center (*Meyendorffina*, *Eclusia-Ilerdorbis*) and pillars. (B) With a marginal hypodermic network, without pillars in the central zone of the chambers, aperture cribrate (amijiellids s. str.), or in one row (*Broeckinella*, etc.). (C) With marginal hypodermic network and cylindrical pillars in the central zone of the chambers (hauraniids). (D) Without marginal partitions s. str., but with radial divisions *extending from the center calcitic infill* to the margin of the chambers (not joining the tectum) as in *Satorina*. (E) Without marginal partitions but with central pillars (*Levantinella*, "*Manghastia*," *Cyclopsinella*). (F) **Biserial** test with *marginal hypodermic network* (cuneolinids). Biseriality is an important Cretaceous innovation among reniform-discoid larger foraminifera, as *Cuneolina* an Early Cretaceous ancestor of *Dicyclina* and *Suraqalatia* (giant test, diam. 7 cm, not shown)

- Yellow surfaces: periods of warming with anoxic conditions, a long period of stress for larger discoid foraminifera with a record of small-size individuals (2–3 mm diam.) and general low biodiversity (icons colored in brown)
- White surfaces: periods of "normal" conditions, medium size (11–13 mm diam.) to very large "monstrous" tests (20–70 mm diam. in dicyclinids), indicating stable and comfortable conditions for vegetative symbiotic life (K-strategist), but fragile and doomed to extinction in a changing environment (Hallock 1985). Icons colored in light green

Note: the small to medium diameters of discoid tests are roughly at scale. The true ratio between very large tests (green) to small tests (brown) is **20:2**, not on the scale in Table 8.2

In Tables 8.2, 8.3, 8.4, and 8.5, numbers 1–11 (brown) refer to main peaks of anoxia (OAE), warming and cooling (blue) events during the Mesozoic (detailed below). For impact on larger foraminifera populations, see also the review of BouDagher (2008)

1. Late Rhaetian, Guex (2016)
2. Triassic-Jurassic boundary, −201 Ma. CAMP Atlantic magmatism and sea-level fall. Hallam and Wignall (1999), etc.
3. Hettangian *planorbis/pacificum* warming event (Guex 2016)
4. Early Sinemurian *semicostatum* warming event
5. Late Pliensbachian (Domerian) *margaritatus* warming and cooling (in blue) events
6. Late Pliensbachian and Early Toarcian *falciferum/exaratum* period and peak of warming (in brown), worldwide T-OAE (−183 Ma). Jenkins (1988), review in Rosales et al. (2018) and Duarte et al. (2019). South Africa Karoo Traps, magmatism
7. Late Tithonian warming event, Oxfordian "black shales," and Kimmeridgian "Orbagnoux bituminous shales" (yellowish brown) are not numbered and considered as local restricted anoxic deposits, like the eutrophic Bathonian-Callovian "*Mytilus* beds" in the Briançonnais realm (black muddy inner platform deposits) rich in imperforate trochospiral larger foraminifera, but without discoid larger morphotypes
8. Early Aptian worldwide anoxic event; OAE 1a or Selli event, Ontong-Java 1. Demise of the Urgonian platform (Föllmi 2008). Barents sea High Arctic magmatism (HALIP)
9. Albian anoxic events (Mid-Cretaceous events). OAE 1b-d (Strasser et al. 2001)
10. Mid- and Late Cenomanian anoxic and drowning events (Mid-Cretaceous events). Late Cenomanian anoxic event OAE 2
11. Maastrichtian warming and cooling events. Deccan Traps, bolide impact on Yucatan and massive extinction. **Shallow water** larger imperforate discoid foraminifera paradoxically not strongly affected. "Lazarus" effect (Wignall and Benton 2000) or recovery in Paleocene-Eocene times

Table 8.3 The **mesoendothyrids** (Mesoendothyridae emend., left and center columns of this table) include morphogenera representing a Paleozoic heritage, through Triassic forms, of rather stable morphotypes (coiling strepto-irregular to planispiral) with **pseudokeriothecal-to-alveolar** structure (symbionts bearing in advanced forms) with a thick wall

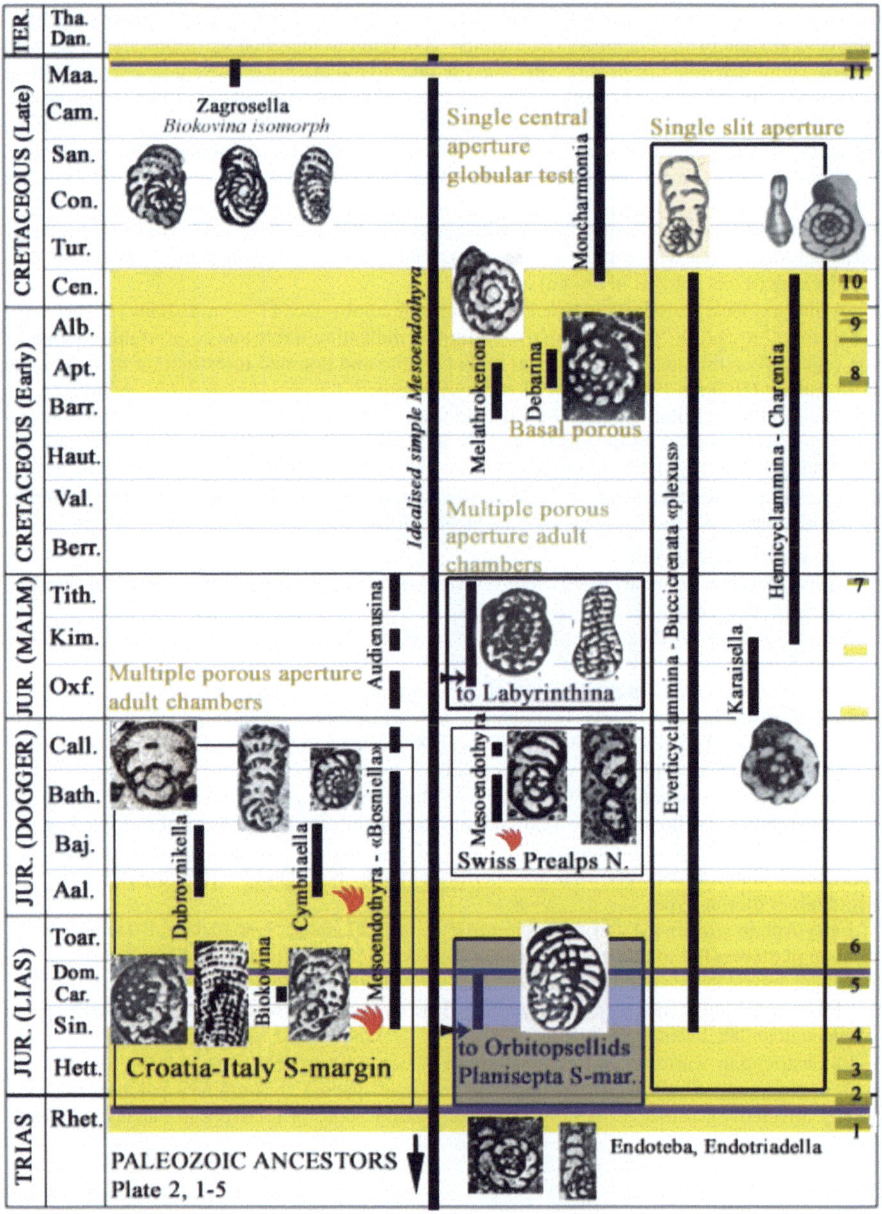

Table 8.3 (continued)

The septa display a conspicuous "spinose, rose thorn" contour, and are tangential to the previous whorl in an equatorial section (see Plate 8.3). The simple basal-to-central opening in early chambers becomes multiple in the adult. The **mesoendothyrids** cross severe anoxic environmental crises (1–11 in brown, yellow band) without being affected, after the main P/T extinction. But some Mesozoic innovations are not known before the Middle Liassic, like multiple porous openings and pillars in the adult stage as recorded in Middle Liassic-to-Dogger and later Maastrichtian levels, following an iterative (somewhat "chaotic") mode of appearance from a simple stock of the *Mesoendothyra* "plexus"

The **everticyclamminids** (right column) with a single-slit aperture (straight or sinuous) were probably issued from a mesoendothyrid ancestor during Early Liassic times. The classic alveolar microstructure of the wall (a Triassic and Liassic innovation) is interpreted as a coarse **pseudokeriotheca** hosting a particular type of larger symbiont (unknown dinoflagellates)

The characters **a–f** defined above are presented in hierarchical order as elements of the classification used in the present study and correspond roughly to the generally accepted classification of lituolids on carbonate platforms by recent authors (Mikhalevitch 2013; Kaminsky 2014).

8.4.3 Defining a Stock of Simple Atavic Persistent Morphotypes

After the main periods of foraminiferal extinctions affecting benthic and planktonic populations during the Mesozoic, the apparition of rare, smaller, morphologically simpler taxa, from which new clades could evolve, has been classically observed by various authors (e.g. Arriaga et al. 2016). These atavic forms (survivors) were partly identified in the Early Jurassic (with a problematic taxonomic status) and briefly discussed by Septfontaine (1984b, 1988) and some siphovalvulinid species introduced by BouDagher and Bosence (2007). They are considered as good candidates at the origin of the different morphogroups considered below (Tables 8.2, 8.3, 8.4, and 8.5) with a very long stratigraphic extension through the Mesozoic and the Tertiary. To be cited are smaller endothyrid-like (*Mesoendothyra*), smaller and rare "praevalvulinids*", abundant siphovalvulinids (*Siphovalvulina*), etc., all persistent atavic morphotypes representing a genetical plexus or "reservoir" from which successive advanced homeomorphic genera diverge iteratively. The origin of hauranids is more problematic. Globally the model of "retrograde evolution" proposed by Guex (2016) for the planktonic rotaliporiids during anoxic Middle Cretaceous extinction crises was not observed among Mesozoic larger benthic foraminifera (see also Arriaga et al. (2016) for the end-Cenomanian crisis), due probably to their particular complexity and intolerance to environmental changes as already stated.

Table 8.4 Mesozoic **valvulinids** (“**praevalvulinids***” and **chrysalidinids**) display trochospiral morphotypes with three or more chambers per whorl, a pseudokeriothecal wall (in advanced larger forms), and a valvular “tooth plate” simple or modified in the central umbilical zone, with conical pillars associated to pores of a cribrate aperture through the tooth plate (advanced forms)

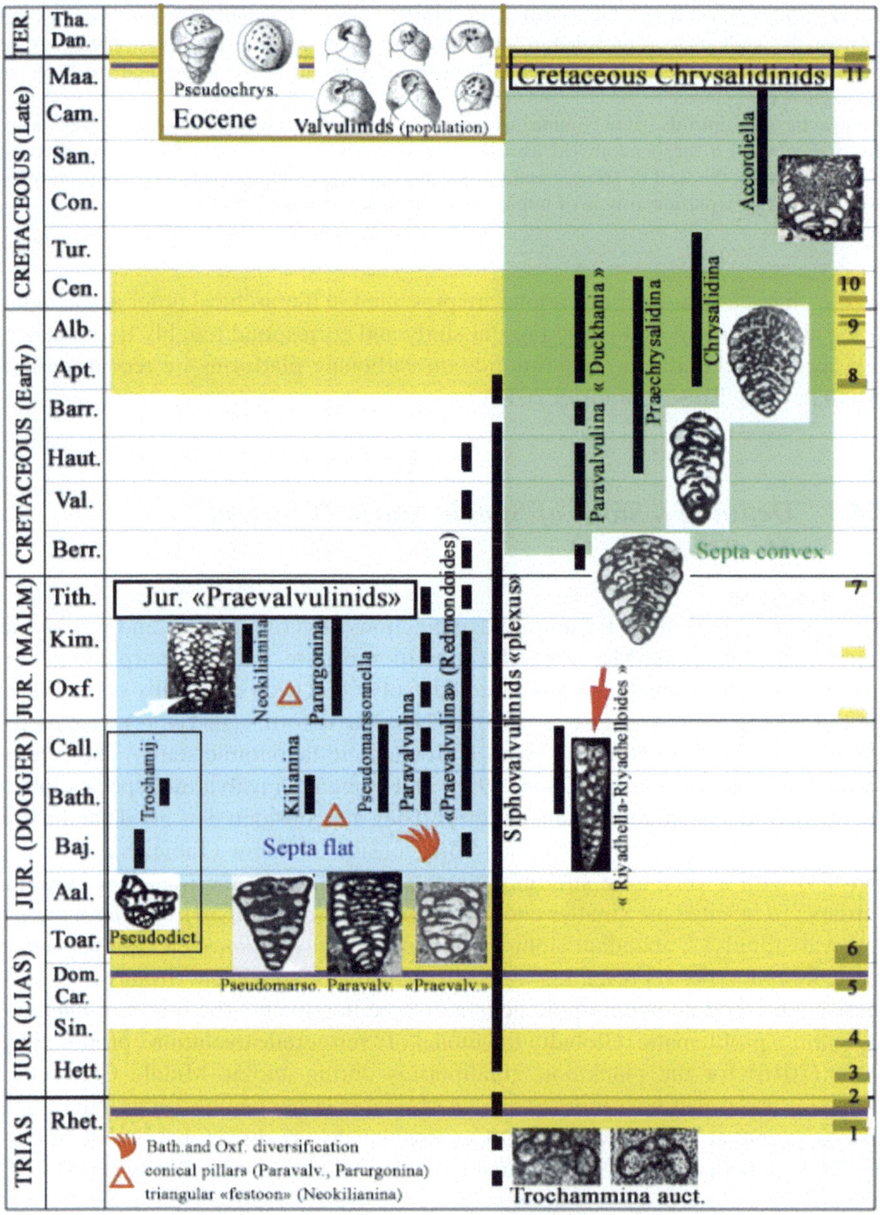

(continued)

Table 8.4 (continued)

The morphotypes include (left of the central **siphovalvulinid** "plexus" on plate) (1) the Jurassic **praevalvulinids∗** (septa flat) and to the right and (2) the Cretaceous **chrysalidinids** (septa convex) with transitions at the J/K limit. The **siphovalvulinid** "plexus" is composed of ancestral simple forms with a central siphon or hollow columella supposed to produce a poorly developed "tooth plate" (after deformation of the columella wall) omnipresent through the Mesozoic (representing a constant "genetic reservoir") from which several complex descends (see also Plate 8.5, **pfenderinids**) are issued following an iterative mode till the Tertiary (Eocene to recent). Red arrow: siphon in the last chambers of "*Riyadhella-oïdes*" morphotype. White arrow: early **praevalvulinid∗** ontogenetic stage in *Neokilianina*

Note apertural (local ecophenotypic) variations of the "tooth plate" in an Eocene synchronous population (top of the plate, from Neumann 1967) as commonly observed in some Jurassic stocks of "**praevalvulinids∗**"

8.5 Impact Appraisal of External Abiotic Factors on Iterative Phyletic Morphogenesis and Individual Morphological Trends

8.5.1 General

The relations between main abiotic events and marine invertebrate population turnovers or extinctions are still unclear (e.g., for the K/T extinction, Mac Leod et al. 1997; Abramovich et al. 1998; Archibald et al. 2010 for Globotruncanids) at a global scale, and the impact of smaller cyclic environmental changes on a morphogenetical trend in a single bioseries (e.g., the orbitopsellid model Fig. 1 discussed by Septfontaine 1984b) seems insignificant. Extensive studies on the subject at a larger scale conclude variable importance of environmental factors (Font et al. 2016) during the major crises on mode, tempo, and synchronism of extinctions, mainly volcanism and main peaks of anoxia (1–11, Tables 8.2, 8.3, 8.4, and 8.5). For example changes in biotic assemblages indicate a marine ecosystem crisis *prior* to the oxygen depletion for P/T anoxia in Norway (Dustira et al. 2013). However, Burger (2019) links the major P/T extinction with massive combustion of coal responsible for global warming and ocean acidification during Siberian Traps volcanism. For other authors, the onset of eruptions slightly post-dates the main phase of faunal disorders during several extinction periods (e.g., Wignall 2001), and the link between LIP's and extinctions remains enigmatic. The Early Jurassic extinction event (Late Pliensbachian-Toarcian OAE) was not synchronous throughout the Tethyan region (Wignall et al. 2006; Metodiev et al. 2012), and this Jurassic extinction may have been a regional crisis rather than a more severe global catastrophe (Wignall and Racki 2006). More precisely equating the timing of LIP eruptions to extinctions is generally difficult (Van de Schrootbrugge and Wignall 2016). The Deccan volcanism is dismissed as much older than the K/T extinction which is considered as a multicausal (warming and cooling, etc.) long-term scenario and not a single one (Keller et al. 2010; Punekar et al. 2015). On the other hand, many faunal turnovers or extinctions, including benthic (larger) foraminifera, began already in

Table 8.5 Repartition of trochospiral **pfenderinid** morphotypes and associated forms in the Mesozoic, with morphological innovations (*Italic*)

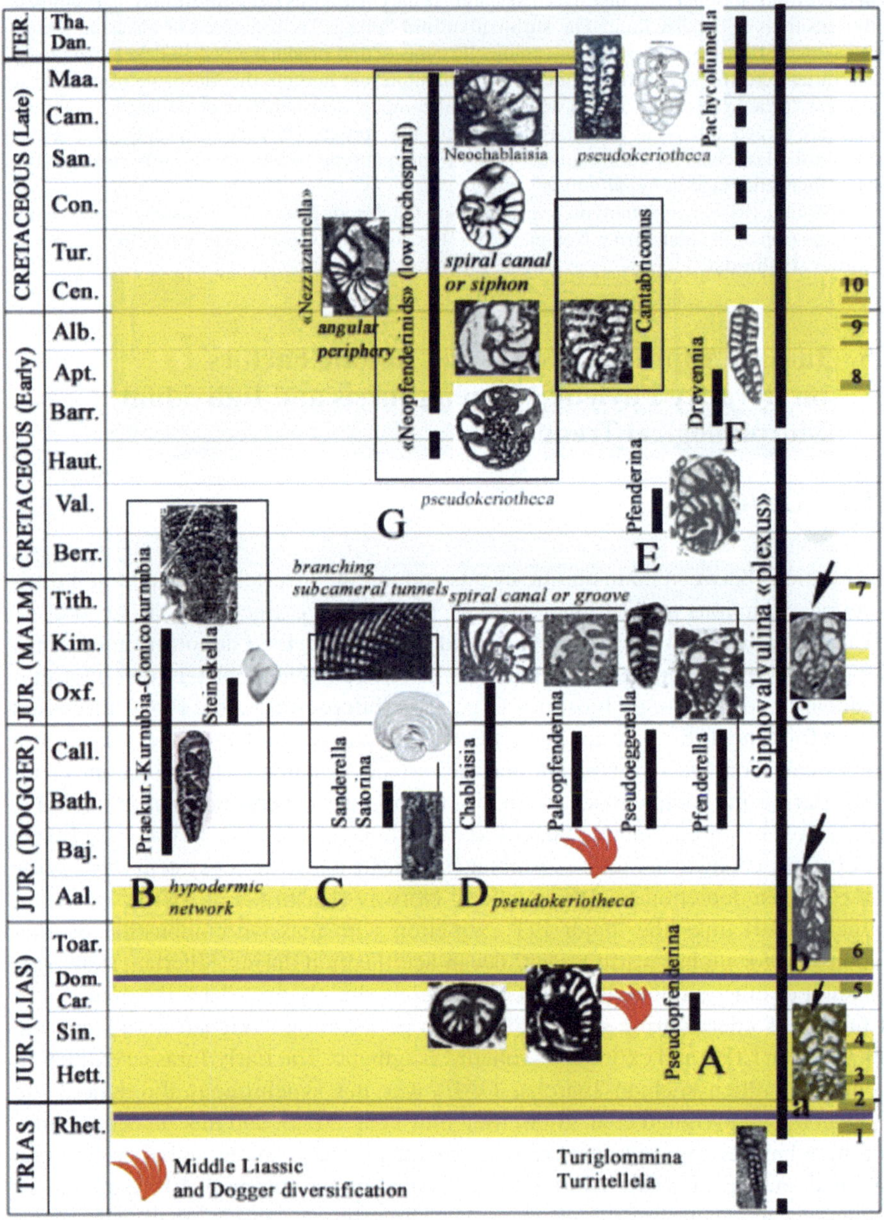

The first appearance of the group, characterized by a central columella (filled central siphon, arrows a, b, c extreme right of the table) and calcitic infill between pillars with later resorption (***groove, canal***), is in the Middle Liassic (A), after a period of climatic changes (warming and cooling) from

(continued)

Table 8.5 (continued)

the Late Triassic to the Early Sinemurian (1–4). A second diversification period (B–D) is in the Late Bajocian-Bathonian to Callovian interval, as for the valvulinids, after the Toarcian anoxic event (6), continuing in the Early Dogger. Advanced forms show an uncoiled "orbitoliniform" cylindrical-to-conical adult stage (B) as *Conicopfenderina* (not shown, see Plates 5–25) or *Conicokurnubia*. The genus *Satorina* shows an exceptional development of arched adult chambers with ***multiple subcameral tunnels*** (C). The **pfenderinids s. str.** are absent from the Kimmeridgian-Tithonian interval, characterized by the **kurnubiids** and advanced *Conicokurnubia* with a ***hypodermic network*** (B), probably issued from a *Siphovalvulina* stock. A new apparition of the **pfenderinids s. str.**, without a direct link with Jurassic morphotypes, occurs in the Valanginian with the well-known *Pfenderina neocomiensis* (E) of large size, issued probably from the ancestral stock, like the younger variant *Drevennia* (F). Other Cretaceous morphotypes with a low trochospire and a spiral canal or an open umbilicus (sometimes filled by pillars) are considered as modifications of the *Siphovalvulina* plexus with simpler morphology. The genus *Pachycolumella* is a Maastrichtian-Paleocene newcomer, homeomorph of *Radoicicina* (Liassic, not shown) with a *Siphovalvulina*-like morphology

the latest Maastrichtian (Mac Leod et al. 1997), and many other faunal taxa, including benthic foraminifera (Table 8.4 and 8.5), survived the K/T crises with minor diversity changes. The Newell's regression hypothesis is tenable for the end Triassic extinction but not for other crises; Globotruncanid demise within the Maastrichtian in several steps occurs far below the K/T crisis (Hallam and Wignall 1999). One of the largest continental flood basalts in the Toarcian is associated with only a minor crisis, so then why do some LIP's trigger major extinction events while others do not? (Bond and Wignall 2014). In Morocco, the Liassic microfaunal turnover occurs earlier than the Toarcian, during the Domerian (margaritatus zone, Table 8.6). A more general review with references for abiotic events affecting the biosphere is presented by Font et al. (2016).

8.5.2 Possible Influence of Environmental Changes and Major Events on Morphogenesis

In normal conditions, between main anoxic events (OAE, 1–11 on Tables 8.2, 8.3, 8.4, and 8.5), members of some ancient species become very large (a tenth of a centimeter in extreme cases) and complex compared with modern forms, and follow roughly the schematic trend displayed in Fig. 8.2. This could be an indication that Mesozoic carbonate platforms, their size and paleogeography, represent more stable and oligotrophic conditions in low latitude than in modern oceans (Hallock 1985). In these oligotrophic conditions larger symbiotic bearing foraminifera (K-strategists) will be favored, but paradoxically at the same time, their high specialization (resulting from coevolution) dooms them to extinction. Indeed larger foraminifera are very sensitive to abrupt changes of environments like sea-level variations and general drowning of the platforms; vast areas of warm shallow seas were isolated from any input of nutrients from land or the deep sea (Hallock 1982, 1985). Then massive extinctions are triggered by environmental changes that are too rapid for many species to respond either by migration or by physiological (morphological) adaptation (Herman 1981; Hallock 1985) but, on the other hand, foraminiferal associations (as the orbitopsellids[3]) support gradual and local environmental changes as recorded in metric parasequences (point C discussed below; Table 8.6).

Evolutionary diversification (adaptive radiation) occurs in rather constant conditions on carbonate platforms through time. Major diversification (and innovations) of larger foraminifera (at a generic level) occurs at different periods, starting with the Early Liassic, after the Triassic extinction, later during Middle and Late Jurassic times after the Late Liassic anoxic crisis (Septfontaine et al. 1991, Table 8.6) and during the Early and Middle Cretaceous at a higher taxonomic rank (Ross 1974; BouDagher 2008) with some innovations (Table 8.1). But strangely enough, several Mesozoic isomorphic morphotypes of hauraniids, valvulinids, and pfenderinids (discussed below) cross the K/T crisis and are represented in the Early Tertiary. The general repartition through time of main morphotypes (genera) of microgranular larger foraminifera of the Mesozoic is resumed in Tables 8.1, 8.2, 8.3, 8.4, and 8.5 and clearly illustrates an iterative mode of evolution as already discussed by Septfontaine (1981, Septfontaine 1988).

General Shape and Coiling (Table 8.1)

The shape of the test reflects adaptation to different types of environments characterized chiefly by water energy, attachment or free living, sediment granulometry (Hohenegger 1994 for porcelaneous recent equivalent), and variation on carbonate platforms from coarse reefal sand to muddy lagoonal sediment. The test shape is also an expression of the chamber volume variations (increasing through time in bioseries), probably related to symbiotic coevolution. On the other hand, the episodic (but widespread through time) transportation of tests by currents or hurricanes may locally complicate the interpretation of fossil associations displaced in variable foreign environments onshore (lagoon or tidal flat, Davaud and Septfontaine (1995) in the Recent; Septfontaine (1985) in the Liassic of Morocco; Septfontaine (unpublished) for transportation of floated fusulinid tests in supratidal environments, Permian olistolites, Hawasina nappe, Oman). Then caution must be taken when interpreting morphology related to environmental conditions for fossil groups in an adaptative (epigenetic) perspective.

At a larger scale, global events like anoxic crises (OAE) drive to general extinctions and leave free ancient ecological niches; they play a major role in the replacement of ancient lituolid clades by new ones (population turnovers), as is the case after the T/J, the Late Pliensbachian (Table 8.6), and the Late Cenomanian events (Arriaga et al., 2016). These events give rise in the aftermath to new morphotypes

[3] Names of suprageneric taxa are informal (morphogroup termination -ids) considering (1) general shape and mode of coiling (plani-plani/strepto-trochospiral) and (2) type of wall microstructure or marginal architecture hosting symbionts and are, provisionally, considered as different convergent, isomorphic repetitive characters in various clades.

belonging to different clades. Among lituolids, these advanced populations are issued from simple pioneer forms persistent on the carbonate platforms, with planispiral (later discoidal tests) or trochospiral (later cylindrical to conical tests) coiling (Table 8.1). Some specific test shapes like the fusiform-to-globular alveolinids or globular lituolids (Barkerinids, not represented in Table 8.1) are curiously absent from the Triassic (except *Triasina*) to the Early Cretaceous. Globular Barkerinids are not related to a major climatic crisis period during the Berriasian (Schlagintweit et al. 2019), but rather possibly start from an adaptation to local conditions. The apparition of globular alveolinids in the Aptian (BouDagher 2008) may be linked to environmental changes, as the Aptian begins a period of general anoxia (OAE1a or Selli event) starting during the Mid-Cretaceous OAE 1b-d events. On the contrary, discoid, larger lituolids are omnipresent from the Liassic hereafter, and are a novelty issued from the aftermath T/J extinction event. The discoid test shape is persistent till the Early Tertiary, shared by some porcelaneous counterparts appearing in the Middle Cretaceous, and still living in recent shallow waters. Biserial fan shape and discoidal cuneolinids represent a major Early Cretaceous (Barremian) innovation apparently not linked to a global environmental change. Trochospiral cylindrical-conical complex lituolids belong also to the new incomers after the T/J crisis colonizing free niches on the carbonate platforms (BouDagher and Bosence 2002, 2007). Following the K/T extinction, several representatives of fan shape to discoidal morphotypes (survivors) are still present in the Paleocene, isomorphs of Mesozoic genera (Table 8.2), as well as trochospiral siphovalvulinids and valvulinids.

Obviously, after the T/J extinction events, driving to a new morphogenetic revolution, the morphological changes and innovations of chambers and the test shape of LIF rather reflect the product of local conditions or the action of internal factors linked to physiology (and symbiosis) of the cell. The general trend shape transformations in one bioseries (as shown in Fig. 8.2 for discoid lituolids) are also a good indicator of evolutionary steps and trends (covering roughly one or two stages), with a general tendency to increase the volume of the adult chambers by chambers overlapping together with an increase of the density or number of openings (pores) communicating to the exterior on the apertural face (Hottinger 1967), a general rule among larger discoid LIF, planispiral mesoendothyrids, and trochospiral valvulinids and pfenderinids or porcelaneous genera.

Internal Microstructures in Discoidal Tests (Table 8.2)

Most planispiral-to-discoid genera grouped artificially in Table 8.2 (morphogroups A–F) follow iteratively the general trend displayed in Fig. 8.2, characterized by an increase of test size together with an increase of the chamber volume by overlapping in the equatorial plane interpreted as an hypermorphosis, with ontogenetic truncation of the early planispiral nepionic stage through time (Brasier 1988). The discoid LIF's are of polyphyletic origin, and the supposed parentage with ancestral planispiral Jurassic mesoendothyrids (for the group A: *Orbitopsella-Orbitammina-Labyrinthina*) or hauraniids (for groups B–C: *Alzonella-Timidonella*) seems reasonable. The internal architecture of the test (see part 4-1, above) consists essentially of various subdivisions of the marginal zone of the chambers by regular radial vertical partitions (for group A), or irregular vertical and horizontal lamellae (for groups B–C–F), including the static protoplasm (Hottinger 1967). The central zone is occupied by pillars and calcitic infills associated with the porous aperture, including the dynamic protoplasm. It is supposed that both types of marginal subdivisions (convergent in independent clades A–F) define cavities containing relatively large symbiotic algae, as their recent porcelaneous counterparts (see Plate 8.1). The absence of a pseudokeriotheca in the wall in both cases could mean a replacement (during the evolution of Early Liassic ancestors mesoendothyrids and hauraniids) of symbiotic bacterial filaments hosted in the pseudokeriothecal wall (Vachard et al. 2004) by larger symbionts (possibly dinoflagellates) in the marginal zone of the chambers only. Conversely, the advanced reniform to discoidal "paleomayncinids" (see Table 8.6) display a *pseudokeriothecal-to-alveolar* microstructure within the wall, without internal microstructures in the chambers, which confirm the incompatibility between the presence of radial vertical partitions or hypodermic networks on the one hand, and the presence of a pseudokeriotheca or alveolae in the wall of a single individual on the other hand. This fact is also confirmed in trochospiral morphogroups (Tables 8.4 and 8.5).

However, it should be noted that some morphogroups of planispiral-discoid LIF (group E, Table 8.2), together with the phylogenetically non-related genera *Levantinella*, *Cyclolina*, *Flabellocyclolina*, and *Cyclopsinella*, are devoid of any exoskeletal (marginal) subdivisions, implying that symbionts lived probably free within the cytoplasm. In group D (*Satorina*), subdivisions of the marginal zone are issued from the center calcitic mass (not from the wall), and the genus could be considered as a pfenderinid (Table 8.5).

The internal structural innovations in the margin or the center of the chambers (like the "crosswise oblique stolon system" at the origin of vertical triangular radial partitions) appear rather randomly within different clades (A–D, F) during periods of relative environmental stability, independently of any perturbations, probably linked to genetic or symbiotic factors (coevolution). Obviously, main abiotic events do not influence the construction of a single morphotype, but rather act as a revealer of the best morphological tenable solution after a major global crisis, creating a major extinction among ILF in a changing environment. It is the case for rare morphotypes of hauraniids like the genera *Vania* or "*Broeckinella*" (homeomorphic of the genus *Alzonella* in the Middle Jurassic (Table 8.2, group B)) which cross the K/T limit and survive in the Paleocene.

Mesoendothyrids (Plate 8.2 and Table 8.3)

Some advanced mesoendothyrid genera (Plate 8.2; Table 8.3) like *Biokovina, Cymbriaella, Dubrovnikella* (Early to Middle Jurassic), or *Zagrosella* (Late Cretaceous) display clearly a heterochronic ontogenetic evolution, which (as for the orbitopsellids) can be interpreted as a recapitulation of characters (Haeckel). The introduction of innovations like multiple large porous apertures, pillars, and enlargement of uniserial-cylindrical or flattened adult chambers is interpreted as a peramorphosis (palingenesis of Haeckel) in terms of phyletic extension. The first chambers of Jurassic advanced forms diverging from the genus *Mesoendothyra* Dain retain the first steps of ancestral morphology of Paleozoic, Triassic or Liassic (Bassoullet et al. 2001) Endothyrids (*Endoteba*, *Endotriadella*). The fossil record in different areas of the Tethyan carbonate platforms, north and south (Briançonnais realm, Dinarids, Italy, etc.), points to a repetitive morphogenesis of these advanced more complicated forms (*Biokovina*-type and other close Middle Jurassic morphotypes) appearing at certain favorable periods (in terms of nutrients, light, etc.), or is limited to local favorable microenvironments. Major anoxic events or sea-level variations, etc. have no direct influence on these evolutionary steps. Clearly, new advanced taxa flourish between anoxic periods, e.g., the Orbitopsellids (*Lituosepta-Orbitopsella-Planisepta* in the Middle Liassic, and *Paleomayncina* Table 8.6) with morphological transitions (Septfontaine 1988) and the bioseries *Mesoendothyra-Biokovina*. Besides, the later genus *Labyrinthina* (with a *Lituosepta* morphology) appears in the Late Jurassic, with a characteristic large planispiral nepionic stage. It is important to consider the large time gap (19 Ma, Toarcian to Callovian), which separates these two almost perfect homeomorphic genera, absent in thoroughly investigated carbonate deposits of Middle Jurassic age all around the Tethyan realm by many authors, as already stated by us (Septfontaine 1981). Still, this morphogenetic redundancy is not linked to any known abiotic event.

Paradoxically, secondary characters, like the "spinose" septal shape in equatorial section, are stable all along Mesozoic times. This "spinose, rose thorn" trait (a Late Paleozoic and Triassic heritage) disappears probably at the K/T limit, and is not recorded during Tertiary times. The combination of characters, the thick wall with pseudokeriotheca-to-alveolae (with transitions, see Plate 8.3)—strepto to planispiral coiling becoming uncoiled-cylindrical in the adult stage—basal-to-central-to-multiple-areal apertures, and finally the "spinose" septa in the equatorial section, permit definition of the mesoendothyrids. The relative importance (weight) of these different characters remains debatable from a taxonomic perspective.

The main Early Jurassic abiotic events triggering extinction and renewal of foraminifera populations is the T/J (CAMP, Atlantic magmatism) anoxic crisis combined with a marine regression (Wignall 2001). However, some simple (small) morphological types like some endothyrids-mesoendothyrids (Plate 8.2 (1–9)) survive these major environmental perturbations. They represent a primitive stock of forms from which larger and more complex populations diverge sporadically. The first adaptive radiation (red flame symbol) is recorded in the Pliensbachian, starting in the Late Sinemurian after a period of warming events (Table 8.3, yellow band, 1–4, red flame symbol), with the first complex morphotype *Lituosepta* (radial partitions), including the "*Planisepta*" flat morphotype (discussed by Fugagnoli and Bassi (2015)), which is a true valid genus as a concurrent to *Orbitopsella* in the Carixian, becoming the only survivor in the Domerian (biozone D, see Table 8.6). *Biokovina*, a large and complex variant of the genus *Mesoendothyra* ("*Bosniella*") was also recorded during this period of favorable environmental conditions (white band in Table 8.3). They all disappeared during the Late Domerian, as victims of a

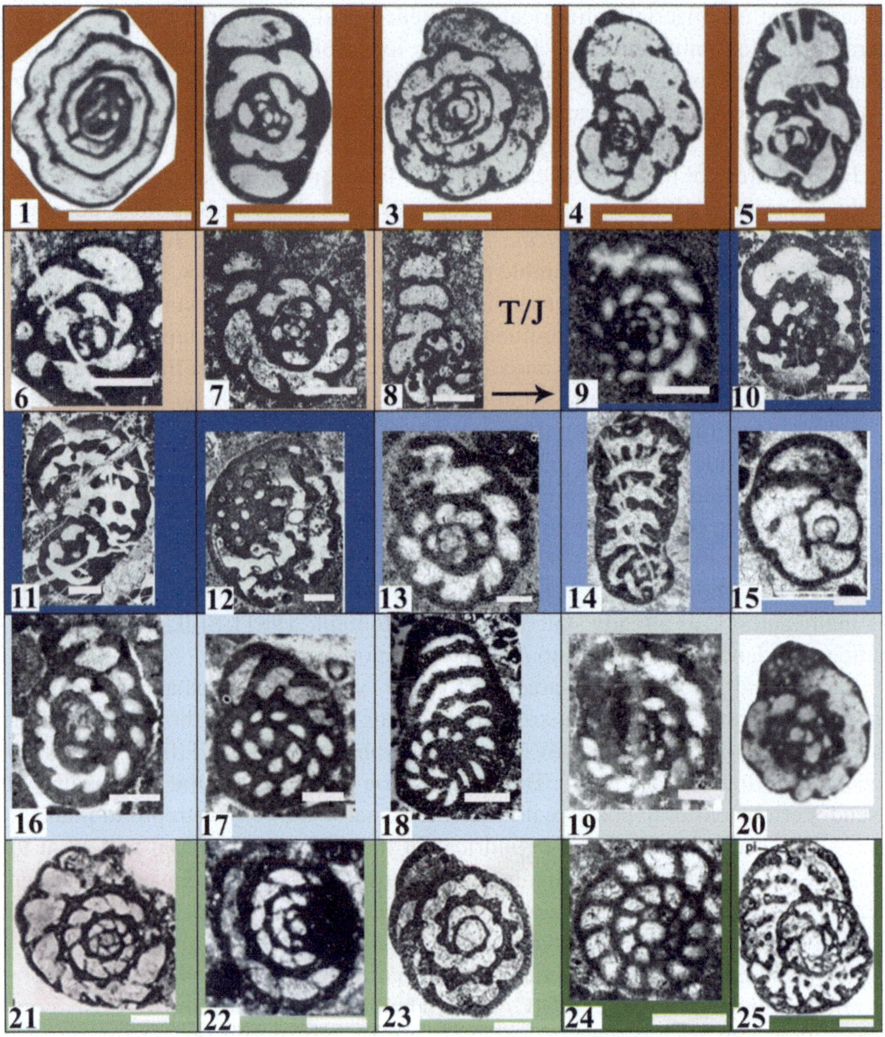

Plate 8.2 Comparison of different "crisis resilient" simple morphotypes among **endothyrids–mesoendothyrids**. More complicated Mesozoic advanced morphotypes (commonly of larger size) are displayed (11–25). They all share some characters of the **endothyrids** (microstructure of the wall, septa shape "rose thorn," etc., see text) including primitive ancestral forms in the Late Paleozoic (Tournayellidae) to advanced descents in the Mesozoic around the genera *Mesoendothyra* and *Everticyclammina* (Table 8.3). The transition from pseudokeriotheca to an alveolar wall is obvious in advanced forms like *Cymbriaella* or *Dubrovnikella*

Scale bar: 0.250 mm

(1–5. Brown: Late Paleozoic, excerpt from Loeblich and Tappan 1988)

1. *Costayella* (Tournayellidae)
2. *Endochernella* (Tournayellidae)
3. *Eblanaia* (Tournayellidae)

(continued)

4. *Rectoavesnella* (Tournayellidae) morphotype *Everticyclammina* (Mesozoic isomorph)
5. *Haplophragmella* (Endothyridae)
(6–8. Light brown: Triassic, excerpt from Zaninetti et al. 1972)
6. *Endoteba* (Endotebidae)
7, 8. *Endotriadella* (Endothyracea)
(9–12. Dark blue: Liassic, 9: excerpt from Septfontaine 1984b; 10–12: from Gusic 1977)
9. *Mesoendothyra* sp. (Mesoendothyridae)
10. *Bosniella* (advanced Mesoendothyridae)
11. *Biokovina* (advanced Mesoendothyridae)
12. *Biokovina* (advanced Mesoendothyridae)
(13–15 Middle blue: Dogger; 13: excerpt from Azeredo 1999; 14, 15: from Schlagintweit and Velić 2011)
13. *Mesoendothyra* (Mesoendothyridae)
14. *Cymbriaella* (advanced Mesoendothyridae, alveolae)
15. *Dubrovnikella* (advanced Mesoendothyridae, alveolae)
(16–18 Light blue: Bathonian; excerpt from Furrer and Septfontaine 1977)
16. *Mesoendothyra* (advanced Mesoendothyridae)
17. *Mesoendothyra* (advanced Mesoendothyridae, planispiral, thick wall)
18. *Mesoendothyra* (advanced Mesoendothyridae, uncoiled adult stage with multiple apertures; the *Biokovina* stage with interseptal pillars is not developed)
(19, 20 Light grey: Malm; 19 excerpt from Tasli 2001; 20 excerpt from Loeblich and Tappan 1988)
19. *Karaisella* = *Charentia* (Mesoendothyrid)
20. *Karaisella* = *Everticyclammina* (Everticyclamminid)
(21–23 Light green: Early Cretaceous; 21 excerpt from Hottinger 1967; 22 excerpt from Neumann 1967; 23 excerpt from Brönnimann and Conrad 1967)
21. *Charentia* (Mesoendothyrid)
22. *Hemicyclammina* (Mesoendothyrid)
23. *Melathrokerion* (Mesoendothyrid)
(24, 25 Dark green: Late Cretaceous; 24 excerpt from Sari et al. 2009; 25 excerpt from Schlagintweit and Rashidi 2017b)
24. *Moncharmontia* or *Fleuryana* (Mesoendothyrid)
25. *Zagrosella*, pi = pillar (Mesoendothyrid)

new environmental crisis (T-OAE 5–6). First simple everticyclamminids with general characters of mesoendothyrids but a slit-like aperture issue in the Lower Liassic persist through the all Mesozoic.

A second period of diversification (Middle Jurassic) can be observed following the environmental crisis of Toarcian-to-Aalenian times on the Apulian plate, Italy, and the Dinarids (Gusic 1977; Velic 2007) and in the Briançonnais realm. However, it is not known if these remarkable steps in morphological diversification are local adaptive variations (increasing the symbiotic culture in the "greenhouse farm" of symbiotic microorganisms), or have a larger repartition through the Tethyan carbonate platforms during favorable environmental conditions (time estimated at one or two stages). In spite of a bad fossil record, we still believe that the second hypothesis is the correct one.

Diversification is low during the Late Jurassic. The basic simple ancestor morphotype *Mesoendothyra* is still commonly represented, with the advanced genus *Labyrinthina* displaying a *Lituosepta* type of internal microstructure (homeomorphy in the same clade). This is not a "retrograde evolution" sensu Guex (2016), as the last orbitopsellids were already extinct 20 Ma ago, but the reappearance of a complicated morphotype in the mesoendothyrid lineage.

During the Cretaceous, mesoendothyrids are well represented by various genera, like *Melathrokerion* and *Debarina* (modified porous aperture) along with the everticyclamminids (with coarse alveolae and a single-slit aperture). Both clades share the same basic characters, general or secondary, as described for *Mesoendothyra*. The relation between environmental climatic crisis as the Mid-Cretaceous events (8–10 in Table 8.3) is not clear as the *Mesoendothyra* morphotype is a rather ubiquitous and easily adaptive form in different hostile environments till the Late Cretaceous. The genus *Zagrosella* is a probable isomorph of *Biokovina* according to Schlagintweit and Rashidi (2017b). This last mesoendothyrid is not affected by the anoxic crisis but the end-Cretaceous events (cooling and warming, etc., Punekar et al. 2015) mark the demise of this otherwise tolerant morphological group.

Valvulinids (Plate 8.3 and Table 8.4)

This morphogroup of trochospiral foraminifera is characterized by a triserial arrangement of the chambers, with a conspicuous horizontal septal flap or "valvular tooth plate[4]" covering partly the umbilicus and the interiomarginal (basal) slit aperture, an Early Dogger innovation. The complications (pillars) associated with the "tooth plate" or septal flap and the basal or areal multiple porous apertures can form a "trematophore-like" plate. The position and number of pores on or around the plate, also called a "sieve plate" by Loeblich and Tappan (1964), could be interpreted as an example of epigenesis as these variations were observed in a single synchronous population (commonly observed by the author in Middle Jurassic assemblages and recorded in the Eocene according to Neumann (1967), Fig. 150, see top of Table 8.4). These variations in a single population could reflect an adaptation to different types of microenvironments. The complications of the apertural face (in advanced morphotypes) have also a phylogenetical meaning, representing an iterative evolutionary trend independent of any major abiotic event.

The valvulinids are absent on the Triassic and Liassic carbonate platforms. Instead trochospiral triserial-to-multiserial microgranular morphotypes (also bearing a pseudokeriotheca) are represented by the siphovalvulinids-pfenderinids (Table 8.5). The ancestor group of the later valvulinids, abundantly recorded in Middle Jurassic to Cretaceous and Tertiary beds, is not known. A morphological relationship between septal flap and central columellar structure of the siphovalvulinids is possible. This would imply a structural deformation of the adaxial side of the hollow vertical columella of Liassic siphovalvulinids into a horizontal septal flap covering the umbilic. Such a morphological trend is not yet clearly documented (work in progress).

[4] Not equivalent to the vertical half tube or "tooth plate" of Hofker (1963).

The first Middle-to-Late Jurassic valvulinids (praevalvulinids*[5]) and Cretaceous chrysalidinids were already thoroughly studied by Banner et al. (1991). It is important to note that usually secondary characters, like septal convexity, which could at first glance be attributed to simple ecophenotypic variations among a synchronous population, can be an evolutionary signature and trend as well, allowing distinguishing Jurassic praevalvulinids* (flat to concave to slightly convex septa) and advanced Cretaceous isomorphic chrysalidinids (strongly convex septa).

The genus *Cantabriconus* (Schlagintweit and Rashidi 2017a) is an Early Cretaceous (Aptian) isomorph of the *Paravalvulina-Parurgonina* "plexus" belonging to the valvulinids. This later morphotype displays a horizontal "valvular tooth plate" modified in an apertural (uniserial) porous plate with pillars in adult chambers. It is, however, tentatively placed with the pfenderinids (Table 8.5) because of the early nepionic ontogenetic stage showing a large open umbilicus and distal (retro) deformation of the septa forming a central tube or straight columella.

Pfenderinids (Plate 8.4 and Table 8.5)

Triassic ancestral forms of the pfenderinids (Siphovalvulinidae) morphotypes and valvulinids (praevalvulinids) may be of the *Turriglomina-Turritellella* type (undivided twisted tube) or the three-chambered *Trochammina* type; these isomorphic taxa display a central columella or siphon. After the end Triassic anoxic crisis and global extinction, the evolutionary link between the first Siphovalvulinidae (Gale et al. 2018) and more advanced pfenderinids (and probably the later valvulinids) seems obvious considering the highly plastic variability of the deformed septa, forming a variable open half-columella (hollow or pillars filled) or umbilical siphon (Plate 8.5). The vertical adaxial part of the chamber's wall (against the spiral axis) appears in the axial section (Plate 8.4, (18, 19); Table 8.5 (A–C)), directed in a backward position against the previous whorl and coated to it. This disposition of the successive chamber's wall leaves a clear central umbilical columella (siphon), which can be filled lately by pillars or calcitic material. Another possible modification (convergent to the valvulinid architecture) is the hypothetical transformation of the vertical wall of the columella (adaxial wall of the chambers) into consecutive horizontal septal flaps arranged in successive apertural porous septal plates joined by pillars plus secreted calcitic infill. The massive columella so formed can be secondarily dissolute, producing lately a groove or spiral canal in advanced pfenderinids of the Middle Jurassic and the Cretaceous.

Thus it may be difficult to differentiate the closely related valvulinids and pfenderinids in early evolutionary stages (Jurassic); the pfenderinids may be of polyphyletic origin. Only advanced forms show, respectively, the typical characters of each morphogroup. The presence of a siphon or a spiral canal or grooves at the surface of the columella may serve as a direct way for the expulsion of larger size

[5] Asterisks point to informal terms like "praevalvulinids" including the Jurassic valvulinids as illustrated in Table 8.4 and Plate 8.4, a proposed new taxonomic term (work in progress).

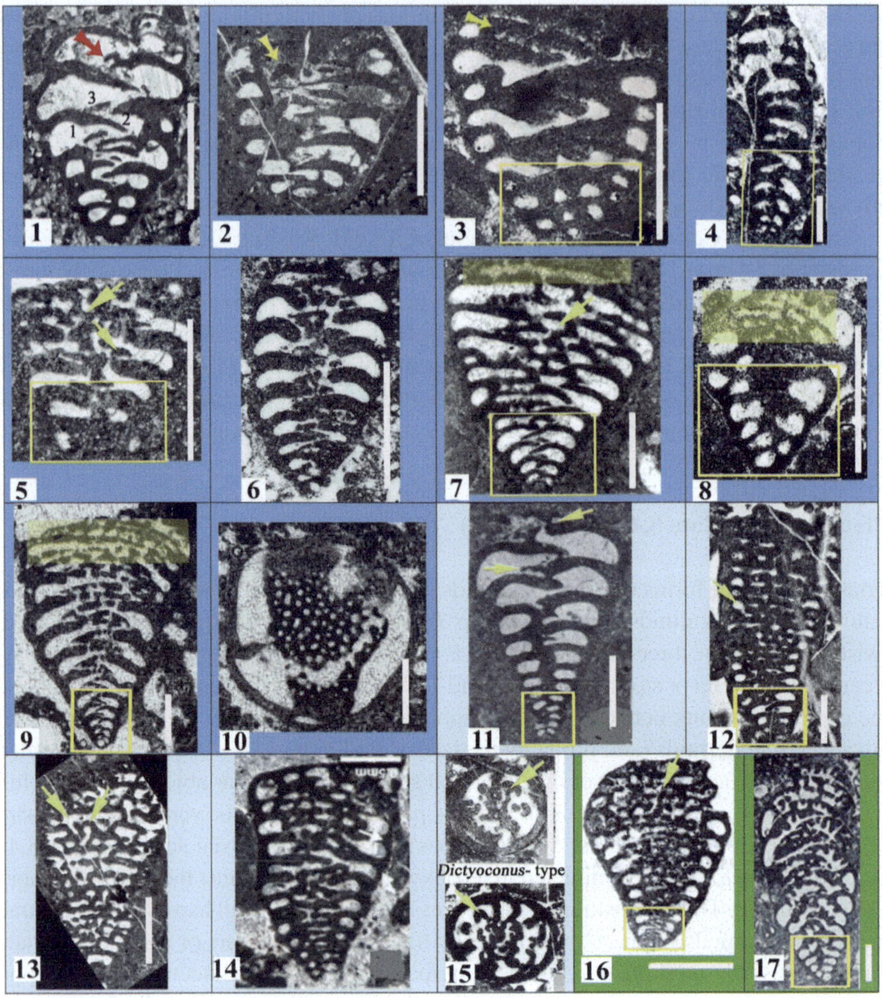

Plate 8.3 **Valvulinids s.l.**, trochospiral-triserial morphotypes (1–17) characterized by a medium size (axial length 1.5–2 mm) and a tooth plate with modifications (including Jurassic "**praealvulinids∗**", Cretaceous **chrysalidinids**, and Tertiary **valvulinids**). A "crosswise oblique" stolon system is developed in advanced forms of "**praevalvulinids∗**", a Bathonian innovation after the Early Jurassic anoxic crisis (Table 8.4). Note the Middle and Late Jurassic diversification of highly variable morphological populations (in space and time) of "**praevalvulinids∗**" with flat to slightly convex septa (1–15, middle and light blue) and Middle Cretaceous **chrysalidinids** with arched strongly convex septa (16–17, green). The early ontogenetic stage of more advanced forms showing ancestor "**praevalvulinid∗**" type of coiling is visible in the yellow rectangle (triserial test). Advanced taxa develop hemi-conical pillars (5–15) and multiple porous apertures in the center of adult chambers

Scale bar: ca 0.5 mm

1–2. *Praevalvulina* (Redmondoides) lugeoni*. Note some deformations of the tooth plate (arrows) in the last three chambers

(continued)

3. *Pseudomarssonella* with numerous (4–5) chambers in the early stage (yellow rect.) and throughout
4. Praevalvulinid with triserial early stage (yellow rect.), becoming uncoiled uniserial in the adult with multiple cribrate aperture
5. *Paravalvulina* with conical pillars and the first known "crosswise oblique" stolon system in the Jurassic (yellow arrows, innovation)
6. *Paravalvulina*
(1–6, excerpt from Septfontaine 1988; 10, from Altiner and Septfontaine 1979)
7–14. Phylogenetically advanced morphotypes attributed to the genera *Paravalvulina* or *Pseudomarssonella* (8) with the transformation of the tooth plate in a very thin apertural plate covering completely the last chambers (7–9, yellow zone) in a uniserial arrangement. Some morphotypes (11) show a regular, simple morphology (triserial throughout) through the all Mesozoic with a tendency of complications (pillars, pores) in the last chambers (yellow arrows). These complications can appear already in the very first chambers (13–14) as an ontogenetic acceleration
(7 excerpt from Azeredo 1999; 8 excerpt from Septfontaine 1981; 9, 10, excerpt from Velic 2007; 11, excerpt from Ramalho 2015; 12, 13 excerpt from Septfontaine 1981; 14, excerpt from Ples et al. 2015)
15. *Parurgonina* with hemi-conical pillars
(excerpt from Septfontaine 1988)
16. *Paravalvulina* ("*Dukhania*") or *Accordiella*; the new genus *Cantabriconus* (Schlagintweit and Rashidi 2017a), not figured here, belong to the same "plexus" of forms. The early stage shows already advanced traits (multiple apertures, yellow rect.)
(excerpt from Jones 2014)
17. *Chrysalidina* with typical arched septa. Juvenile part of the test: Valvulinid ancestor (yellow rect.)
(excerpt from Neumann 1967)

gamonts by the dynamic protoplasm flux (Hottinger 2006) and the construction of a new chamber.

Pfenderinids (advanced *Pseudo*-to-*Paleopfenderina* morphotypes) are absent from the Late Jurassic; the genus *Chablaisia* is the only representative in Oxfordian times with a spiral canal. Moreover, the trochospiral kurnubiids with a thick and opaque central columella (considered herein as a modification of the central siphon of *Siphovalvulina* in a porous plate with pillars and calcitic infill) are conspicuous and widespread along the margins of the Tethyan domain during late Jurassic times. The new clade appears earlier in the Middle Jurassic, with the small form *Praekurnubia* ancestor of the Late Jurassic larger *Kurnubia* and *Conicokurnubia* bioseries. The whole group shows an original innovation with the presence of a hypodermic network in the marginal zone of the chambers, *the first known example in trochospirally arranged tests* in the Jurassic, shared with the genus *Pseudodictyopsella* (Septfontaine and De Matos 1998) in the Bajocian of Oman and the Dinarids (Velic 2007). This novelty among trochospiral LIF is interpreted herein as tentatives (at least two) of a new association of LIF with larger symbiotic algae (dinoflagellates) succeeding to the probable symbiosis of pfenderinids-valvulinids

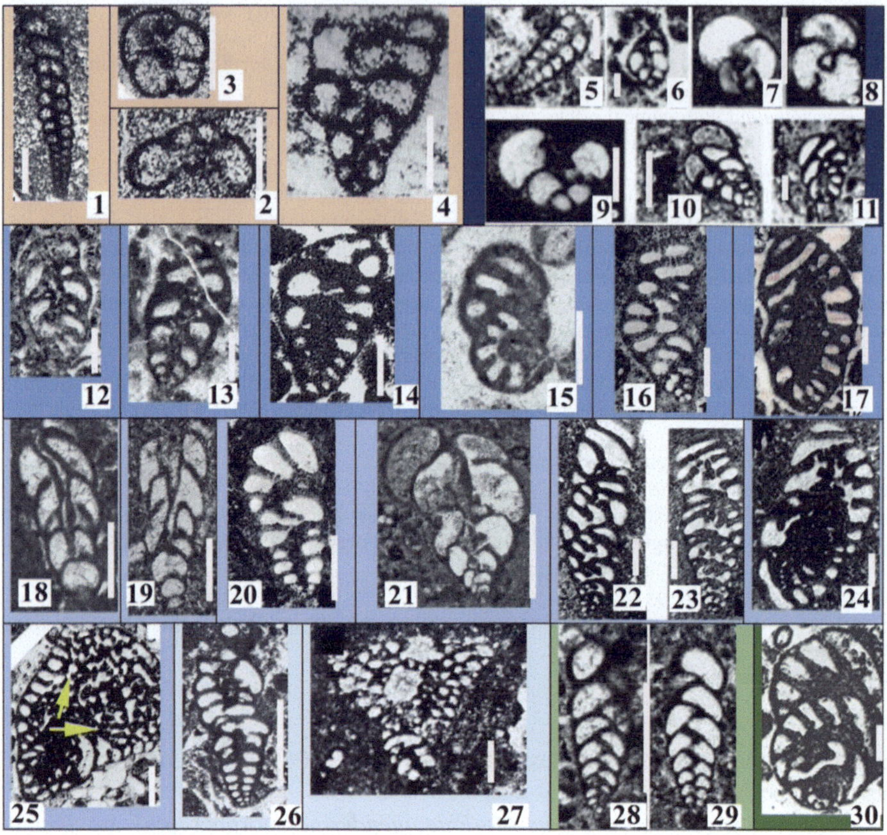

Plate 8.4 Siphovalvulinids to pfenderinids s.l. (*Pseudopfenderina-Paleopfenderina-Pfenderina* and allied Cretaceous forms, see Table 8.5); a very polymorphic group of trochospirally coiled (triserial to multiserial) small tests with a tendency to uncoil in the adult stage. The presence of a **pseudokeriotheca** is visible in some advanced genera with a thick wall, like the genus *Paleopfenderina*, but doubtful in other thin-walled morphotypes. In spite of great disparities in shape and number of chambers per whorl, the group shares a conspicuous general microstructure in the center of the chambers, well recognized by authors, but sometimes highly modified: a central siphon in primitive morphotypes (6–10) well developed in advanced forms (18, 19), which can be filled by secreted calcitic material and pillars forming an indistinct columellar mass (14, 15, 17), which can be secondarily resorbed in a groove or spiral tunnel in advanced genera (30). *The presence of small primitive* ***siphovalvulinid*** *morphotypes throughout the Mesozoic points to a possible example of a resilient simple morphotype (opportunist and possibly asymbiotic) against environmental crises in shallow water environments*

Scale bar: ca 0.25 mm

Light brown: Triassic

1. "*Turritellella–Turriglommina*" (undivided tube, with doubtful central siphon)
2. *Trochammina*
3. *Trochammina*
4. "*Tetrataxis-Paleolituonella*"

(1–4, excerpt from Zaninetti 1976)

Dark blue: Early Liassic

(continued)

5. *Siphovalvulina* var
6. *Siphovalvulina* var
7. *Siphovalvulina* var
8. *Siphovalvulina* var
9. *Siphovalvulina* var
10. *Siphovalvulina* var
11. aff. *Pseudopfenderina*

(5–11, excerpt from BouDagher and Bosence 2002)
Middle blue: Middle Liassic
12. aff. *Pseudopfenderina*
13. aff. *Pseudopfenderina*
14. aff. *Pseudopfenderina*
15. *Pseudopfenderina* (central button)
16. Advanced *Siphovalvulina* to *Pseudopfenderina*; uncoiled adult stage with porous aperture
17. *Pseudopfenderina* (typical axial section)

(12–17, excerpt from Septfontaine 1981, Septfontaine 1984b)
Light blue: Dogger (Bathonian)
18, 19. *Siphovalvulina* (elongated three-chambered morphotype, "half-moon" chambers with deformed backfolded septal tooth plate forming a partially closed central siphon)
20. Advanced *Siphovalvulina* morphotype (early chambers triserial) becoming multiserial in the adult with a porous aperture like (16)
21. Advanced *Siphovalvulina* var. (with resorption of the central siphon)
22, 23. *Paleopfenderina* var. (associated with more typical forms in the same thin section)
24. *Paleopfenderina* (typical subaxial section with some pillars visible and pseudokeriotheca)
25. Advanced *Paleopfenderina (Conicopfenderina)* with uncoiled discoid adult chambers, central pillars with a tendency to develop a "crosswise-oblique" stolon system (arrows) in a general orbitoliniform test
(18–20, excerpt from Septfontaine 1977, 1981; 21, excerpt from Granier et al. 2016; 22–24, excerpt from Septfontaine 1978)
Light blue-grey: Malm
26. Advanced *Siphovalvulina* (triserial throughout, with abnormal adult stage and central pillars, porous aperture)
27. "*Neokilianina*" (showing a *Siphovalvulina* early ontogenetic stage; later chambers divided into numerous arched chamberlets in a "crosswise-oblique" system; orbitoliniform test)
(27, excerpt from Ramalho 2015)
Green: Cretaceous
28, 29. Elongated typical *Siphovalvulina* var. (*Belorussiella* auct., with "half-moon-shaped" triserial chambers; the Late Cretaceous specimen is identical to the Jurassic morphotypes)
30. Classical advanced *Pfenderina neocomiensis* (with a central spiral tunnel formed by resorption of the central calcitic mass (or pillars); Early Cretaceous)
(28, 29, excerpt from Kobayashi and Wernli 2014; 30 excerpt from Loeblich and Tappan 1964)

with bacterial filamentous microorganisms in the pseudokeriothecal wall, as shown by Vachard et al. (2004) for the analogous keriotheca of fusulinids. The whole group disappears during the Kimmeridgian.

Smaller forms belonging to the siphovalvulinid ancestral genetic stock are still present in the Late Jurassic, and may be ancestral morphotypes of the genus *Neokilianina* forming its early ontogenetic stage (Ramalho 2015 and our Plates 5–27), without an apparent central passage or canal in later whorls.

True pfenderinids reappear in the Lower Cretaceous with the characteristic *Pfenderina neocomiensis* in the Berriasian (homeomorphic of *Paleopfenderina*) and

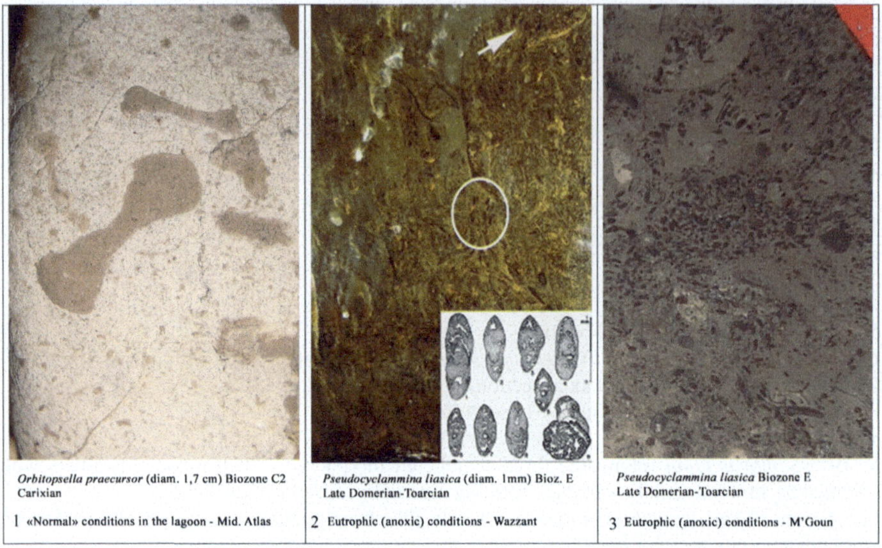

Plate 8.5 The Late Sinemurian to Domerian Aganane formation has registered the Late Domerian-Toarcian anoxic crisis visible on top, in the last tenth of meters (2–3) in a shallow water dark eutrophic carbonate facies arranged in metric parasequences. Specialized small-size and thin-shell dwellers are lithiotids and other Bivalvia (megalodontids etc., in biostromes), Gastropoda, and dwarf foraminifera (*Haurania* and *Pseudocyclammina* ("*Paleocyclammina*"). The macro- and microorganism populations are adapted to a eutrophic environment characterized by an olive-grey to black muddy (wackestone to packstone) facies. These unfavorable conditions (for larger orbitopsellids) contrast with the "normal," more oxygenated facies colored in clear yellow to light grey as shown in (1), the surface of an orbitopsellid bed of Carixian age
See also Table 8.6 for the stratigraphical setting
Axial sections of *P. liasica* visible on photo 2 (circle and arrow); drawings from Septfontaine (1984b); photos 1 and 2 same magnification; diam. *Orbitopsella*: 1.7 cm; photo 3 slightly enlarged

other low trochospiral morphotypes during the Barremian attributed to *P. globosa.* Later in Middle Cretaceous times, these low trochospiral forms give rise to classical morphotypes attributed to "*Nezzazatinella spp.*" or "*Valvulammina picardi*" by different authors. They all possess a central hollow columella with a spiral canal or a straight siphonal structure. It is important to note here that (apparently) secondary characters in these taxa, as the shape of chambers or the axial contour (rounded or angular) of the test, can be of prime importance for biostratigraphy. Thus angular periphery (in axial sections) of the "*Nezzazatinella*-type" can be considered an epigenetic innovation (ecophenotypic) during Middle Cretaceous times, unknown in the Jurassic and the Early Cretaceous, becoming later genetically fixed and conspicuous till the Maastrichtian (Septfontaine et al. 2019). Moreover, some advanced morphotypes (*Cantabriconus* with a siphovalvulinid early ontogenetic stage) in the Middle Cretaceous show uniserial chambers in the adult stage with an orbitoliniform test. The last pfenderinids s.l. (Pfenderinoidea) are diversified in the Maastrichtian and the Paleocene, with several genera and species as *Pachycolumella* spp., bearing a central hollow columella (Septfontaine et al. 2019) and the genus "*Nezzazatinella*" with an angular periphery, associated with ancestral siphovalvulinids. These Middle to Late Cretaceous and Paleocene representatives are grouped herein within two families: the Siphovalvulinidae (Gale et al. 2018), and the informal family "Neopfenderinidae∗". The new genus *Pachycolumella* (homeomorph of the Liassic genus *Radoicicina*, Gale et al. 2018) is placed in the Siphovalvulinidae, and is represented by its two species *elongata* and *acuta* in the Maastrichtian, but the former (*elongata*) did not cross the K/T limit. However, *P. acuta* (a survivor) is abundant and morphologically diversified in Paleocene levels in Turkey (Sirel, 2015) and Iran.

8.5.3 At a Shorter Time Scale, a Case History: The "Orbitopsellid" Model (Fig. 8.2; Table 8.6; Plate 8.5)

The orbitopsellids (*Lituosepta-Orbitopsella-Planisepta*) on the Liassic carbonate platform of the central High Atlas in Morocco (Hottinger 1967; du Dresnay 1977; Septfontaine 1985) represent probably the best continuous record of morphological steps in a bioseries of larger foraminifera, here in the Aganane formation (Late Sinemurian to Pliensbachian) studied by Jenny (1988) and Crevello (1991). Other carbonate platforms around the Mesozoic Neotethys (e.g., Arabia, Oman, Adria, the Apennines) contain orbitopsellids as well. These platforms have a long history from the Triassic to the Cretaceous (Septfontaine and De Matos 1998; Cobianchi and Picotti 2001; Velic 2007; Husinec and Read 2007; Tisljar and Velic 1993; Ciocchini et al. 2008), recording hundreds of parasequences to third-order sequences in formations of thousands of meters thickness. However, the microfossil record in Liassic beds is often scarce or concentrated to some horizons only, due to facies variations and discontinuities.

In Morocco, the demise of the genus *Orbitopsella* (LAD of *O. dubari* or *Cyclorbitopsella*) occurs during the Late Carixian in a still constant (but cyclical) environment built of metric "shallowing upward" parasequences (Aganane formation, Jenny 1988). The first-order sedimentary sequences (Fig. 8.3) or elementary sequences of Strasser (2016) are related to the 20 kyr precession cycles (in the allocyclic model). In this context, the morphological steps displayed in the *Lituosepta-Orbitopsella* bioseries (Fig. 8.2) can be identified in terms of size variations (test and proloculus diameter increase) and internal morphological complications increase. After a bed-by-bed study in different locations (till the final extinction of the bioseries in a constant environment), a total of 24 stations (in green in Table 8.6) deliver abundant populations of *Orbitopsella,* permitting a rough statistical estimation of the mean diameter. A break in the mean diameter was clearly observed after a long period of stagnation, suggesting a punctuated mode of evolution of the mean size. This break is not related to exterior environmental changes, but recorded in the constant environmental setting characteristic of the Aganane formation (Fig. 8.3).

Table 8.6 Growth and demise of "**orbitopsellids s.l.**" (**orbitopsellids s.str.**, and the *Paleomayncina* clade) and main OAE environmental crisis provoking a faunal turnover during the Middle Domerian (followed by the main T-OAE event in the *falciferum-levisoni* zone), which is consistent with the conclusions of Duarte et al. (2019) in Iberia for deeper water foraminifera)

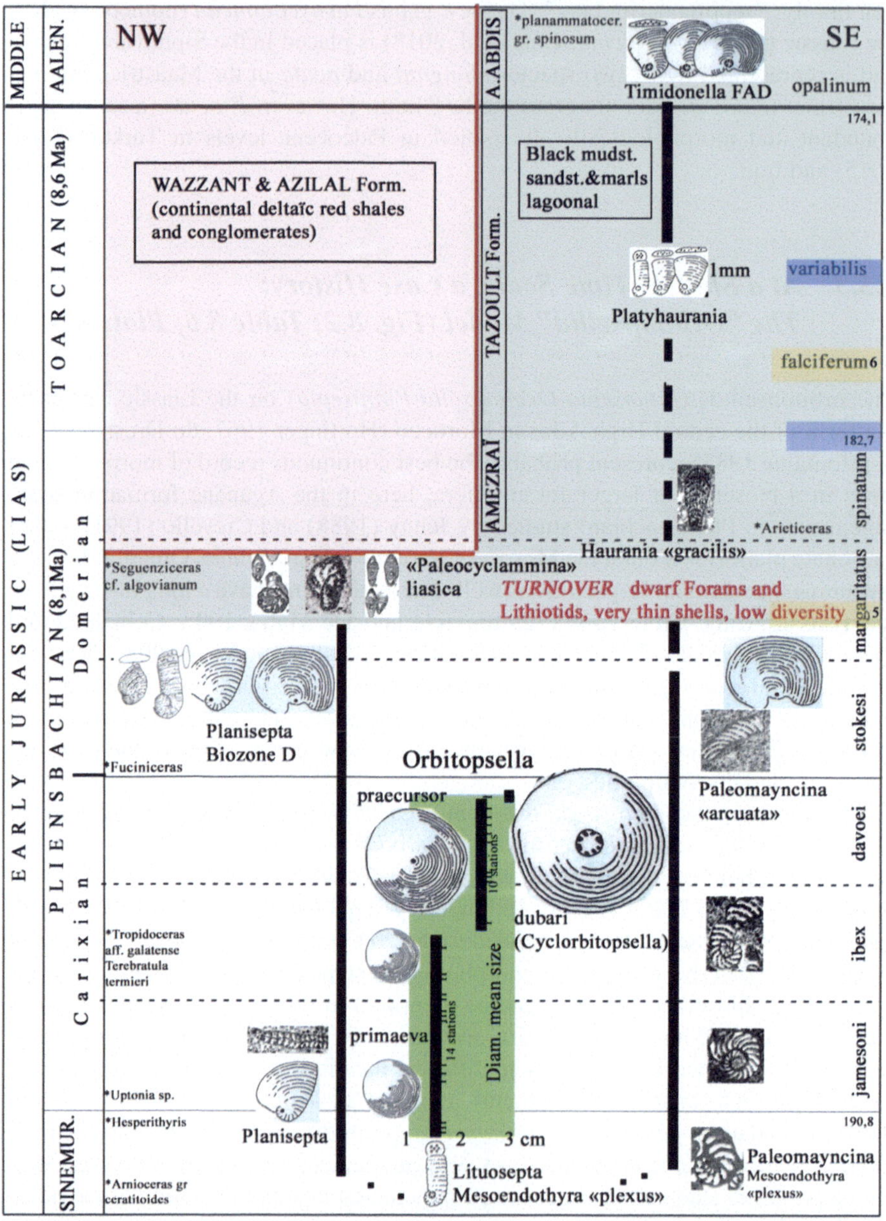

The situation in the central High Atlas and Middle Atlas of Morocco (R'nim, Adoumaz, and Koumch sections) is figured in green; the fine horizontal dashed lines in green space: 14 sampling stations with *O. primaeva* and 10 stations with *O. praecursor*). The table shows the general stratigraphical setting and ammonite discoveries and control. Discussion in text

The elementary cycles (shallow subtidal to inter- and supratidal) have no direct influence on the populations (or morphological variations) of foraminifera, which obviously find a "refuge" in a submerged area of the platform, migrating probably near the outer shelf during marine low-stand periods (Septfontaine 1985). This situation will last until the Early Domerian before the anoxic crisis registered on top of the Aganane formation (Plate 8.5).

Thus the rapid disappearance of large orbitopsellids during the Late Carixian in a more stable environment point rather to (1) an internal biological/genetical cause that may be related to a deterioration of the symbiotic relation becoming lethal for the host (Lefèvre et al. 2010), or the suppression of the sexual reproduction phase reducing morphological diversity (Hallock 1985), and (2) a mechanical cause related to the large diameter of the test which can no more afford a constant stability on the ground against storm or tidal current turbulences. We postulate that the attachment by pseudopods through the porous apertures situated on the peripheral side of the otherwise imperforate test offers a too weak anchorage for large shells with a diameter up to 2.5–3 cm. However the smaller genus *Planisepta* (an isomorph of *Orbitopsella* already present as morphovariant of *Lituosepta* in the Late Sinemurian) persists, defining the D biozone (Septfontaine 1984b, 1988). But this new bioseries (test contour flabelliform to reniform in the last levels) could not reach the ultimate discoidal steps (d, e, Fig. 8.2), impeded by the Middle Domerian anoxic crisis, creating a drastic microfaunal turnover. Thus the LAD of the last orbitopsellid *Planisepta* (together with the LAD of the sister group *Paleomayncina*, biozone D in the Tethyan realm) is clearly due to a major climatic event and microfaunal turnover on top of the Aganane formation in the central High Atlas of Morocco, with deposition of olive-grey to black shallow water bioclastic limestones indicating anoxia (Plate 8.5). Dwarf or rare lithiotid bioherms are widespread in the Tethyan domain till Oman and the Himalayas (Wignall et al. 2006). During the long crisis period and aftermath (alternation of warm, 5–6 in tables, or cold anoxic events from the top Pliensbachian and the Toarcian to the Aaleno-Bajocian), a new microfaunal assemblage characterized by very small size (millimetric) and simple interior microstructure inhabited the eutrophic lagoons, *Haurania* "*gracilis*∗" and "*Paleocyclammina*∗" *liasica*, defining the biozone E, and later the very small discoid "fragile" genus *Timidonella* (biozone F) in the Aalenian-Bajocian. This foraminiferal turnover during the Middle Domerian (followed by the main T-OAE event in the *falciferum-levisoni* zone) is consistent with the conclusions of Duarte et al. (2019) in Iberia for deeper water foraminifera. The Late Liassic anoxic crisis was also recognized above the orbitopsellid LAD in Mallorca (Rosales et al. 2018). A similar situation with a drastic microfaunal faunal turnover following a climatic crisis event (anoxic OAE-2), at the Cenomanian-Turonian boundary, is described by Arriaga et al. (2016) in the Apennines (Italy). The survivor taxa are very small ("Lilliput effect"), with a simple architecture after the demise of larger discoid foraminifera.

8.6 Conclusions

The variable mode of life and strategy of reproduction among foraminifera is a key for the understanding of the brutal response (demise) of LIF to severe environmental crises. When compared to the r-strategist planktonic foraminifera, with a high degree of variability and adaptation during anoxic periods (related to rapid reproduction rate), the larger benthic K-strategists with a slow reproduction rate impede rapid morphological changes and adaptation to new environmental conditions. Their highly specialized relation with specific endosymbionts (coevolution with symbiotic algae) conditions the internal architecture of the test in the marginal zone (size and shape of chamberlets) and the shape and size of the chamber itself (crescentic or cylindrical, to increase the volume of the "greenhouse" farm). Thus the ILFs are totally "padlocked" in their morphological characters. Such a situation is responsible for an unstable and fragile equilibrium in the changing environment of the carbonate platforms. Each lineage follows an analogous (parallel) irreversible way increasing the chamber volume. Contrary to planktonics, they are exposed to near-surface changes and perturbations (climatic, temperature, and sea-level variations, atmosphere composition, etc.) in the photic zone. Moreover, they are bottom dependent, with a vagile mode of life on the surface of an algal film (or algae), by contrast with the free-living planktonics in the water mass. Thus they have a very limited possibility to quit their biotope for better conditions in the photic zone during environmental stress periods. Moreover, the large size of discoid forms (gigantism) at the end of a bioseries may be inconvenient in terms of stability against currents and storms, provoking their extinction.

The mode of evolution appears definitely iterative for all groups considered here (Tables 8.1, 8.2, 8.3, 8.4, and 8.5). More advanced morphotypes in each clade appear several times (with slight modifications) throughout the Mesozoic, together with an increase of biodiversity during periods of favorable environmental conditions (Late Sinemurian to Early Domerian–Bajocian to Callovian–Late Jurassic to Early Cretaceous–Late Cretaceous), that is, between two main environmental crisis periods (anoxic events, global sea-level changes, etc.). More investigations are needed to assert this conclusion for specific morphogroups. Advanced morphotypes diverge from a stock of simple, primitive, smaller forms with long stratigraphical range and large adaptive possibilities (morphological plasticity) for each clade considered, acting as a "genetic memory." The large size and specialized architectural complications of advanced forms are obstacles to later morphological variations. The concept of "retrograde evolution" introduced by Guex (2016) seems to be a rather limited strategy present only in morphologically simpler planktonic microorganisms (planktonic foraminifera, Radiolaria, etc.) and some benthic forms (hyaline perforate foraminifera) affecting essentially secondary characters like ornamentation (e.g., keeled globotruncanids back to a smooth ancestor during stressful environmental periods), and uvigerinids, according to depth (ecophenotypes). These characters, considered epigenetic by Guex (2016), are absent in the unkeeled imperforate larger foraminifera (ILF) with a smooth test surface.

The influence of abiotic events like anoxia concerns especially population composition as aftermath of extinction events. The new population is characterized by small, fragile forms, as stated after the Pliensbachian-Toarcian or the Late Cenomanian anoxic crisis. However, some simpler morphotypes like endothyrids and larger survivor morphotypes of lituolids (e.g., the hauraniids *Broeckinella* or *Pachycolumella*, a siphovalvulinid) cross the severe P/T crisis for the former, or the K/T limit for the latter. This particularity remains unexplained.

Acknowledgments I am grateful to Professor John S. Torday for his help to improve the English of the last version of the manuscript.

References

Abramovich S, Almogi-Labin A, Benjamini C (1998) Decline of the Maastrichtian pelagic ecosystem based on planktonic foraminifera assemblage change: implication for the terminal Cretaceous faunal crisis. Geology 26(1):63–66

Altiner D, Septfontaine M (1979) Micropaléontologie, stratigraphie et environnement de déposition d'une série Jurassique à faciès de plate-forme de la région de Pinarbasi (Taurus Oriental, Turquie). Rev Micropaléontol 22(1):3–18

Archibald J-D, Clemens W et al (2010) Cretaceous extinctions: multiple causes. Science 328:973

Arriaga M-A, Frija G, Parente M, Caus E (2016) Benthic foraminifera in the aftermath of the Cenomanian-Turonian boundary extinction event in the carbonate platform facies of the southern Apennines (Italy). J Foraminiferal Res 46(1):9–24

Azeredo A-C (1999) Micropaleontologial studies on Middle Jurassic carbonate successions from the Lusitanian Basin (Portugal). Comun Inst Geol e Mineiro Portugal 86:59–84

Banner F-T, Simmons M-D, Whitaker J-E (1991) The Mesozoic chrysalidinidae (Foraminifera, Textulariacea) of the middle East: the Redmond (Aramco) taxa and their relatives. Bull Br Mus Nat Hist (Geol) 47(2):101–152

Bassoullet J-P (1997) Les grands foraminifères. In: Cariou, Hantzpergue (eds) Biostratigraphie du Jurassique Ouest Européen, Elf Mém. 17, p 293–304

Bassoullet J-P, Chabrier G, Fourcade E (1976) Données compémentaires sur la morphologie, la structure interne et la position stratigraphique de *Timidonella sarda* (Foraminifère, Lituolidae). Rev Micropaléontol 19(1):3–18

Bassoullet J-P, Boutakiout M, Vachard D (2001) D'éventuels foraminifères Fusulinina au Mésozoïque, à propos de la description d'*Endotriadella ifranensis* n.sp. dans le Lias du Maroc Comptes Rendus de l'Académie des Sciences, Série IIA. Earth Planet Sci 332(7):473–478

Bond D, Wignall P (2014) Large igneous provinces and mass extinctions: an update. Geol Soc Am Spec Paper 505:1–27

BouDagher-Fadel M, Bosence W.-J (2002) Early Jurassic benthic foraminiferal diversification and biozones in shallow-marine carbonates of western Tethys. Senckenberg lethaea 87(1):1–39

Bosence D, Procter E, AurelL M, Belkahla A, BouDagher-Fadel M, Casaglia F, Cirilli S, Mehdie M, Nieto L, Rey J, Scherreiks R, Soussi M, Waltham D (2009) A dominant tectonic signal in high-frequency, peritidal carbonate cycles?—a regional analysis of Liassic platforms from western Tethys. J Sediment Res 79:389–415

BouDagher M (2008) Evolution and geological significance of larger benthic foraminifera: the Mesozoic larger benthic foraminifera, the Jurassic. Dev Palaeontol Stratigr 21:544, Elsevier Ed

BouDagher M, Bosence D-W-J (2007) Early Jurassic benthic foraminiferal diversification and biozones in shallow-marine carbonates of western Tethys. Senckenb Lethaea 87(1):1–39

Brasier M-D (1988) Foraminiferal extinction and ecological collapse during global biological events. In: Larwood GP (ed) Extinction and survival in the fossil record, The systematic association spec. vol. 34. Oxford Science Publ., Oxford, pp 37–64
Brönnimann P, Conrad M (1967) Cinquième note sur les Foraminifères du Crétacé inférieur de la région genevoise. *Melathrokerion valserinensis*, n.gen., n.sp., un Foraminifère nouveau du Barrémien à faciès urgonien dans le Jura français. CR Séances SPHN Genève NS 1(3):129–151
Burger B (2019) What caused Earth's largest mass extinction event? New evidence from the Permian-Triassic boundary in northeastern Utah. Global Planet Change, submitted pre-print manuscript, 59 p
Chaline J (1987) Paléontologie des vertébrés. Dunod Ed, Paris, 178 p
Cherchi A, Schroeder R (1975) *Eclusia decastroy* n. sp. (Lituolidae, foram.) del Barremiano della Sardegnia nord-occidentale. Soc Pal Ital Boll Modena 14(1):66–74
Ciocchini M, Chioccini RA, Didaskalou P, Potetti M (2008) Ricerche micropaleontologiche e biostratigrafiche sul Mesozoico della piattaforma carbonatica laziale-abruzzese (Italia centrale). Mem descr della Carta geol d'Italia 84:1–65
Cobianchi M, Picotti V (2001) Sedimentary and biological response to sea-level and palaeoceanographic changes of a lower-middle Jurassic Tethyan platform margin (Southern Alps, Italy). Palaeogeogr Palaeoclimatol Palaeoecol 169:219–244
Crevello P (1991) High-frequency carbonate cycles and stacking patterns: interplay of orbital forcing and subsidence on lower Jurassic rift platforms, High Atlas, Morocco. In: Franseen et al. (eds), Sedimentary modelling: computer simulations and methods for improved parameter definition, Kansas Geological Survey, 223, p 207–230
Cushmann J-A (1933) Foraminifera, their classification and economic use. Norwood Press Mass USA:349 p
Darwin C (1859) The origin of species. John Murray London: 490 p
Davaud E, Septfontaine M (1995) Post-mortem onshore transportation of epiphytic foraminifera: recent example from the Tunisian coastline. J Sediment Res 65(1A):136–142
du Dresnay R (1977) Le milieu récifal fossile du Jurassique inférieur (Lias) dans le domaine des chaînes atlasiques du Maroc. 2d Symposium international sur les Coraux et récifs coralliens fossiles, Paris 1975. Mém. BRGM no 89, p 296–312
Duarte, et al. (2019) The Toarcian oceanic anoxic event at Peniche. An exercise in integrated stratigraphy—Stop 1.3. In: 2nd International Workshop on Toarcian Anoxic Event, Coimbra 2018, Portugal. IGCP Project 655. Field Trip Guidebook, p 33–54
Dustira A-M, Wignall P-B, Joachimsky M, Blomeier D, Hartkopf-Fröder C, Bond D (2013) Gradual onset of anoxia across the Permian-Triassic boundary in Svalbard, Norway. Palaeogeogr Palaeoclimatol Palaeoecol 374:303–313
Fleury J-J (2018) Rhapydioninidés du Campanien-Maastrichtien en région méditerranéenne: les genres *Murciella*, *Sigalveolina* n. gen. et *Cyclopseudedomia*. Carnets de Géologie 18(11):233–280
Föllmi K-B (2008) A synchronous, middle Early Aptian age for the demise of the Helvetic Urgonian platform related to the unfolding oceanic anoxic event 1a ("Selli event") Comment on the article "Sur la présence de grands foraminifères d'âge aptien supérieur dans l'Urgonien de la Nappe du Widhorn (Suisse centrale). Note préliminaire " by R.Schroeder, K.Schenk, A.Cherchi & B.Schwitzer, Revue de Paléobiologie, 2007, 665-669. Revue de Paléobiologie Genève 27(2): 461–468
Font E, Adatte T, Planke S, Svensen H, Kürschner W-M (2016) Impact, volcanism, global changes and mass extinction. Palaeogeogr Palaeoclimatol Palaeoecol 441(1):1–3
Fugagnoli A, Bassi D (2015) Taxonomic and biostratigraphic reassessment of *Lituosepta recoarensis* CATI, 1959 (Foraminifera, Lituolacea). J Foraminiferal Res 45(4):402–412
Furrer U, Septfontaine M (1977) Nouvelles données biostratigraphiques (à l'aide des foraminifères) dans le Dogger à facies briançonnais des Préalpes médianes romandes. Eclogae Geol Helv 70(3):717–737
Gale L, Barattolo F, Rettori R (2018) Morphometric approach to determination of lower Jurassic siphovalvulinid foraminifera. Res Paleontol Stratigr 124(2):265–282

Gould S-J (1977) Ontogeny and phylogeny. The Belknap Press, Harvard Univ. Press, Cambridge, 501 p
Granier B, Toland C, Gèze R, Azar D, Maksoud S (2016) Some steps toward a new story for the Jurassic-Cretaceous transition in Mount Lebanon. Carnets Geol 16(8):247–269
Guex J (2003) A generalization of Cope's rule. Bull de la Soc Geol France 174(5):449–452
Guex J (2016) Retrograde evolution during major extinction crisis. Springer, Heidelberg, 75 p
Gusic I, Velic I (1978) *Lituolipora polymorpha* n.gen, n sp. (Foraminiferida Lituolacea ?) from the Middle Liassic of the Outer Dinarids in Croatia and the establishment of a new family Lituoliporidae. Geol Vjesn Inst Geol istraz 30(1):73–93
Gusic Y (1977) A new foraminiferal family, Biokovinidae, from the Jurassic of the Dinarids and its phylogenetic relationships. Palaeont Jugosl 18:7–31
Hallam A, Wignall P (1999) Mass extinctions and sea level changes. Earth Sci Rev 48:217–250
Hallock P (1982) Evolution and extinction in larger foraminifera. Proc 3d North Am Paleontol Conv 1:221–225
Hallock P (1985) Why are larger foraminifera large? Paleobiology 11(2):195–208
Hennig W (1966) Phylogenetic systematics. Univ. of Illinois Press, Urbana, 263 p
Herman Y (1981) Causes of massive biotic extinctions and explosive evolutionary diversification through Phanerozoic time. Geology 9:104–108
Hofker J (1963) Studies on the genus *Orbitolina* (Foraminiferida). Thèse Leiden, Leiden
Hohenegger J (1994) Distribution of living larger foraminifera NW of Seseko-Jima, Okinawa, Japan. Mar Ecol 15(3/4):291–334
Hottinger L (1967) Foraminifères imperforés du Mésozoïque marocain. Notes Mém Serv Géol Maroc 209:1–168
Hottinger L (2006) The "face" of benthic foraminifera. Boll Soc Paleontol Italiana 45(1):75–89
Hottinger L, Caus E (1982) Marginoporiform structure in *Ilerdorbis decussatus* n. gen., n.sp. a Senonian agglutinated discoidal Foraminifer. Eclogae Geol Helv 75(3):807–819
Husinec A, Read JF (2007) The Late Jurassic Tithonian, a greenhouse phase in the Middle-Early Cretaceous "cool" mode: evidence from the cyclic Adriatic Platform, Croatia. Sedimentology 54:317–337
Jenkins H-C (1988) The Early Toarcian (Jurassic) anoxic event. Stratigraphic, sedimentary and geochemical evidence. Am J Sci 288:101–151
Jenny J (1988) Carte géologique du Maroc au 1:100.000 Azilal/Haut Atlas central—Mémoire explicatif. Notes et Mém du Service géol du Maroc 339:1–104
Jones R-W (2014) Foraminifera and their Applications. Cambridge Univ Press: 391p
Kaminsky M-A (2014) The year 2000 classification of the agglutinated foraminifera. In: Bubik M, Kaminsky M-A (eds) 2004 Proc. 6th Int. Workshop on Agglutinated foram. Grzybowski Found. Spec Publ, 8, p 237–255
Keller G, Adatte T, Pardo A, Bajpai S, Khosla A, Samant B (2010) Cretaceous extinctions: evidence overlooked. Sci Lett 328:974–975
Kobayashi F, Wernli R (2014) Early Cretaceous (Berriasian to Valanginian) foraminifers from the Torinosu limestone at the type locality of Sakawa, Shikoku,Japan. Revue de Paléobiologie Genève 33(1):67–78
Lecointre G, Le Guyader H (2001) Classification phylogénétique du vivant. Belin Ed, Paris, 560 p
Lee J-J, McEnery M-E, Kahn E-G (1979) Symbiosis and the evolution of larger foraminifera. Micropaleontology 25:118–140
Lefèvre T, Renaud F, Selosse M-A, Thomas F (2010) Evolution des interactions entre espèces: systèmes symbiotiques. In: Biologie évolutive. de boeck Ed., p 556–594
Loeblich A.-R, Tappan H (1964) Treatise on Invertebrate Paleontology, Part C, Protista 2 (vol. 1-2). Geol Soc of Amer, Univ Kansas Press
Loeblich AR, Tappan H (1988) Foraminiferal genera and their classification. Springer Science Ed., New York. (softcover reprint)
Mac Leod N, Rawson PF, Forey PL, Banner FT, BouDagher-Fadel MK, Brown PR, Burnett JA, Chambers P, Culver S, Evans SE, Jeffery C, Kaminsky MA, Lord AR, Milner AC, Milner

AR, Morris E, Owen B, Rosen BR, Smith AB, Taylor PD, Urquhart E, Young JR (1997) The Cretaceous-Tertiary biotic transition. J Geol Soc London 154:265–292

MacGillavry H-J (1978) Foraminifera and parallel evolution. How or why? Geol en Mijnb 57(3):385–394

Metodiev LS, Savov IP, Gröcke D-R, Wignall P-B et al (2012) Upper Pliensbachian-Toarcian palaeoenvironmental perturbations—belemnite isotopes Bulgaria. Solid Earth Discuss 4:315–361

Mikhalevitch V-I (2013) New insight into the systematics and evolution of the foraminifera. Micropaleontology 59(6):493–527

Neumann M (1967) Manuel de micropaléontologie des foraminifères. Gauthier-Villars, Paris, 298 p

Ples G, Bucur I, Pâcurariu A (2015) Foraminiferal assemblages and facies associations in the Upper Jurassic carbonates from Ardeu unit (Metaliferi Mountains, Romania). Acta Pal Rom 11(2):43–57

Punekar J, Keller G, Khozyem H-M, Adatte T, Font E, Spangenberg J (2015) A multi-proxy approach to decode the end-cretaceous mass extinction. Palaeogeogr Palaeoclimatol Palaeoecol 441(1):116–136

Ramalho M (2015) Stratigraphic Micropaleontology of the Upper Jurassic neritic formations of Portugal and its Tethyan context . 1 The Algarve basin. Memorias Geologicas Lab Nac de Energia e Geologia Lisboa 35: 110 p

Richardson SL (2006) Endosymbiont-bleaching in epiphytic populations of *Sorites dominicensis*. Symbiosis 42:1–15

Richardson SL (2009) An overview of symbiont bleaching in the epiphytic foraminifera Sorites dominicensis. In: Lang et al. (ed) Proc. of the Smithsonian Marine Science Symposium, p 429–436

Richardson SL, Rützler K (1999) Bacterial endosymbionts in the agglutinating foraminifera *Spiculidendron corallicolum* Rützler and Richardson, 1996. Symbiosis 26:299–312

Rigaud S, Vachard D, Martini R (2014) Agglutinated versus microgranular foraminifers: end of a paradigm. J Syst Palaeontol 1–20

Rosales I, Barnolas A, Goy A, Sevillano A, Armendáriz M, López-Garcia J-M (2018) Isotope records (C-O-Sr) of late Pliensbachian-early Toarcian environmental perturbations in the westernmost Tethys (Majorca Island, Spain). Palaeogeogr Palaeoclimatol Palaeoecol 497:168–185

Ross C-A (1974) Evolutionary and ecological significance of large, calcareous Foraminiferida (Protozoa). Great Barrier Reef Proc. 2d Int. Coral Reef Symp., vol. 1, p 327–333

Sari B, Tasli K, Özer S (2009) Benthonic Foraminiferal Biostratigraphy of the Upper Cretaceous (Middle Cenomanian-Coniacian) Sequences of the Bey Daglari Carbonate Platform, Western Taurides, Turkey. Turk J Earth Sci 18:393–425

Schlagintweit F, Velic I (2011) New and poorly known Middle Jurassic larger benthic foraminifera from the Karst Dinarides of Croatia. Geol Croatica 64(2):81–99

Schlagintweit F, Rashidi K (2017a) *Persiacyclammina maastrichtiana* n.gen., n. sp., a new larger benthic foraminifer from the Maastrichtian of Iran. Acta Palaeont Romaniae 13(1):15–23

Schlagintweit F, Rashidi K (2017b) *Zagrosella rigaudi* n. gen., n. sp., a new biokovinoidean foraminifer from the Maastrichtian of Iran. Acta Palaeont Romaniae 13(1):3–13

Schlagintweit F, Septfontaine M, Rashidi K (2019) *Pseudochablaisia subglobosa* gen. et sp. nov., a new pfenderinid foraminifera from the Upper Cretaceous of Iran. Cretaceous Research 100:105–113

Septfontaine M (1971) *Eclusia moutyi* gen. et sp. nov., un foraminifère nouveau du Valanginien du Jura méridional. Arch Sci Genève 24(2):285–298

Septfontaine M (1977) Niveaux à foraminifères (Pfenderininae et Valvulininae) dans le Dogger des Préalpes médianes du Chablais occidental (Haute-Savoie, France). Eclogae geol Helv 70(3):599–625

Septfontaine M (1978) *Chablaisia* nov. gen., un nouveau genre de foraminifère du Jurassique briançonnais (nappe des Préalpes médianes. Arch Sci Genève 31(1):39–50

Septfontaine M (1981) Les Foraminifères imperforés des milieux de plate-forme au Mésozoïque: détermination pratique, interprétation phylogénétique et utilisation biostratigraphique. Rev Micropaleontol 23(3/4):169–203

Septfontaine M (1984a) Le Dogger des Préalpes suisses et françaises-stratigraphie, évolution paléogéographique et paléotectonique. Mém Soc Helv Sci Nat 97:121

Septfontaine M (1984b) Biozonation (by means of imperforate foraminifera) of the liassic carbonate inner platform of the haut atlas (Morocco). Rev Micropaléontol 27(3):209–229

Septfontaine M (1985) Depositional environments and associated foraminifera (lituolids) in the Middle Liassic carbonate platform of Morocco. Rev Micropaléontol 28(4):265–289

Septfontaine M (1988) Vers une classification évolutive des Lituolidés (Foraminifères) jurassiques en milieu de plates-formes carbonatées. Rev Paléobiol vol. spéc. 2, Benthos'86, 229–256

Septfontaine M, De Matos E (1998) *Pseudodictyopsella jurassica* n. gen., n. sp. A new foraminifera from the Early Middle Jurassic of the Musandam Peninsula, N-Oman mountains; sedimentological and stratigraphical context. Rev Micropaléontol 41(1):71–87

Septfontaine M, Arnaud-Vanneau A, Bassoulet J-P, Gusic Y, Ramalho M, Velic I (1991) Les foraminifères imperforés des plates-formes carbonatées jurassiques: état des connaissances et perspectives d'avenir. Bull Soc Vaud Sci Nat 80(3):255–277

Septfontaine M, Schlagintweit F, Rashidi K (2019) *Pachycolumella* nov. gen., shallow-water benthic imperforate foraminifera and its species from the Maastrichtian and Paleocene of Iran. Micropaleontology 65(2):145–160

Sirel E (2015) Reference sections and key localities of the Paleogene stage and discussion C-T, P-E and E-O boundaries by the very shallow water foraminifera in Turkey. Ankara University Yayinlari, 461:171p

Song Y, Black G, Lipps H (1994) Morphological optimization in the largest living foraminifera: implication from finite element analysis. Palaeobiology 20(1):14–26

Strasser A (2016) Hiatuses and condensations: an estimation of time lost on a shallow carbonate platform. Deposit Rec J IAS (2015) 1(2):91–117

Strasser A, Caron M, Gjermeni M (2001) The Aptian, Albian and Cenomanian of rotter Sattel, Romandes Prealps, Switzerland: a high-resolution record of oceanographic changes. Cretaceous Res 22:173–199

Tasli K (2001) Benthic Foraminifera of the Upper Jurassic Platform Carbonate Sequence in the Aydincik (Icel) Area, Central Taurides, S Turkey. Geologia Croatica 54(1):1–13

Tisljar J, Velic I (1993) Upper Jurassic (Malm) shallow-water carbonates in the western Gorski Kotar area: facies and depositional environments (western Croatia). Geol Croatica 46(2):263–279

Vachard D, Munnecke A, Servais T (2004) New SEM observations of keriothecal walls: implications for the evolution of Fusulinida. J Foraminiferal Res 34(3):232–242

Van de Schrootbrugge B, Wignall PB (2016) A tale of two extinctions: converging end Permian and end Triassic scenarios. Geol Mag Cambridge 153:1–23

Velic I (2007) Stratigraphy and paleobiogeography of Mesozoic Benthic Foraminifera of the Karst Dinarids (SE Europa). Geol Croatica 60(1):1–113

Wignall P (2001) Large igneous provinces and mass extinctions. Earth Sci Rev 53(1–2):1–33

Wignall P, Benton M.-J (2000) Discussion on Lazarus taxa and fossil abundances at times of biotic crisis. J Geol Soc London 157:511–512

Wignall P, Racki G (2006) Mass extinctions, volcanism and anoxia: comparisons of the end Permian and Early Jurassic events. ISC 2006. In: 17th International Sedimentological Congress, Fukuoka, Japan. Abstract

Wignall PB, Hallam A, Newton RJ, Sha JG, Reeves E, Mattioli E, Crowley S (2006) An eastern (Tibetan) record of the Early Jurassic (Toarcian) mass extinction event. Geobiology 4:179–190

Zaninetti L (1976) Les Foraminifères du Trias—Essai de synthèse et corrélation entre les domaines mésogéens européen et asiatique. Riv Ital Pal 82(1):1–258

Zaninetti L, Brönnimann P, Baud A (1972) Microfacies particuliers et Foraminifères nouveaux de l'Anisien supérieur de la coupe du Rothorn (Préalpes médianes rigides, Diemtigtal, Suisse). Mitt Ges Geol Bergbaustud Innsbruck 21:465–498

Chapter 9
Morphological Deformation of Foraminiferal Tests Caused by Intertidal Oil Spills (Black Tides)

Marie-Thérèse Vénec-Peyré, Annachiara Bartolini, Michele Weber, and Jere H. Lipps

Abstract The causes of morphological foraminiferal test deformations (chamber malformations, errors in directional growth) have been extensively studied for a long time. Such deformations are found in areas of pollution, including oil spills, or "black tides" in tidal regions. Nevertheless, questions have arisen on whether or not deformed foraminifera are a bioindicator in such ecological catastrophes.

This study compares, for the first time, the impact of the *Amoco Cadiz* (1978) and *Erika* (1999) oil spills on foraminifera in two close areas, the Bay of Morlaix and Bay of Bourgneuf, NW France. In these similar environments (mud flats of tidal areas) with similar faunas (characterized by the foraminifera *Haynesina germanica*), the foraminiferal response to oil pollution was different. During the 13 months following the *Amoco Cadiz* black tide (1978), foraminiferal deformations increased (from 3 to 21%) while after the *Erika* black tide (1999) their ratios did not exceed those observed in non-polluted areas (3%). In addition, turbellarians (flatworms) increased their attachment of egg capsules to foraminiferal tests in *Amoco Cadiz* black tide areas.

We review and compare the causes of such differences in the response of foraminifera to oil spills by examining the main characteristics of both black tides (magnitude of the spills, the physical and chemical properties of discharged oils, their concentration in the studied sediments, as well as the characteristics of the spill context). This analysis shows that both studies on deformed foraminifera, far from

---

M.-T. Vénec-Peyré (✉)
Département Origines et Evolution, Muséum National d'Histoire Naturelle, Paris, France
e-mail: marie-therese.venec-peyre@mnhn.fr

A. Bartolini · M. Weber
Département Origines et Evolution, Muséum National d'Histoire Naturelle, CR2P UMR 7207 MNHN CNRS SU, Paris, France
e-mail: bartolini@mnhn.fr

J. H. Lipps
Museum of Paleontology, University of California, Berkeley, CA, USA
e-mail: jlipps@berkeley.edu

J. Guex et al. (eds.), *Morphogenesis, Environmental Stress and Reverse Evolution*, https://doi.org/10.1007/978-3-030-47279-5_9

contradicting each other, only emphasize the differences between the characteristics of *Amoco Cadiz* and *Erika* oil spills. Deformed foraminifera are good indicators of the severity of oil spill impact and the rate of recovery after such a disaster.

Keywords Foraminifera · Pollution · Environmental stress · Black tides · Morphological deformations

9.1 Introduction

More than 200 years ago, morphological deformations of foraminiferal tests were recognized by Soldani (1789), and later by d'Orbigny in the "Planches Inédites" finished in 1826 (Vénec-Peyré 2005) and Parker et al. (1865) who published illustrations showing the great morphological diversity of these microorganisms. Interest in deformations as indicators of stressed environments appeared much later. After the mid-twentieth century, numerous authors noted the occurrence of abnormal deformed foraminifera and their relationship to variations in natural environmental parameters leading to unfavorable growth conditions. Later, some authors observed the effects of anthropogenic pollution on foraminifera (see reviews of the huge number of papers by Saraswat et al. 2004 and Nigam et al. 2006). Over the period from 1959 to 1979, authors focused mainly on domestic pollution relating the effect of sewage outfalls on population dynamics (foraminiferal abundances and species diversity). Now, foraminifera are recognized as useful indicators of the health of modern and past environments because of their high sensitivity and rapid reproduction which permit their fast response to small changes in physicochemical parameters. Their great numbers and their wide geographic distribution in most marine environments allow meaningful statistical analyses of sensitive marine environments.

In the 1980s, interest moved towards the effect of industrial pollutants. Besides the background pollution caused by human activities (agriculture, industrial, and urban development), accidental catastrophic oil spills are also main sources of pollution impacting the marine environment. They are defined as "violent" spillages "concentrated in a specific area, surpassing the natural assimilation capacities of the surrounding environment" (CEDRE 2007). The most tragic events resulted from the sabotage of oil wells, pipelines, or oil tankers during wars. The most frequent, however, were caused by accidental explosions or accidents on oil drilling platforms or shipwrecks due to collision or grounding (CEDRE 2007). The world's largest accidental marine spill was the Deepwater Horizon in 2010 in the Gulf of Mexico (USA).

If these spills are limited in space and time, they create "black tides," defined as "major oil slicks which threaten to reach and pollute the shore" (Zilberberg 1998). Since the 1960s, more than 160 black tides occurred in the world, attracting huge media coverage with shocking pictures of desolation of birds in the intertidal zones. Scientifically, most publications detailing the consequences of black tides on ecosystems concern macrofauna, such as birds and marine mammals as well as the economically important organisms like fishes, mollusks, and arthropods (ITOPF 2014), but only a few studies examined impacts on foraminifera. Among those, field studies mainly focused on species diversity and standing stocks (Morvan et al. 2004; Schwing et al. 2015). Abnormal foraminifera were documented for the first time by Vénec-Peyré (1981) in sediments contaminated by black tides as a result of the wreck and spill of *Amoco Cadiz* in Northern Brittany, France, in 1978. After the wreck of *Erika* tanker in Southern Brittany in 1999, the resulting black tide had little impact on foraminiferal tests observed in field samples, although the potential toxicity of *Erika* oil in culture experiments was confirmed (Morvan et al. 2004). Nor was a significant impact on test morphology seen in experimental *Erika* oil-treated mesocosms (Ernst et al. 2006). Such effects in the vicinity of offshore petroleum platforms have also not been observed (Locklin and Maddocks 1982; Murray 1985). As test deformations were not always observed in oil-polluted areas and because several interrelated parameters may induce deformed tests even in non-polluted sites submitted to other stresses (anoxia, variations of salinity, pH, etc.), oil spill impacts on foraminiferal morphology were considered speculative (Murray 1985; Stouff et al. 1999; Le Cadre et al. 2003). More recently, the *Deepwater Horizon* pollution impacted foraminifera by population declines in deep marine environments (Schwing et al. 2015); but, in marshes in Louisiana and Mississippi, that were differentially oiled, deformed foraminifera were only observed in well-oiled areas (Brunner et al. 2013). Deformed specimens were also observed in black tide-contaminated marshes on the west coast of Taean Peninsula, Korea (Lee et al. 2014), and on the northern coast of the Bohai Sea, China (Lei et al. 2015). In both cases, the deformations of foraminiferal tests mostly occur in more heavily oiled nearshore environments.

In this chapter, focused on foraminiferal test deformations in oil-polluted environments, we review the main characteristics of the *Amoco Cadiz* and *Erika* black tides in order to determine the causes of the differences observed in the response of foraminifera to these two oil spills which impacted the coasts of Brittany (NW France).

9.2 Black Tides in Brittany (NW France)

The coasts of Brittany in NW France have been subjected to repeated oil spills due to shipwrecks (Fig. 9.1). This area remains threatened by such accidents due to the important maritime traffic with more than 14,000 vessels each year transporting dangerous products (among them 2300 million tons of oil) (CEDRE 2007). More than 20 oil spills have occurred in this area since 1967. Among them, the two most important resulted from the grounding of the supertankers *Torrey Canyon* on 18 March 1967 and *Amoco Cadiz* on 16 March 1978, which released 123,000 and 227,000 tons of oil into the sea, respectively (CEDRE 2007; Spooner 1978).

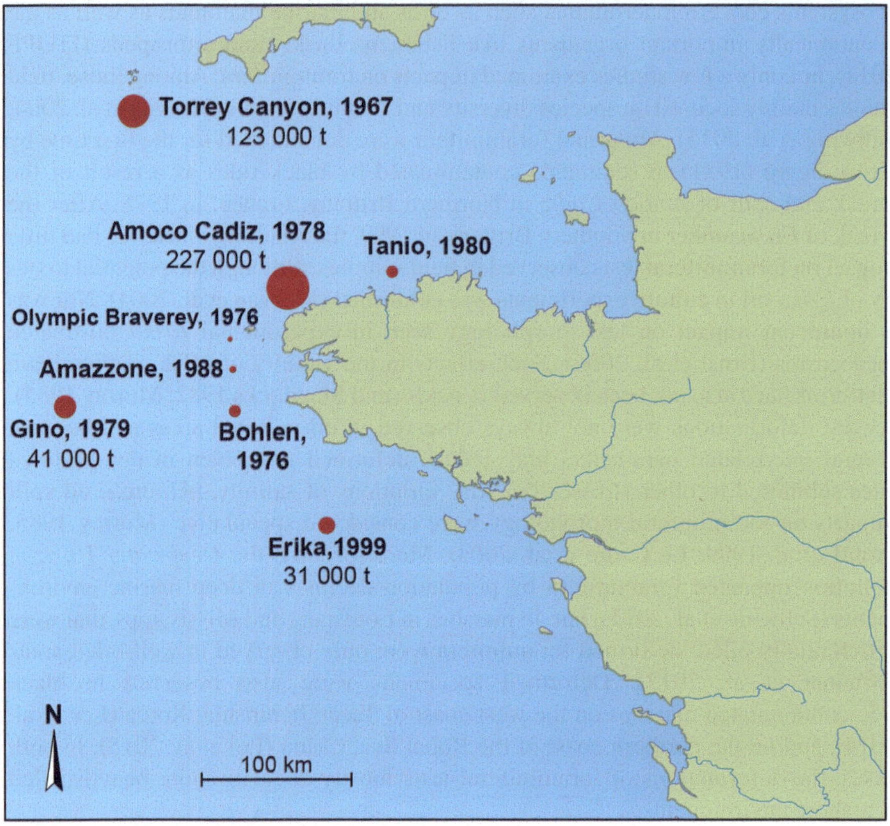

Fig. 9.1 Main oil spills along the coasts of Brittany (NW France) between 1967 and 1999 (from https://wwz.cedre.fr/ressources/accidentologie)

Those were followed in magnitude by the spills from the wrecks of *Gino* and *Erika* tankers on April 1979 and December 1999, respectively, loaded with 41,000 tons and 31,000 tons of oil. The accidents of the tankers *Boehlen* (October 1976), *Tanio* (March 1980), *Amazzone* (January 1988), and *Olympic Bravery* (January 1976), respectively, released less than 7000 tons of oil (Fig. 9.1).

9.3 *Amoco Cadiz* Black Tide and Foraminiferal Data

In 1978, the wreck of the *Amoco Cadiz* supertanker north of Finistere Department, Brittany, France, caused the most impactful oil spill recorded up to that time (Fig. 9.2). Scientists formed a Franco-American joint cooperative effort to examine the effects of the spill, coordinated by the Centre National pour l'Exploitation des Océans (CNEXO) and the U.S. National Oceanic and Atmospheric Administration

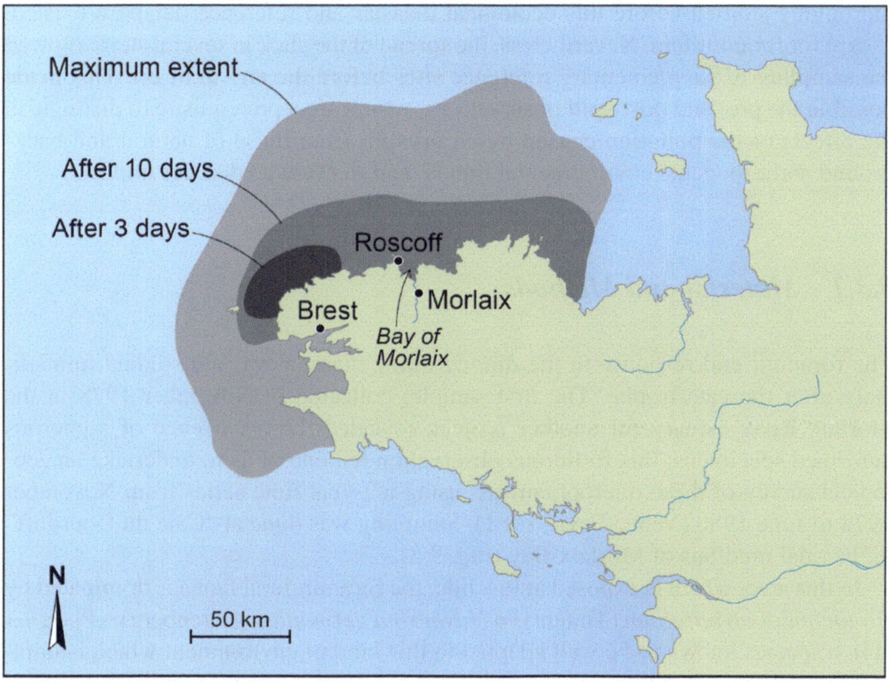

Fig. 9.2 The gradual extent of the *Amoco Cadiz* oil slick (from CEDRE, https://wwz.cedre.fr/Ressources/Outils-pedagogiques, modified)

(NOAA). Seventeen scientific teams studied the ecological impact of this black tide. This oil spill remains one of the most comprehensively studied and a reference for scientific and juridical impacts (CEDRE 2007) for several reasons: its magnitude, a stepwise expansion of the oil slick in several steps, and its location which favored quick and thorough involvement of scientists.

Amoco Cadiz supertanker went aground near the rocks of Portsall (48°35 N, 4°43 W), about 2 nautical miles (3.6 km) off the shore on 16 March 1978. It broke into two sections on the north coast of Brittany, France, on 17 March. The ship carried 223,000 tons of crude oil (light Iranian and light Arabian, with similar physicochemical characteristics) and 4000 tons of bunker fuel. All of that oil spilled at sea (Marchand et al. 1982; CEDRE 2007).

Bad weather conditions and prevailing winds and currents quickly drove much of the oil ashore. The slick spread over the sea in several steps over a period of 2 weeks (Fig. 9.2). It reached the Bay of Morlaix on 21 March, 4 days after the wreck (Boucher 1980). In the end, the resulting black tide contaminated about 360 km of coastline extending over 60 km offshore, eventually polluting coastal rocks, bays, marshes, and estuaries, as well as offshore subsurface areas. Due to the proximity of several research centers (Marine Station of Roscoff, CNEXO [today IFREMER], and University of Brest), scientists reacted very quickly to the accident. Most organisms of the biological communities on these Brittany coasts had already been thoroughly studied before this ecological disaster and reference database existed, except for foraminifera. Nevertheless, the spread of the slick in several steps allowed the sampling of supplementary reference sites before the arrival of oil. This made possible the pre- and post-spill observations, which are a prerequisite to distinguish the effects of the pollution caused by an oil spill from those of natural and background anthropogenic environmental inputs, and to evaluate disaster impacts.

9.3.1 Material and Methods

The foraminiferal response to the *Amoco Cadiz* disaster was not studied immediately after the catastrophe. The first sample, collected in November 1978 in the Morlaix River estuary for another project, revealed the occurrence of numerous deformed specimens; this fortuitous observation led one of us to undertake an ecological survey of these microorganisms using a 2-year time series from November 1978 to June 1980 (Vénec-Peyré 1981). Sampling was done at "Cale du Dourduff" in the tidal mudflats of Morlaix Bay (Fig. 9.3).

In this area, which is exposed at low tide, the foraminiferal fauna is dominated by *Protoelphidium paralium* (Tintant) (= *Haynesina germanica* (Ehrenberg)) (Plate 9.1 (1)), a species known to be well adapted to this kind of environment which is unfavorable to biocalcification. In the ecological survey performed from 1978 to 1980, the majority of the samples were collected using cores from sites with undisturbed surfaces. From 1978 to 1980, the cores, with a cross-section area of 5.7 cm^2, were sliced in 0.5 cm intervals. The upper 0.5 cm mainly contained living foraminifera.

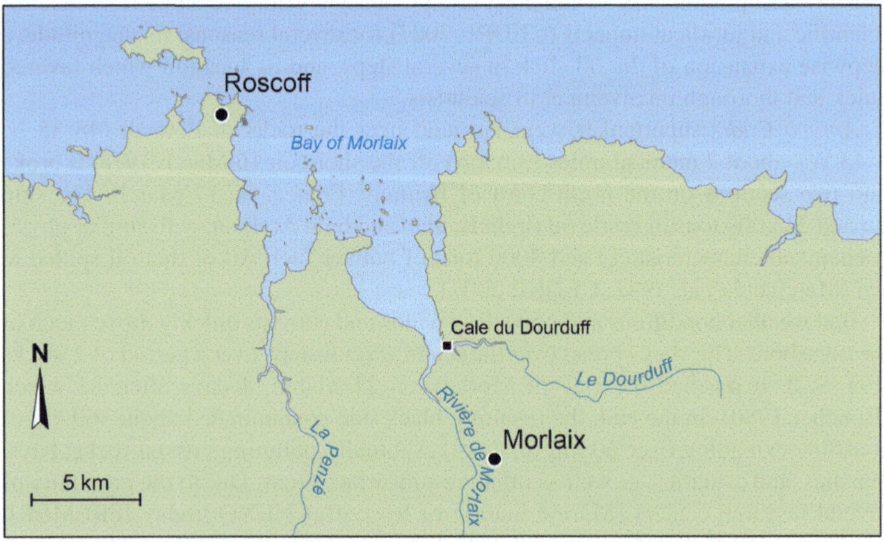

Fig. 9.3 Morlaix River estuary and location of the sampling area (Cale du Dourduff)

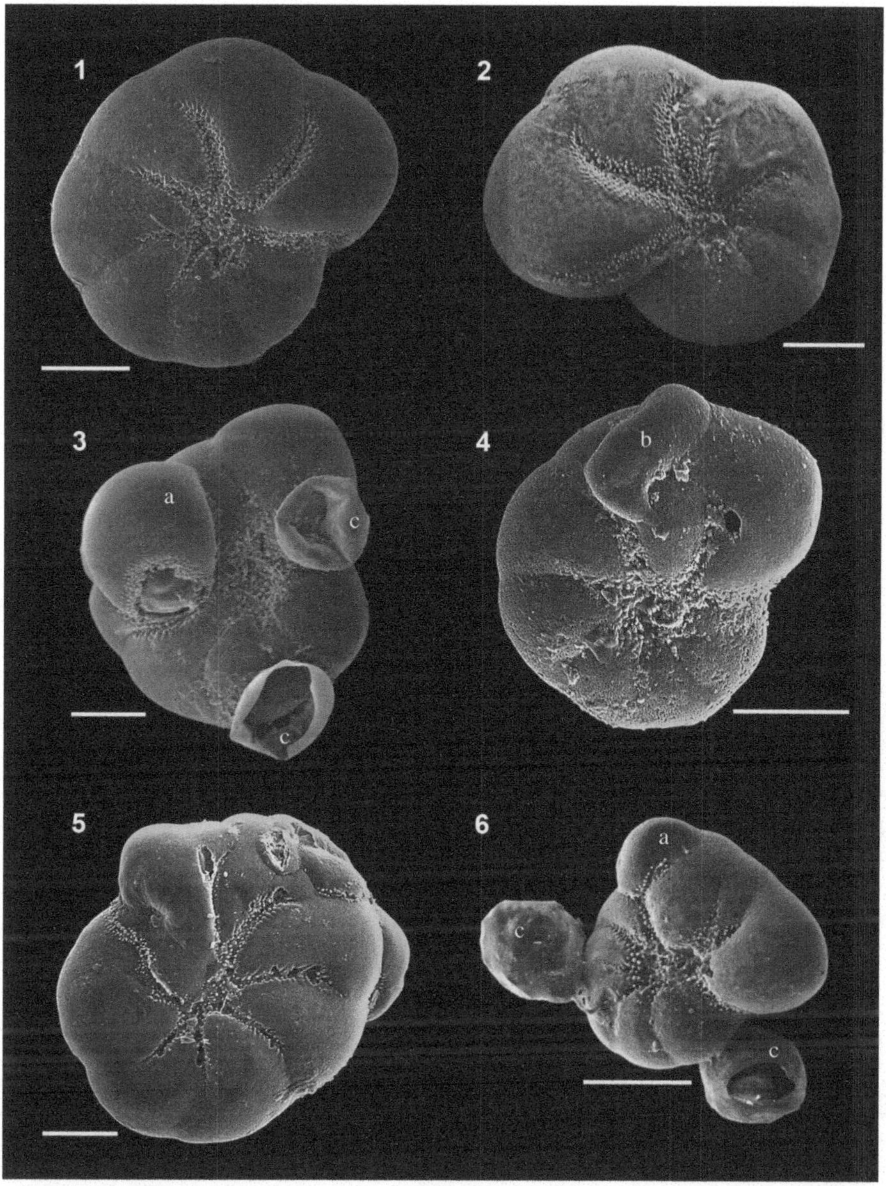

Plate 9.1 SEM microphotographs of *Haynesina germanica* from Cale du Dourduff in the Morlaix River estuary (NW France). (**1**). A normal specimen sampled in 2010. (**2–6**) Abnormal specimens collected between 8 and 13 months after the Amoco Cadiz oil spill. (**2**) Occurrence of undersized chamber. (**3** and **6**) Deformations of tests. Additional chambers (**a**), calcified protrusions (**b**), and presence of turbellarian egg capsules (**c**). (**4** and **5**) Calcified protrusions on the tests. Scale bar = 100 μm

In addition, one sample was scraped off on the surface in April 1979, in order to obtain additional sediment. In April 2010, two samples about 1 m apart (Fig. 9.4) were collected by core, and the upper first centimeter of each core was examined. All the collected sediments were immediately fixed in alcohol. Then, they were washed on 63 micron mesh sieves. Observations and picking of foraminifera were made on the wet sediment and the living specimens were detected by the natural green color of their cytoplasm.

Fig. 9.4 Collecting sites in the Morlaix River estuary. Upper: The general area of the estuary at Cale du Dourduff. Lower: Sampling site in the mud on the left side of the anchored sailboat

The counts performed in 1978 used a sediment volume of 2.86 cm^3, while those of 2010 used a volume of 10 cm^3. Thus, the total number of specimens reported in Table 9.1 cannot be interpreted in terms of changes in the foraminiferal densities. To understand fluctuations through time, the seasonal and natural variations related to the species' life cycle must be known, and, for each period, several samples should be collected simultaneously in the same area due to the patchy distribution of these benthic microorganisms.

9.3.2 Observation of Anomalies

No immediate effect of the black tide on the foraminiferal density could be determined because the study started 8 months after the spillage. The most striking observations, made in the first sample collected in November 1978, were different kinds of abnormalities in the living specimens as well as the high number of flatworm (turbellarian) egg capsules attached to foraminiferal tests, which are the dominant solid substrate in the muddy sediments. Some foraminiferal tests had undersized chambers interfering with the regularity of the spire, which indicates episodic slow growth (Plate 9.1 (2)). Others were deformed by additional chambers (Plate 9.1 (3, 6)), or calcified protrusions such as knobs or swellings (Plate 9.1 (4, 5) and Plate 9.2 (7)). Calcification imperfections on the chamber surface, as well as hiatuses in crystallization inside the test microstructure, were also observed (Plate 9.2 (1, 2, 8)). On some specimens, the surface appears to be creased (Plate 9.2 (1)), suggesting a softening of the wall that may be linked to decalcification. The percentages of these different morphological deformities in each sample collected during the ecological survey are given in Table 9.1. This table groups the results published in 1981 (concerning the period November 1978–April 1979) and unpublished data (related to the period May 1979–June 1980 and reported in Vénec-Peyré 1982).

The living fauna collected between 8 and 13 months after the wreck (from 13 November 1978 to 19 April 1979) was severely impacted: up to 21% of the foraminifera showed slow growth, up to 8% were deformed by additional chambers or protrusions, and up to 50% had attached turbellarian egg capsules (Table 9.1). Between 14 and 24 months following the spill (13 May 1979 to 12 June 1980), these values significantly decreased.

In the thanatocoenosis of the same samples, the empty tests exhibited similar abnormalities but at lower percentages (Table 9.1). Although their percentages were slightly different from those observed in the biocenosis, they showed a similar trend. Moreover, several individuals were entirely decalcified, and only their organic skeletons were observed in these samples. Variations observed in the proportions of deformed and "epiparasitized" specimens during the survey suggested a close relationship with an episode of strong environmental stress.

Table 9.1 Quantitative data related to the different abnormalities observed on *Haynesina germanica* collected in the tidal mudflats of "Cale du Dourduff" (Morlaix Bay) during the survey 1978–1980 and in 2010. The total number of specimens reported in this table cannot be interpreted in terms of changes in the foraminiferal densities (cf. Sect. 9.3.1). *Sample was collected before the oil reached Morlaix Bay. **Sample scrapped off on the surface

	Biocenosis				Thanatocoenosis			
Date	Specimen nb	%Growth lowering	%Calcifi. defaults	%Parasited ind.	Specimen nb	%Growth lowering	%Calcifi. defaults	%Parasited ind.
*19/03/1978**					*223*	*5*	*0.8*	*4*
13/11/1978	64	21	7	15.5	70	21	3	3.5
12/12/1978	41	21	8	14.5	65	21	5	9
15/1/1979	25	20	8	16	76	16	6	17
19/3/1979	31	16	3	35	74	23	7	0
19/4/1979	10	15	3	50	113	16	5	12
*19/4/1979***	*224*	*14.5*	*3*	*50*	*64*	*11*	*4.5*	*12*
13/5/1979	46	4.5	0	30	81	5	5	17
19/6/1979	75	3	0	20	90	4	2	4.5
20/9/1979	42	4	0	9.5	174	3	6	8
22/10/1979	26	4	0	15	102	8	3	5
18/2/1980	24	4	0	0	106	6	2	6
15/4/1980	34	3	0	5				
15/5/1980	22	0	0	4				
12/6/1980	26	0	0	2	52	4	2	6
30/04/2010 (1)	267	2	3	15				
30/04/2010 (2)	201	3	3	10				

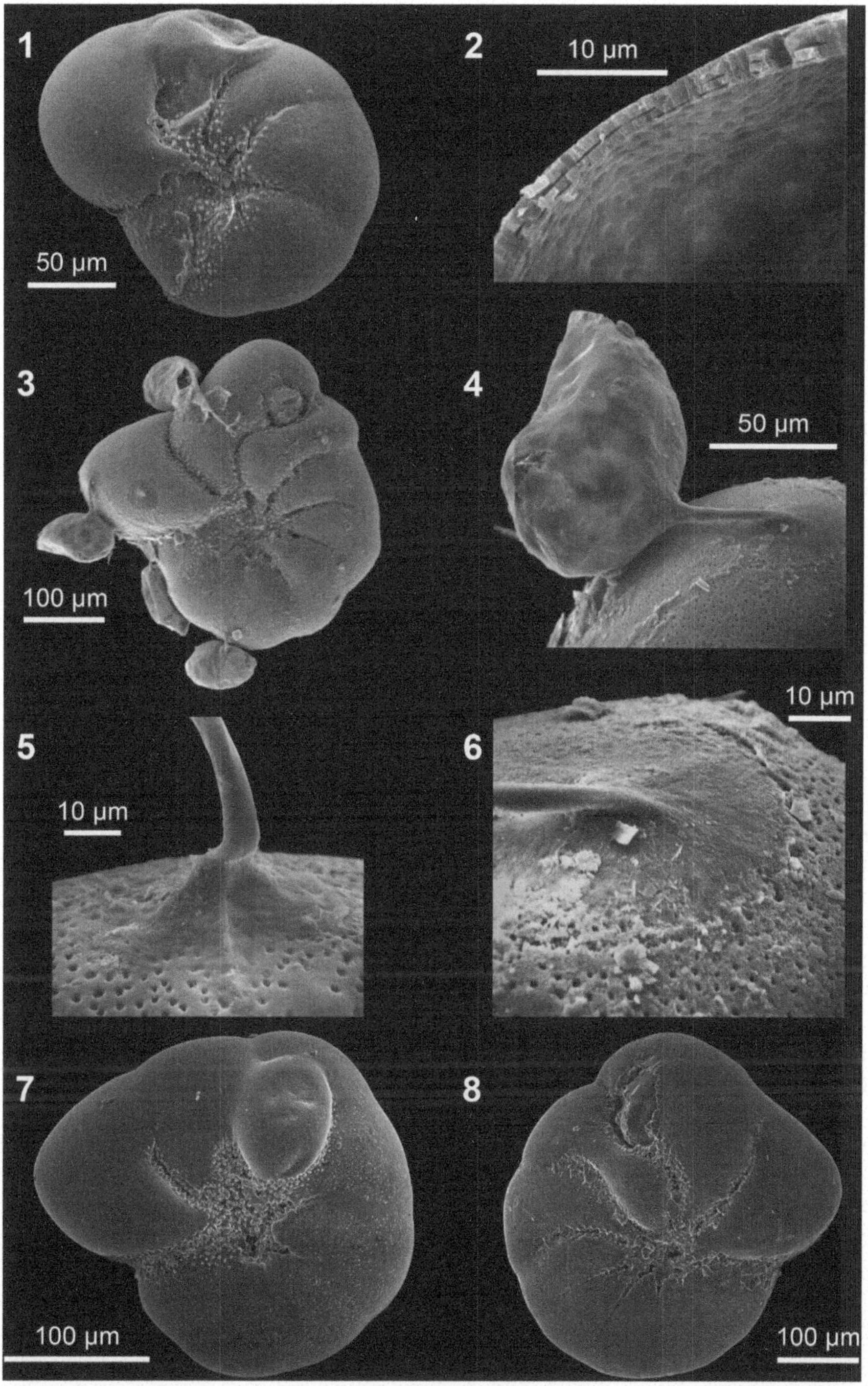

Plate 9.2 SEM microphotographs of *Haynesina germanica* from Cale du Dourduff in the Morlaix River estuary (NW France) collected between 8 and 13 months after the Amoco Cadiz oil spill. (**1**) Defects in the calcification of the penultimate chamber. (**2**) Hiatuses in the calcification of a test wall. (**3–4** and **6**) Abnormal specimen epiparasitized with four turbellarian egg capsules and enlargements of the last chamber showing the site of capsule attachment. (**5**) Enlargement showing capsule attachment on another specimen. (**7, 8**) Deformed specimens collected in 2010

9.3.3 Comparisons of Pre- and Post-spill Data

Morphological Abnormalities

Several causes could trigger morphological deformities. In the past, these deformities have been attributed to physical damage, highly fluctuating environmental parameters (salinity, amount of organic material, oxygen content, pH), anthropogenic pollution (urban runoff, heavy metals, hydrocarbons, PCBs, etc.), or any combination thereof (Nigam et al. 2006). In our studied tidal area, salinity variations as well as local human pollution could be potential causes to explain the observed deformations, but they are not sufficient to interpret the strong temporal fluctuations noted during the survey. The most drastic environmental changes in this area, over the 2 years of monitoring, were induced by the *Amoco Cadiz* black tide. By chance, due to the vigilance of the biologists of the marine laboratory of Roscoff who took care to collect sediments just after the wreck and before the first arrival of oil in this intertidal location, a comparison could be made between pre- and post-spill samples. In an unpolluted sample collected on 19 March 1978 in the same area, the dead microfauna also displayed anomalies but with very low numbers, quite similar to those found after June 1979; less than 5% of specimens exhibited growth irregularities, and less than 1% were deformed.

In the Bay of Morlaix, highly impacted by the *Amoco Cadiz* oil spill, the average hydrocarbon concentrations reached around 311 ppm in surface sediments in July 1978 (4 months after the spillage) and 172 ppm in February 1979 (Marchand 1981; Marchand et al. 1982). In the surface sediments of Cale du Dourduff, where the foraminifera were collected, the aliphatic hydrocarbon concentrations fluctuated around 300 ppm between May 1978 and February 1979 (with a peak of 1086 ppm in July 1978), and decreased below 74 ppm in March 1979. Before the pollution, they were less than 50 ppm. The aromatic hydrocarbons, highly toxic, increased from 93 ppm in May 1978 to values ranging from 126 to 251 ppm between July 1978 and January 1979, and decreased to 55 ppm in February and March 1979 (Boucher, personal communication in Vénec-Peyré 1981).

Due to the similar fluctuations of the percentages of abnormalities and the levels of pollution, Vénec-Peyré (1981) concluded that the *Amoco Cadiz* oil spill negatively impacted the foraminiferal fauna by affecting the process of calcification. She pointed out the difficulty distinguishing a potential effect of the oil toxicity from a possible impact of oiling acting upon other environmental parameters. The foraminiferal fauna recovered their pre-spill characteristics between 14 and 24 months after the wreck. In a general way, recovery of all the taxonomic groups was observed over a period of time ranging from 2 to 10 years following the *Amoco Cadiz* spill (Laubier 2005).

Flatworm Egg Capsules Attached to Foraminiferal Tests

Some flatworms (turbellarians; Platyhelminthes) form egg capsules as part of their reproductive processes and attach them to hard substrates (Shinn 1985, 1993; Crespo-González et al. 2005). The egg capsules were from either free-living or parasitic, unidentified flatworms, informally grouped as turbellarians (see Tyler et al. 2006–2019). In our samples, egg capsules of these flatworms were attached to many of the foraminiferal tests (Plate 9.1 (3, 6) and Plate 9.2 (3–6)). These egg capsules, considered as epibionts on foraminifera, were previously observed on foraminifera found in shallow waters (Jepps 1942; Boltovskoy 1963, 1965; Boltovskoy and Wright 1976). Such egg capsules also were reported on subtidal foraminifera around offshore petroleum production platforms on the Louisiana shelf in the Gulf of Mexico (Locklin and Maddocks 1982). Although egg capsules formed in cultures may have various patterns on their surfaces (Shinn 1985, 1993), those on our foraminifera were smooth although dissected and distorted. The turbellarians attach the egg capsules by a stalk to the substrate and these are present on the foraminifera.

Higher numbers of foraminifera epi-parasitized by egg capsules in samples collected between November 1978 and June 1979 are probably not directly related to the deleterious effect of oil pollution. According to Marchand (1981), turbellarians and copepods were the groups of meiofauna that were most affected by the spill in the Morlaix River Bay, exhibiting a reduction in their density. Marchand also reported that 7 months after the accident, this decline of the meiofaunal community was followed by an increase of the whole meiofauna (but unfortunately without specific data on turbellarians). This period of meiofaunal recovery coincides with the beginning of the foraminiferal survey. Thus, the higher percentages of epi-parasitized foraminifera observed in Cale du Dourduff may correspond to the progressive meiofaunal recovery and would result, by that fact, from a side effect of the oil contamination. In turn, then, their decrease between June 1979 and June 1980 could correspond to a return to more normal environmental conditions and a stabilization of the turbellarian population. Due to the lack of information on the turbellarian life cycle and population dynamics, this explanation remains hypothetical.

In any case, the capsules are firmly attached to their host by stalks that induce deformations of the foraminiferal test wall. Marks of etching are observed (Plate 9.2 (4, 6)), and in other cases, the attachment creates protuberances (Plate 9.2 (5)) which may heighten when calcitic lamellae form new chambers to cover previous ones. Then, when the capsules fall off, these attachment marks leave scars on the foraminiferal test (Plate 9.2 (4)).

9.3.4 Thirty-Two Years Later

In April 2010, 32 years after the spill, new samples were taken at Cale du Dourduff. The foraminiferal fauna was dominated by the same species. The same test deformations (occurrence of smaller chambers interfering with the coiling regularity, defects in calcification) were observed, but with percentages lower than 3%, similar to the pre-oil spill levels (Table 9.1). The abnormalities observed before pollution and after May 1979 could result from pollutants (oil, toxic chemicals, etc.) entering the Morlaix River estuary from daily activities such as boating and agriculture.

9.4 *Erika* Black Tide and Foraminiferal Data

The Maltese oil tanker *Erika* split up in two parts and sank at a depth of 120 m and position of 47°09 N, 4°15 W, south of Finistere Department, Brittany, France, on 12 December 1999 about 30 nautical miles (56 km) from the nearest shore (Fig. 9.5). A period of violent storms and high winds blew the oil slick ashore along about 400 km of shoreline that was polluted to different degrees. Of the 31,000 tons of heavy fuel oil (fuel oil n°6, according to international nomenclature) transported by the vessel, about 19,000 tons were released into the sea (Laubier et al. 2004).

Field samples were collected within the 21 months following the spill in the mudflats of the southern Bay of Bourgneuf (Vendée, W. France) (Morvan et al. 2004), where the foraminiferal fauna was dominated by nonionids. A series of mesocosm experiments using sediments and foraminifera from the mudflats of Bourgneuf Bay were also run by Ernst et al. (2006).

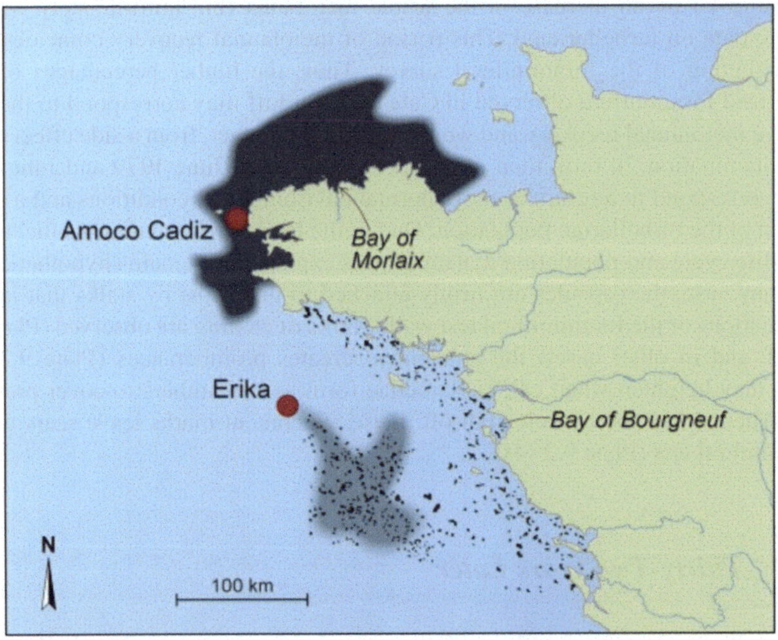

Fig. 9.5 The areas, respectively, impacted by Amoco Cadiz and Erika oil spills

After the *Erika* oil spill, morphological abnormalities recorded in the field samples were lower than 3% during the monitoring (Morvan et al. 2004); these authors did not report pre-spill reference data, but they observed that their data "never exceed those recorded in unpolluted environments" and "did not indicate a significant impact of the oil spill" (Morvan et al. 2004). They did not give the oil concentrations at the site studied. Nevertheless, they demonstrated the potential toxicity of *Erika* oil in cultures of *Ammonia tepida*. All the individuals raised in cultures, which also contained 3 mg of oil per 100 mL of seawater, were normal, whereas a concentration of pollutant higher than 5.5 mg per 100 mL of seawater produced more than 40% deformed tests. Ernst et al. (2006) reported few deformed tests both in control and *Erika* oil-treated mesocosms with quite similar relative frequency. They indicated that, in their experiments, foraminifera "have been subjected to much smaller quantities of fuel" than the doses of 5.5 mg of oil per 100 mL of water. Moreover, these authors did not observe significant numerical increase of foraminifera with attached turbellarian egg capsules in their experiment and field survey.

9.5 Discussion

If few studies dealt with black tide effects on foraminifera, their impacts on other organisms (mostly macroorganisms) were extensively studied (Laubier 2005; CEDRE 2007). These studies showed that ecosystem response to oil pollution differs from one spill to another, and that the responses of organisms differ from one taxonomic group to another as well. As with the other organisms, the consequences of *Amoco Cadiz* and *Erika* oil spills on foraminifera were different: 18–29% deformed living specimens versus <3%, respectively, reported during the months following each disaster, as well as increase of epibionts only observed after *Amoco Cadiz* disaster.

Similarities between the *Amoco Cadiz* and *Erika* spills include the following: both spills impacted the North West French coastlines during a period of adverse weather; both studies were conducted in low-energy areas (bays); sampling was done in tidal mudflats; and both foraminiferal faunas were dominated by *Haynesina germanica*. This species is opportunistic, and hence well adapted to tidal environments where anoxia occurs and pH and salinity fluctuate under normal conditions. Although the foraminifera survive the additional oil pollution, their physiology may be disturbed and the calcification of their tests may be abnormal. On the contrary, in both cases, the amount and physicochemical characteristics of the oils, as well as the characteristics of the spill itself, are quite different (Table 9.2), which could control the mechanisms through which black tides may impact the ecosystems (oil toxicity and physical smothering of biotopes).

9.5.1 Chemical Composition of Oils

The toxicity of a pollutant depends upon its chemical composition, the quantity of spilled oil, the tolerance threshold above which its concentration could affect the specimens in their environment, and its bioavailability. The light *Amoco Cadiz* oil

Table 9.2 Comparison between the general characteristics of *Amoco Cadiz* and *Erika* oil spills and their respective impact on foraminifera

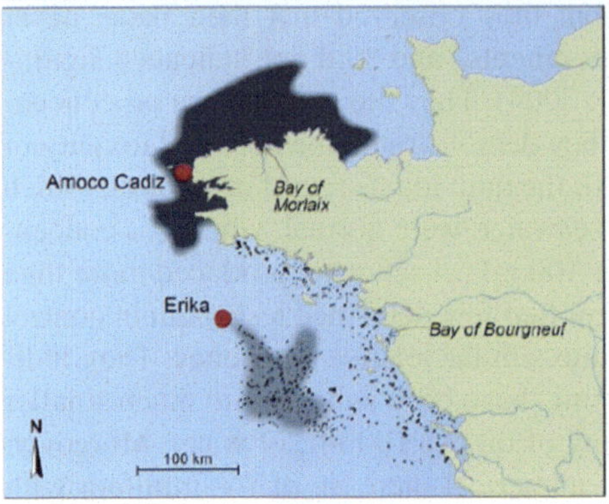

		Amoco Cadiz wreck	*Erika* wreck
General characteristics	Date	March 1978	December 1999
	Meteorological conditions	Storms	Storms
	Distance to the coast	2 nautical miles	30 nautical miles
	Spillage	227,000 tons	19,000 tons
	Polluted coastlines	360 km	400 km
	Density	0.8-0.85	>0.95
	Viscosity	Soluble ($v = 8$ cSt)	Highly viscous ($v = 20{,}000$ cSt)
	Dispersal mechanisms	Slicks, chocolate mousse	Pellets
	% Aliphatic hydrocarbons, biodegradable	45–47%	22–30%
	% Aromatic hydrocarbons (toxic)	31–34%	42–50%
	% Resins (toxic)	16–17%	31%
	% Asphaltenes (toxic)	4–5%	5%
	Sulfur content	2.35%	2.28%
	Nickel	14–16.5 ppm	45 ppm
	Vanadium	45–60 ppm	82.7 ppm
Post-spill characteristics in the sampling area	Sampling area	Intertidal area	Intertidal area
	Dominant foraminifera	Nonionids	Nonionids
	Impact on foraminiferal density	Not investigated	Probable
	Deformed foraminifera in field samples	18–29%	<3%

(continued)

Table 9.2 (continued)

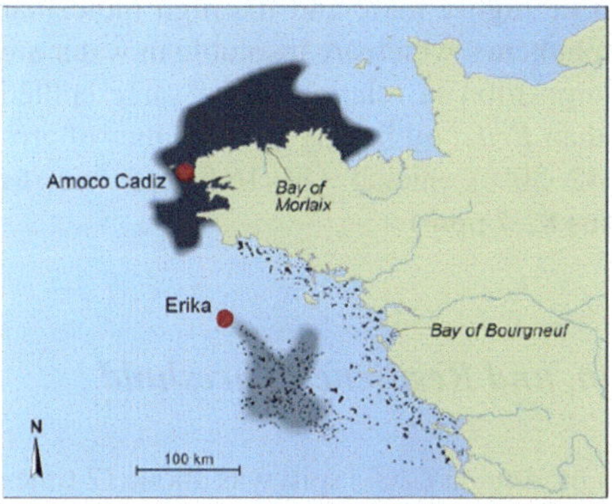

		Amoco Cadiz wreck	*Erika* wreck
	Deformed foraminifera in cultures	Not investigated	>40% only in cultures contaminated with Erika oil (>5.5 mg/100 mL of seawater)
	Turbellarian egg capsules	Increase	No increase
	Pollutant content (at the sampled sites)	Aliphatic hydrocarbons: 74–1086 ppm Aromatic hydrocarbons: 54–251 ppm	Not indicated
	Complete ecosystem recovery (Laubier 2005)	10 years	1 year

initially contained 45–47% saturated aliphatic hydrocarbons or alkanes, 31–34% aromatic hydrocarbons, 16–17% polar compounds (resins), and 4–5% asphaltic compounds, which include nickel (14–16.5 ppm), vanadium (45–60 ppm), and sulfur contents (2.35%) (Ducreux 1982). The aromatic fraction included a large number of polycyclic hydrocarbons (PAHs) such as naphthalenes (dominant), phenanthrenes, dibenzothiophenes, and naphthobenzothiophenes and monocyclic aromatic hydrocarbons such as benzenes (Atlas et al. 1981; Boehm 1982). Over the course of a year, the saturated hydrocarbons as well as light aromatic compounds progressively disappeared, whereas heavy aromatics, resins, and asphaltenes persisted. Moreover, the processes of decontamination were slower in the muddy sediments (Ducreux 1982).

The heavy *Erika* oil was initially composed of 22–30% saturated aliphatic hydrocarbons or alkanes, 42–50% aromatic hydrocarbons (with a dominance of naphthalenes, phenanthrenes, chrysenes), 31% resins, 5% asphaltenes, and 2.28% sulfur (Tiercelin et al. 2000; Baars 2002; see a complete list of aromatic compounds in Tronczynski et al. 2004). This oil was rich in nickel (45 ppm) and vanadium (82.7 ppm) (Tiercelin et al. 2000; source TotalFina).

The alkanes are mildly toxic and biodegradable by bacteria. The light, unsaturated aromatic hydrocarbons are considered to be highly toxic and the high molecular PAHs are carcinogenic. The resins and asphaltenes which are insoluble in water are very toxic. Thus, based on its initial composition in relation to the same spilled volume, *Amoco Cadiz* oil was less toxic than *Erika* oil having lower values of aromatic hydrocarbons (31–34% versus 42–50%), nickel (14–16.5 ppm versus 45 ppm), and vanadium (45–60 ppm versus 82.7 ppm).

9.5.2 Amount, Oil Concentration, and Resilience Threshold

The total volume of oil released in the sea by *Amoco Cadiz* spill was about 12 times more than the *Erika* accident (223,000 tons versus 19,000 tons).

The toxicity of a pollutant depends not only upon its chemical composition and the quantity of discharged oils but also upon the tolerance threshold above which its concentration could affect organisms in their environment. The survey conducted in Cale du Dourduff revealed a similar trend between the rate of deformities and oil concentrations (highest percentages of abnormal specimens were observed when the oil concentrations were >300 ppm). This trend, observed at the time scale of this survey, was also observed after the 2011 *Penglai* oil spill in the Bohai Sea, China (Lei et al. 2015), where the percentage of deformed foraminifera increased as the distance between oil spill and sampling sites decreased, with the highest percentages corresponding to the concentration of oil between 239 and 487 μg/g (=239–487 ppm).

The pollutant content in Bourgneuf Bay fouled by black tide from the *Erika* spill was not recorded when surveyed (Morvan et al. 2004). Nevertheless, laboratory cultures confirmed the toxicity of *Erika* oil, showing that a concentration of at least 5.5 mg of pollutant per 100 mL of seawater caused foraminiferal deformities of *Ammonia tepida*. Considering this experimental result, the concentration of oil in Bourgneuf Bay may not have been sufficient to induce foraminiferal deformities. Additional information about *Erika* oil concentration is given by Baars (2002). Assessing the human health risks for tourists in the case of *Erika* spill, Baars (2002) reported: "it is expected that near the beaches the dissolved oil will be diluted with a factor of 100 at least. This would result in a maximum oil concentration at beach level of 50 μg/L" (Baars 2002, p. 64). The 50 μg/L, expected at the beach by Baars, is 1000 times less than the 55 mg/L which caused 40% deformed tests in the cultures performed by Morvan et al. (2004). These estimated values are well below the levels measured in Cale du Dourduff just after the *Amoco Cadiz* event.

9.5.3 Physical Oil Properties, Deposition Pattern, and Weathering

The *Amoco Cadiz* oil was light ($d = 0.8$–0.85) and had a viscosity of 8 cSt (at 21 °C); thus, it was highly volatile and soluble with a high dispersal potential. Due to its solubility, the light *Amoco Cadiz* oil mixed with seawater in a water-in-oil emulsion, also called "chocolate mousse," which increased the volume of pollutant up to five times and facilitated the penetration of oil in all parts of the ecosystem. In Cale du Dourduff, one of the most heavily impacted areas, oil contaminated the first 10 cm of the sediment column. During storms and high tides, remobilization of this oil trapped within sediments continues to generate high levels of pollution (Beslier et al. 1980). On the contrary, the *Erika* oil was heavy ($d \geq 0.95$) and very viscous ($v = 20{,}000$ cSt at 10 °C), with a low evaporation, dissolution, and dispersal potential. The slick was fragmented in a multitude of "pancake"-like tar globs or in pellets which came ashore 10 days after the wreck. The depositional patterns of the two spills differently influenced the smothering of biotopes.

The smothering of sediments by oil pollution can modify the ambient environment in a cascading way (changes in the light penetration, supply of dissolved oxygen and nutrients, food web, pH). Thus, this indirect impact on biocalcification processes cannot be excluded. Nevertheless, as already stipulated by Vénec-Peyré (1981), the determination of the contributions of the direct effects of the pollutant due to its toxicity from that of the indirect impact caused by oiling of biotopes is difficult.

The weathering processes (volatilization, dissolution, photooxidation, biodegradation, etc.), which depend upon the physical properties of the oil, are also controlled by the delay between the time of spillage and the arrival of pollutants on the sampling sites. This time interval is related to the distance between the accident location and the sampling area. The *Amoco Cadiz* wreck sank at a depth of 30 m about 75 km from the site of foraminiferal sampling in Morlaix Bay, while the *Erika* wreck, lying at a depth of about 120 m, is located 175 km from the foraminiferal sampling in Bourgneuf Bay. Consequently, the pollution arrived in Morlaix Bay earlier (4 days after the wreck) than in Bourgneuf Bay (14 days after the wreck) and the oil characteristics evolved differently.

All these interacting parameters related to the physicochemical properties of oil and to the context of both black tides likely impacted differently the evolution of harmfulness of *Amoco Cadiz* and *Erika* pollutants as well as their biological availability. The light *Amoco Cadiz* fuel oil was probably more influential on foraminifera than the viscous *Erika* oil.

Despite some similarities, the differences observed in the multiple factors involved in both oil spills may have created specific conditions that are particular to each of the investigated areas and resulted in a larger physiological impact on foraminifera in the Morlaix estuary than in Bourgneuf Bay. These various conditions may explain the differences observed in the foraminiferal response in both areas.

Although the two studies of the consequences of the *Amoco Cadiz* (Vénec-Peyré 1981) and *Erika* (Morvan et al. 2004) oil spills on foraminifera did not use the same methods nor was information about oil concentration in Bourgneuf Bay sediment available, we suggest the hypothesis that the original characteristics and their changes over time of the oil created different responses in the foraminifera. The use of deformed specimens as pollution indicators has been criticized because too many parameters may induce them, even without industrial pollution. To decrease the parameters, culture experiments are recommended. Cultures of foraminifera are, of course, a powerful tool to study some biological aspects of foraminifera. They can be useful in attempting to constrain the reaction of a species to a given parameter (salinity, temperature, or diverse type of pollutants) under controlled conditions, but a culture or microcosm experiment, whatever the quality, is always a stressed environment, and cannot perfectly recreate the various processes occurring in the polluted natural environments.

This study points out that a single parameter impacting foraminifera cannot be determined in the case of oil spill pollution. A number of factors are involved and any of them or a combination of some of them may impact living foraminifera including the deformation of test formation or modification by the foraminifera, The comparisons of field data collected before and after a special event, which could affect the quality of environment, can provide strong inference in favor of a negative effect of black tides.

9.6 Conclusions

This study compares for the first time two cases of oil spill impacts on foraminiferal morphology and turbellarian attachments to the tests in quite similar environments dominated by nonionids and located on the French NW coasts. The comparison of the percentages of anomalous specimens in Cale du Dourduff before and after the *Amoco Cadiz* disaster provides strong evidence for a negative effect of the oil spills on foraminiferal tests. On the contrary, observations performed on field samples in Bay of Bourgneuf and on oil-treated mesocosms did not demonstrate a clear impact of *Erika* black tide on foraminiferal morphology despite the potential toxicity of the pollutant. The comparison between the magnitude of the spills, the physical and chemical properties of discharged oils, their degradation over time, their concentration in the sediments, as well as the characteristics of the spill context allows a better understanding of the differences in the foraminiferal and turbellarian responses. With their ability to deform in oil-polluted areas, foraminifera constitute a useful monitoring tool to assess the environmental stress caused by catastrophic oil spills under specific conditions. This analysis confirms that the severity of such a disaster depends upon a number of parameters linked, of course, to the characteristics of pollutants but also to their multiple interactions with environmental factors which control the physical and chemical processes involved in hydrocarbon evolution. Both studies far from contradict each other. This analysis shows that the results of both studies directly reflect the differences observed in observable environmental parameters. The anomalous foraminifera are good indicators of the severity of oil spill impact and the rate of recovery after such a disaster.

Acknowledgements The authors are indebted to Alexandre Lethiers (CR2P, Sorbonne Université) for his assistance with the preparation of text figures and plates, and to S. Laroche (MNHN) and O. Boudouma (ISTEP, Sorbonne Université) for SEM photographs. We thank the France-Berkeley Fund of the University of California, Berkeley, for financial support of this project. This is UC Berkeley Museum of Paleontology publication number 2095.

References

Atlas RM, Boehm PD, Calder JA (1981) Chemical and biological weathering of oil from the *Amoco Cadiz* oil spillage, within the littoral zone. Estuar Coast Mar Sci 12:589–608

Baars BJ (2002) The wreckage of the oil tanker '*Erika*'—human health risk assessment of beach cleaning, sunbathing and swimming. Toxicol Lett 128:55–68

Beslier A, Birrien JL, Cabioch L, Larsonneur C, Le Borgne L (1980) La pollution des Baies de Morlaix et de Lannion par les hydrocarbures de l' "*Amoco Cadiz*": Répartition sur les fonds et évolution. Helgoländer Meeresunters 33:209–224

Boehm PD (1982) The *Amoco Cadiz* analytical chemistry program. In: Report of the NOAA-CNEXO join scientific commission, Ecological study of the *Amoco Cadiz* oil spill, p 35–99

Boltovskoy E (1963) Sobre las relaciones entre Foraminiferos y turbelarios. Neotropica 9:29

Boltovskoy E (1965) Los foraminiferos recientes. Editorial Universitaria de Buenos Aires, Argentina, 510 p

Boltovskoy E, Wright R (1976) Recent Foraminifera. W. Junk, La Hague, 515 p

Boucher G (1980) Impact of *Amoco Cadiz* oil spill on intertidal and sublittoral meiofauna. Mar Pollut Bull 11(4):95–101

Brunner CA, Yeager KM, Hatch R, Simpson S, Keim J, Briggs KB, Louchouarn P (2013) Effects of oil from the 2010 Macondo well blowout on marsh foraminifera of Mississippi and Louisiana, USA. Environ Sci Technol 47:9115–9123

Centre de documentation, de recherche et d'expérimentations sur les pollutions accidentelles des eaux (CEDRE) (2007). http://wwz.cedre.fr/en/Resources/Spills/Spills/Amoco-Cadiz

Crespo-González C, Reza-Álvarez RM, Rodríguez-Domínguez H, Soto-Búa M, Iglesias R, Arias-Fernández C, García-Estévez JM (2005) In vitro reproduction of the turbellarian *Urastoma cyprinae* isolated from *Mytilus galloprovincialis*. Mar Biol 147:755–760

Ducreux J (1982) Evolution of the hydrocarbons present in the sediments of the Aber Wrac'h estuary. In: Report of the NOAA-CNEXO join scientific commission, Ecological study of the *Amoco Cadiz* oil spill, p 111–142

Ernst SR, Morvan J, Geslin E, Le Bihan A, Jorissen F (2006) Benthic foraminiferal response to experimentally induced *Erika* oil pollution. Mar Micropaleontol 61:76–93

ITOPF (2014) Effects of oil pollution on the marine environment. Technical Information Paper, 13. http://www.itopf.org/knowledge-resources/documents-guides/document/tip-13-effects-of-oil-pollution-on-the-marine-environment/

Jepps MW (1942) Studies on *Polystomella* Lamarck (Foraminifera). J Mar Biol Assoc U K 25:607–666

Laubier L (2005) Après l'*Erika* et l'*Amoco Cadiz*, la renaissance des écosystèmes. Pour la Science 332:36–42

Laubier L, Le Moigne M, Flammarion P, Thybaud E, Cossa D (2004) The monitoring programme of the ecological and ecotoxicological consequences of the "*Erika*" oil spill. Aquat Living Resour 17:239–241

Le Cadre V, Debenay JP, Lesourd M (2003) Low pH effects on *Ammonia beccarii* test deformation: implications for using test deformations as a pollution indicator. J Foraminifer Res 33:1–9

Lee YG, Kim S, Jeong DU, Lee JS, Wood HJ, Park MW, Kim BH, Son MH, Choi YH (2014) Decalcification of benthic foraminifera due to "*Hebei Spirit*" oil spill, Korea. Mar Pollut Bull 87:276–285

Lei YL, Li TG, Bi H, Cui WL, Song WP, Li JY, Li CC (2015) Responses of benthic foraminifera to the 2011 oil spill in the Bohai Sea, China. Mar Pollut Bull 96:245–260

Locklin JA, Maddocks RF (1982) Recent foraminifera around petroleum production platforms on Southwest Louisiana shelf. Gulf Coast Assoc Geol Soc Trans 32:377–397

Marchand M (1981) *Amoco Cadiz*, Bilan du Colloque sur les conséquences d'une pollution accidentelle par hydrocarbures, Brest-novembre 1979. Publications de CNEXO – Rapports Scientifiques et Techniques 44:1–84

Marchand M, Bodennec G, Caprais JC, Pignet P (1982) The *Amoco Cadiz* oil spill. Distribution and evolution of oil pollution in marine sediments. In: Report of the NOAA-CNEXO join scientific commission, ecological study of the *Amoco Cadiz* oil spill, p 143–157

Morvan J, Le Cadre V, Jorissen F, Debenay JP (2004) Foraminifera as potential bio-indicators of the "*Erika*" oil spill in the Bay of Bourgneuf: field and experimental studies. Aquat Living Resour 17:317–322

Murray JW (1985) Recent foraminifera from the North Sea (Forties and Ekofisk areas) and the continental shelf west of Scotland. J Micropaleontol 4(2):117–125

Nigam R, Saraswat R, Panchang R (2006) Application of foraminifers in ecotoxicology: retrospect, perspect and prospect. Environ Int 32:273–283

Parker WK, Jones TR, Brady HB (1865) On the nomenclature of the foraminifera. Pt. XII. The species enumerated by d'Orbigny in the «Annales des Sciences Naturelles» vol vii. 1826. Ann Mag Nat Hist 3(16):15–41

Saraswat R, Kurtarkar SR, Mazumder A, Nigam R (2004) Foraminifers as indicators of marine pollution: a culture experiment with *Rosalina leei*. Mar Pollut Bull 48:91–96

Schwing PT, Romero IC, Brooks GR, Hastings DW, Larson RA, Hollander DJ (2015) A decline in benthic foraminifera following the Deepwater Horizon Event in the northeastern Gulf of Mexico. PLoS One 10(3):e0120565. https://doi.org/10.1371/journal.pone.0120565

Shinn GL (1985) Reproduction of *Anoplodium hymanae*, a turbellarian flatworm (Neorhabdocoela, Umagillidae) inhabiting the coelom of sea cucumbers; production of egg capsules, and escape of infective stages without evisceration of the host. Biol Bull 169:182–198

Shinn GL (1993) Formation of egg capsules by flatworms (Phylum Platyhelminthes). Trans Am Microsc Soc 112(1):18–34

Soldani AB (1789) Testageographiae ac Zoophythographiae parvae et microscopicae. Rossi F Sienne 1:1–80

Spooner MF (ed) (1978) The *Amoco Cadiz* oil spill. Mar Pollut Bull 9(7), Special edition

Stouff V, Geslin E, Debenay JP, Lesourd M (1999) Origin of morphological abnormalities in *Ammonia* (foraminifera): studies in laboratory and natural environments. J Foraminiferal Res 29:152–170

Tiercelin C, Marchand M, Rousseau C (2000) Spécial accident de l'*Erika*, golfe de Gascogne (Sud-Bretagne), 12 décembre 1999. Bulletin d'Information du CEDRE 13:10–18. http://wwz.cedre.fr/Ressources/Publications/Bulletins-d-information/n-13

Tronczynski J, Munschy C, Héas-Moisan K, Guiot N, Truquet I, Olivier N, Men S, Furaut A (2004) Contamination of the Bay of Biscay by polycyclic aromatic hydrocarbons (PAHs) following the T/V "*Erika*" oil spill. Aquat Living Resour 17:243–259

Tyler S, Schilling S, Hooge M, Bush LF (2006–2019) Turbellarian taxonomic database. Version 1.8 http://turbellaria.umaine.edu

Vénec-Peyré MT (1981) Les Foraminifères et la pollution : étude de la microfaune de la Cale du Dourduff (Embouchure de la Rivière de Morlaix). Cah Biol Mar 22:25–33

Vénec-Peyré MT (1982) Étude de l'influence du milieu sur la distribution, la morphologie et la composition du test des Foraminifères benthiques. Implications paléoécologiques. Thèse de doctorat *és* Sciences, Université Pierre Marie Curie, Paris VI, 260 p

Vénec-Peyré MT (2005) Les Planches Inédites d'Alcide d'Orbigny. A l'aube de la Micropaléontologie. Publications Scientifiques du Muséum, Collection Des Planches et des Mots, Paris, 304 p, 62 planches

Zilberberg LJ (1998) Elsevier's dictionary of marine pollution. Elsevier B.V., Amsterdam, 708 p, English to French Ed

Chapter 10
Environmental Control on Biotic Development in Siberia (Verkhoyansk Region) and Neighbouring Areas During Permian–Triassic Large Igneous Province Activity

Yuri D. Zakharov, Alexander S. Biakov, Micha Horacek, Ruslan V. Kutygin, Evgeny S. Sobolev, and David P. G. Bond

Abstract We propose an updated ammonoid zonation for the Permian–Triassic boundary succession (the lower Nekuchan Formation) in the Verkhoyansk region of Siberia: (1) *Otoceras concavum* zone (uppermost Changhsingian); (2) *Otoceras boreale* zone (lowermost Induan); (3) *Tompophiceras morpheous* zone (lower Induan); and (4) *Wordieoceras decipiens* zone (lower Induan). The *Tompophiceras pascoei* zone, previously defined between the *Otoceras boreale* and *Tompophiceras morpheous* zones, is removed in our scheme. Instead of this the *Tompophiceras pascoei* epibole zone is proposed for the lower part of the *Tompophiceras morpheous* zone. New and previously published nitrogen isotope records are interpreted as responses to climatic fluctuations in the middle to higher palaeolatitudes of

Y. D. Zakharov (✉)
Far Eastern Geological Institute of Russian Academy of Sciences (Far Eastern Branch), Vladivostok, Russia

A. S. Biakov
N.A. Shilo North-East Interdisciplinary Scientific Research Institute, Russian Academy of Sciences (Far Eastern Branch), Magadan, Russia

Kazan (Volga Region) Federal University, Kazan, Russia

M. Horacek
Institute of Lithospheric Research, Vienna University, Vienna, Austria

R. V. Kutygin
Diamond and Precious Metal Geology Institute, Russian Academy of Sciences, Yakutsk, Russia

E. S. Sobolev
Trofimuk Institute of Petroleum Geology and Geophysics, Koptyug Ave. 3, Novosibirsk, Russia

D. P. G. Bond
Department of Geography, Geology and Environment, University of Hull, Hull, UK

J. Guex et al. (eds.), *Morphogenesis, Environmental Stress and Reverse Evolution*, https://doi.org/10.1007/978-3-030-47279-5_10

Northeastern Asia and these suggest a relatively cool climatic regime for the Boreal Superrealm; however the trend towards warming across the Permian–Triassic boundary transition is also seen. The evolutionary development and geographical differentiation of otoceratid ammonoids and associated groups are considered. It is likely that the Boreal Superrealm was their main refugium, where otocerid, dzhulfitid and some other ammonoids survived the major biotic crisis at the end of the Permian. The similarity of ontogenetic development of suture lines of *Otoceras woodwardi* Griesbach and *O. boreale* Spath gives some grounds for suggesting a monophyletic origin of the genus *Otoceras*, having bipolar distribution.

Keywords Permian–Triassic transition · South Verkhoyansk · Molluscs · N-isotopes

10.1 Introduction

The presence of Permian–Triassic (P–T) boundary sequences in Siberia was first documented by Popow (1956, 1958, 1961), who described *Otoceras indigirense* Popow and *O. boreale* Spath from the Indigirka River basin and determined some *Otoceras* from the lower part of the Nekuchan Formation in the South Verkhoyansk region (Setorym River basin) using materials collected by Domokhotov (1960). *Otoceras*-bearing sediments were earlier known only in the Himalayas (Diener 1897) and Greenland (Spath 1930).

The biostratigraphy of the Upper Permian and P–T boundary transition on right bank of the Setorym River, where the Imtachan and Nekuchan formations are well exposed, was initially investigated by Domokhotov (1960) and has since been reported by Korostelev (1972), Arkhipov (1974), Dagys et al. (1984, 1986), and Biakov et al. (2016).

Otoceras boreale Spath from the South Verkhoyansk region was first described by Zakharov (1971) on the basis of his own extensive collection (of more than 150 individual *Otoceras*) from the left bank of the Setorym River. The following biostratigraphic units were recognised from the lower part of the Nekuchan Formation: *Otoceras boreale* beds and overlying "*Vishnuites*" cf. *decipiens* beds (= *Tompophiceras morpheous* zone; Zakharov 2002) on the basis of faunas from the Nikolkin Klyuch and Seryogin Creek sections.

Ammonoids from the *Otoceras*-bearing sequences and overlying units of the Pravyi Suol-1 section have been described by Dagys and Ermakova (1996). Those authors divided the lower part of the Nekuchan Formation into four zones: (1) *Otoceras concavum* (12.0 m thick); (2) *Otoceras boreale* (39.3 m thick); (3) *Tompophiceras pascoei* (13.2 m thick); and (4) *Tompophiceras morpheous* (68.0 m thick) zones. Subsequent investigations (Zakharov 2002; Biakov et al. 2018) suggest that the thicknesses of these units have been significantly overstated by Dagys and Ermakova (1996).

Zakharov (1971, 2002, 2003) has recognised dimorphs in *Otoceras boreale* Spath from the Setorym River basin named "morpha A" (="leaniconchs") and "morpha B" (="faticonchs") (Zakharov 2002). Zakharov (2002) considered these dimorphs to be a strictly corresponding sexual dimorphic pair by analogy with living *Nautilus*. Mature males of living *Nautilus* differ from females in having generally wider (in the apertural region) and larger shells because of presence of the voluminous organ (spadix).

More recently chemostratigraphical (C-isotope) data from P–T boundary sections in the South Verkhoyansk region has been provided by Zakharov et al. (2014, 2015).

In 2016 and 2017, additional expeditions were organised by A.S. Biakov to investigate the bio- and chemostratigraphy of Upper Permian and P–T boundary strata in the Setorym River basin (the Levyi Suol, Pravyi Suol-1, Ustupny and Nikolkin Klyuch sections). The former section has not been investigated previously because until now it was obscured by a debris stream. The latter section has not been visited by geologists during the past 47 years.

This chapter focuses on a revision of data on the stratigraphic distribution of otoceratid and dzhulfitid ammonoids in the P–T boundary strata of the Setorym River basin (South Verkhoyansk region) and the reconstruction of habitats for marine biota existing in the middle and higher palaeolatitudes of Northeast Asia in late Wuchiapingian, Changhsingian and early Induan times based on isotope data. We also evaluate the mass extinction at the end of the Permian and the evolutionary development of otoceratid ammonoids.

Our investigated ammonoid collections are curated at Far Eastern Geological Institute FEB RAN (Vladivostok; collection no. 803) and the Diamond and Precious Metal Geology Institute SB RAN (Yakutsk, collection no. 231).

10.2 Materials and Methods

New macrofossil remains were excavated, prepared and analysed statistically from strata stradding the P–T transition at Pravyi Suol-1, Levyi Suol, Ustupny, Seryogin Creek-2, Nikolkin Klyuch and some other sections in the Setorym River basin (South Verkhoyansk region) for biostratigraphic purposes (Fig. 10.1).

Nitrogen isotope analyses were performed on 144 samples, which were previously analysed for their carbon isotopic composition (Zakharov et al. 2014, 2015). These samples were collected at a spacing of ~0.125–0.14 m in the Imtachan (upper part) and Nekuchan (lower part) formations of the Pravyi Suol-1 section. The samples were ground by a ball mill and decarbonated with 5% HCl on a heated surface. Small aliquots of the samples were weighed into tin capsules and analysed for their N-isotope ratio using a Flash EA (Thermo, Bremen/Germany), connected via a CONFLO IV (Thermo, Bremen/Germany) to a Delta V advantage isotope ratio mass spectrometer (Thermo, Bremen/Germany) in Wieselburg, Austria (HBLFA Francisco–Josephinum). The results are reported in the conventional δ-notation in

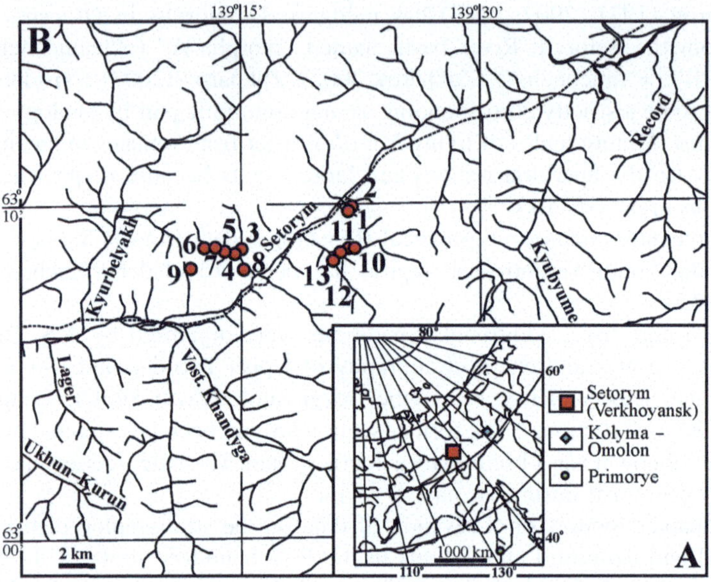

Fig. 10.1 Location map of the investigated area in the Setorym River basin, South Verkhoyansk area. 1, Nikolkin Klyuch (base of section is located at 63°11′56.6″ N, 139°26′35.6″ E); 2, Inessin Klyuch; 3, Levyi Suol (63°08′43.0″ N, 139°14′04.9″ E); 4, Pravyi Suol-1 (63°09′18.5″ N, 139°12′10.9″); 5, Pravyi Suol-2; 6, Pravyi Suol-3; 7, Pravyi Suol-4; 8, Nizhny Suol; 9, Ustupny; 10, Seryogin Creek-1; 11, Seryogin Creek-2; 12, Seryogin Creek-3 (Shagali); 13, Dolgachan

permil (‰) relative to the international N-air standard for nitrogen. Reproducibility of replicate standards is better than ±1‰ for nitrogen (because of the low content of organic matter and corresponding low N content of the samples).

10.3 Results: Biostratigraphy, Nitrogen Isotopes

The studied area (Setorym River basin) is located within the western part of the Verkhoyansk–Kolyma folded area (South Verkhoyansk Zone) adjacent to the eastern edge of the Siberian platform (Parfenov and Kuzmin 2001), where P–T sequences comprise the Imtachan and Nekuchan formations (Domokhotov 1960; Biakov et al. 2016). The upper part of the Imtachan formation (*Intomodesma postevenicum* subzone of the *Intomodesma costatum* zone; Biakov et al. 2016, 2018) in our sections comprises alternations of sandstone, siltstone and sandy siltstone. The exposed part of the overlying Nekuchan Formation in the Pravyi Suol-1 and Levyi Suol sections consists of mudstone (95–100 m thick) with rare layers of sandstone and tuffs.

Following Domokhotov (1960), the P–T boundary was for a long time placed at the Imtachan–Nekuchan formational contact (Zakharov 1971, 1995, 2002; Arkhipov 1974; Korostelev 1972; Dagys et al. 1984, 1986; Dagys and Ermakova 1996). However, chemostratigraphical (C-isotope) data across the P–T boundary transition at the Pravyi Suol-1 section (Zakharov et al. 2014, 2015; Biakov et al. 2018) have demonstrated that the P–T boundary in the South Verkhoyansk region is located within the *Otoceras*-bearing sequences in the lower part of the Nekuchan Formation. In the Pravyi Suol-1 and Levyi Suol sections, the P–T boundary is placed at approximately 6.3 m above the formational contact (Zakharov et al. 2014, 2015).

The following ammonoid zones are recognised in the lower part of the Nekuchan Formation (from oldest to youngest), modified from Domokhotov (1960), Zakharov (1971, 2002), Dagys et al. (1979, 1984, 1986), Dagys and Ermakova (1996), Zakharov et al. (2014) and Biakov et al. (2018): (1) *Otoceras concavum*, (2) *Otoceras boreale*, (3) *Tompophiceras morpheous* and (4) *Wordieoceras decipiens*. The *Tompophiceras pascoei* zone, offered by Dagys and Ermakova (1996), has been rejected by us, because the first representatives of its species–index were collected from the *Otoceras boreale* zone and their later representatives occur in association with *Tompophiceras morpheous* (Popow) in the Nikolkin Klyuch and Seryogin Creek sections, which are characterised by the most abundant mollusc fossils in the region. However, following epibole zones are offered additionally for the lower and upper parts of the *Tompophiceras morpheous* zone in the Nikolkin Klyuch section: *Tompophiceras pascoei* and *Tompophiceras morpheous*, respectively. On the right bank of the Setorym River (e.g. Levyi Suol, Pravyi Suol-1 and Ustupny), representatives of the genus *Tompophiceras* were reliably found only within members 7–9, corresponding to the *Tompophiceras morpheous* zone.

Additional palaeontological data from the P–T boundary beds in our investigated sections from the Setorym River basin are given below.

10.3.1 Nikolkin Klyuch

The Nikolkin Klyuch section (Fig. 10.2) is located on the left bank of the Setorym River, ~19 km upstream from its mouth. This section is accessible by car, as it is only 0.6 km SE of the road, but it is located on a very steep slope in the ravine. Nikolkin Klyuch is one of the most complete P–T boundary sections in Siberia, containing members 1–11 of the Imtachan (upper part) and Nekuchan (lower part) formations (see lithological column in Fig. 10.3). Otoceratid, dzhulfitid and some other ammonoids described by Zakharov (1971, 2002, 2003) are from this section. The recent revision of some ammonoid species from this section and the additional collection of fossil material in 2016 permit clarification and revision of the fauna.

No fossils were found in the upper part of member 1 of the Imtachan formation or in member 2 of the Nekuchan Formation represented by sandstone; otoceratid ammonoids in submember 3a of member 3 of the Nekuchan Formation (4.2 m thick), belonging to the *Otoceras concavum* zone, are quite rare. Samples collected on the left slope of the Nikolkin Klyuch ravine, 1.0 m above the base of member 3, contain *Otoceras* cf. *gracile* Tozer (Fig. 10.4d–h; Plate 10.1 (1, 2)). Some specimens of this species, found in the member 3a (= member 26; Zakharov et al. 2014), were previously erroneously determined as *Otoceras boreale* Spath (morpha A; Zakharov 2002).

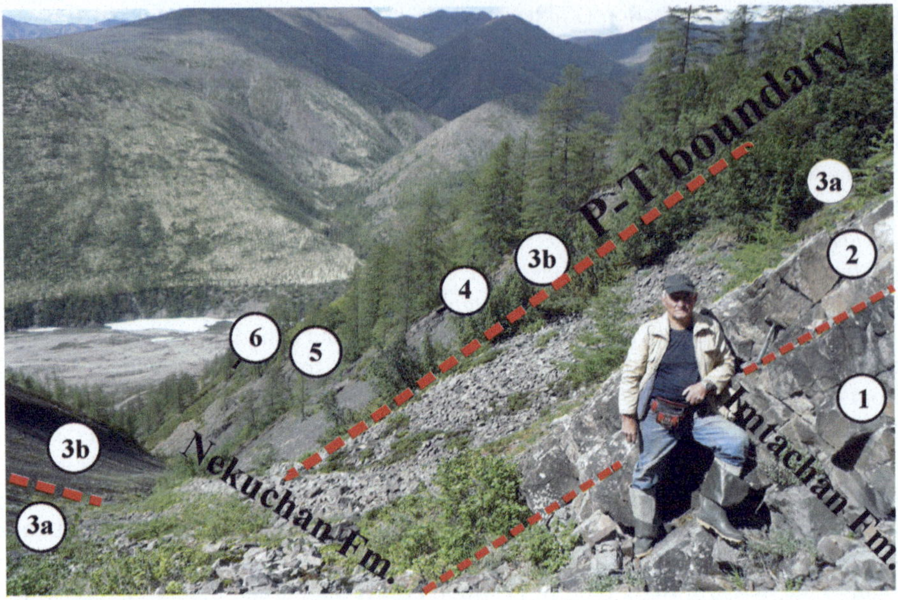

Fig. 10.2 Outcrop photograph showing the location of different units in the Imtachan (uppermost part) and Nekuchan formations (Nikolkin Klyuch section)

Three samples of the *Otoceras concavum* Tozer were found in submember 3a (left slope of the ravine), ~4.0 m above the base of member 3 and from talus (Fig. 10.4a, b; Plates 10.1 (3, 4) and 10.2 (1)). Some specimens of this species, found in the submember 3a (=26; Zakharov et al. 2014), were also previously erroneously determined as *Otoceras boreale* Spath (morpha B; Zakharov 2002).

Submember 3b of member 3, corresponding to the lower part of the *Otoceras boreale* zone, is characterised by both the *O. boreale* morpha A and *O. boreale* morpha B (Figs. 10.4j–n; Plates 10.1 (5) and 10.2 (2)). These are associated with other ammonoid species (e.g. *Aldanoceras* sp.), the nautiloid *Tomponautilus setorymi* Sobolev, the bivalve *Claraia?* sp. and small gastropods and conodonts.

The most abundant collection *Otoceras boreale* originates from 0.3 m-thick layer, situated ~5.2 m above the base of member 3. The conodonts *Hindeoodus typicalis* (Sweet) and *Clarkina* cf. *changhsingensis* Wang et Wang and the earliest representative of the genus *Tompophiceras* (*T.* cf. *pascoei* (Spath)) were discovered in this layer. This 0.3 m-thick layer, which dates to the lower part of the *Otoceras boreale* zone, is proposed as the epibole zone of the same name. A large specimen of *Otoceras boreale* morpha B (Plates 10.2 (5) and 10.2 (2)) was found slightly higher in the sequence, 0.5 m below the top of the member 3.

No fossils have been found in the sandstone belonging to member 4. In overlying mudstone of the lower–middle part of member 5 mollusc remains are extremely rare in the section (only *O. boreale* morpha A has been found somewhat below its top). However, rare ammonoids belonging to *Tompophiceras pacoei* (Spath), *T. morpheous* (Popow) and *Aldanoceras* sp. as well as the bivalve *Claraia?* sp. are common in its upper part, 2–3 m below its top.

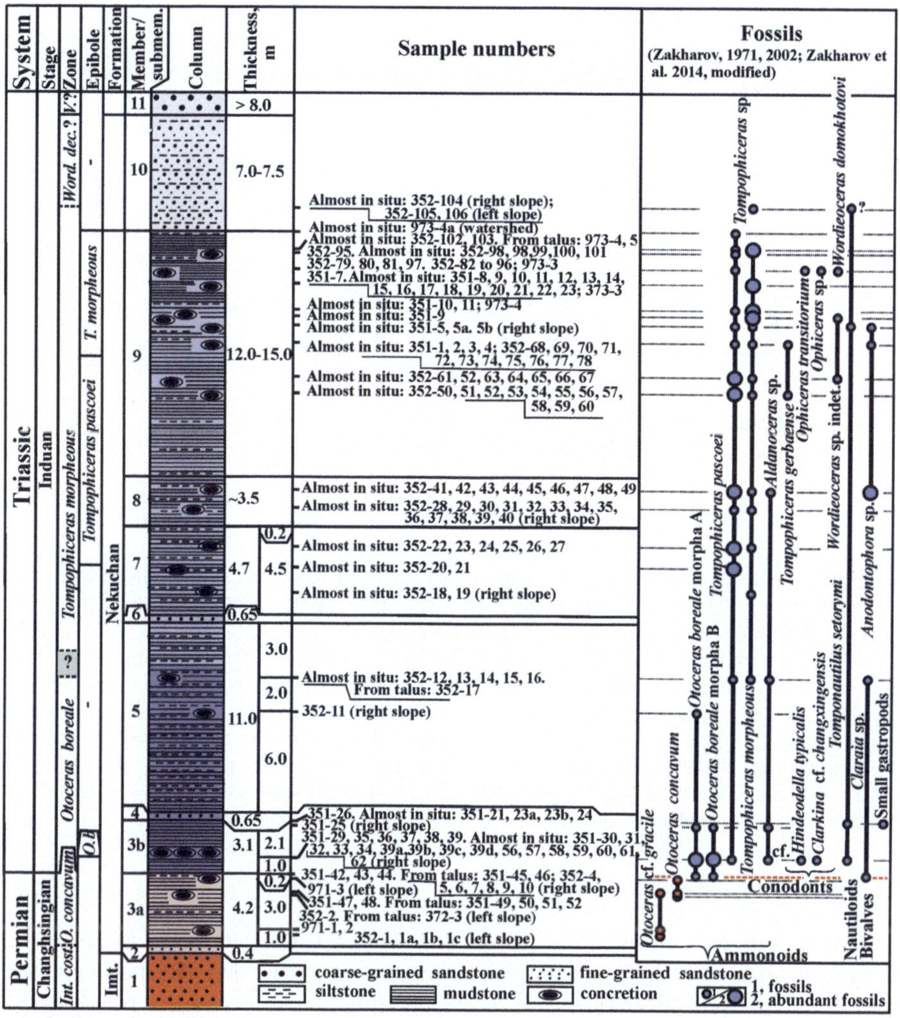

Fig. 10.3 Stratigraphic range chart of important fossils from Permian–Triassic transition of the Nikolkin Klyuch section. Abbreviations: *Int. cost.* Intomodesma costatum, *O. concavum Otoceras concavum*, *Word. dec.? Wordieoceras decipiens*, *V.? Vavilovites sverdrupi*, *O.b. Otoceras boreale* epibole-zone, *Imt* Imtachan, *submem.* submember

No fossils have been found in the sandstone of member 6. Members 7 (upper part), 8 and 9 (lower part) represent a *Tompophiceras pascoei* epibole zone, belonging to the *Tompophiceras morpheous* zone, and are rich in individuals of the ammonoid *Tompophiceras pascoei*, as well as *Tompophiceras morpheous*, *Aldanoceras* sp., *Wordieoceras* sp. indet. and some other fossils.

The upper part of member 9, representing a *Tompophiceras morpheous* epibole zone as a part of the zone of the same name, is rich in *Tompophiceras morpheous*, associated with rarer representatives of *T. pascoei* (Spath), *Wordieoceras domokhotovi* Zakharov, the nautiloid *Tomponautilus setorymi* Sobolev and some other fossils.

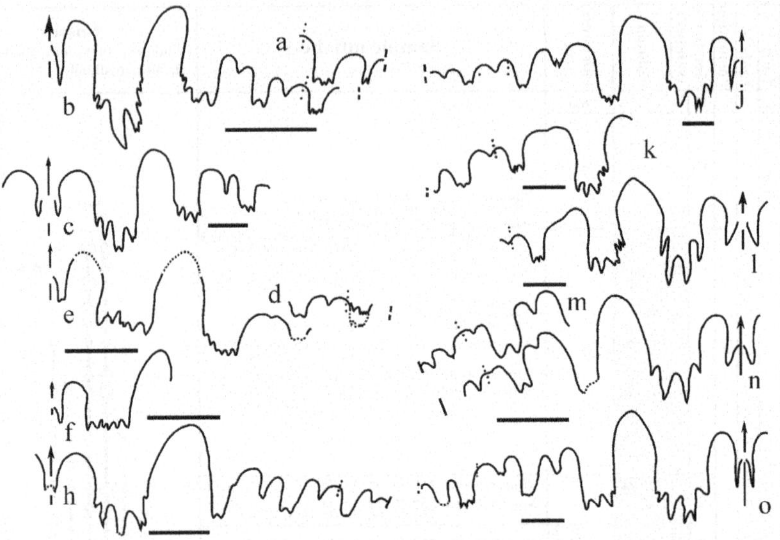

Fig. 10.4 Suture lines of *Otoceras* from the Setorym River basin (a, b), *O. concavum*, FEGI no. 93/803 (field no. 971-30), Nikolkin Klyuch: (a) at $H = 25.5$ mm; (b) at $H = 25.0$ mm; (c) *O. concavum*, DPMGI no. 234/241-1, Ustupny, at $H = 48.0$ mm; (d–h) *O.* cf. *gracile*, Nikolkin Klyuch: (d) FEGI no. 91/803 (field no. 971-2), at $H = 42.5$ mm; (e) FEGI no. 91/803 (field no. 971-2), at $H = 42.3$ mm; (f) FEGI no. 91/803 (field no. 971-2); at $H = 23.9$ mm; (h) FEGI no. 90/803 (field no. 971-1), at $H = 42.2$ mm; (j, l), *O. boreale* morpha B, Nikolkin Klyuch: (j) FEGI no. 31/802 (field no. 351-25), at $H = 76.5$ mm; (k) FEGI no. 80/803 (field no. 351-59), at $H = 56.0$ mm; (l) FEGI no. 80/803 (field no. 351-59), at $H = 51.0$ mm; (m–o) *O. boreale* morpha A: (m) FEGI no. 3/803 (field no. 351-59), at $H = 35.5$ mm, Nikolkin Klyuch; (n) same sample, $H = 35.2$ mm; (o) FEGI no. 151/803 (field no. 970-1), at $H = 58.1$ mm, Levyi Suol. Scale bar 5 mm

The lower part of member 10 is characterised by very rare fossils (e.g. *Tompophiceras* sp.). No fossils were discovered in overlying deposits in this section (lower–middle part of member 10), corresponding apparently to the *Wordioceras decipiens* zone, known for the Lekeer Creek area, Tompo River basin (Dagys and Ermakova 1996).

10.3.2 Inessin Klyuch

The Inessin Klyuch section, located in a neighbouring ravine, 0.4 km upstream from the Nikolkin Klyuch section, is similarly exposed and has comparable member thicknesses. Limited time was spent collecting fossils from this newly discovered section. Therefore, only *Otoceras boreale* Spath was discovered in the upper part of member 3, corresponding to the lower part of the *Otoceras boreale* zone (R.V. Kutygin's coll.). *Tompophiceras pascoei*, collected from the uppermost part of member 9, cor-

Plate 10.1 (1, 2) *Otoceras* cf. *gracile* Tozer: (1) FEGI no. 90/803 (field no. 971-1); (2) DVGI no. 91/803 (field. no. 971-2); Nikolkin Klyuch; Nekuchan Formation, submember 3a, *Otoceras concavum* zone. (3, 4) *O. concavum* Tozer: (3) FEGI no. 92/803 (field no. 351–49); (4) 93/803 (field no. 971-3); Nikolkin Klyuch; Nekuchan Formation, submember 3a, *Otoceras concavum* zone; (5) *Otoceras boreale* Spath, morpha B, FEGI no. 31/803 (field no. 351-25); Nikolkin Klyuch; Nekuchan Formation, member 3b, *Otoceras boreale* zone

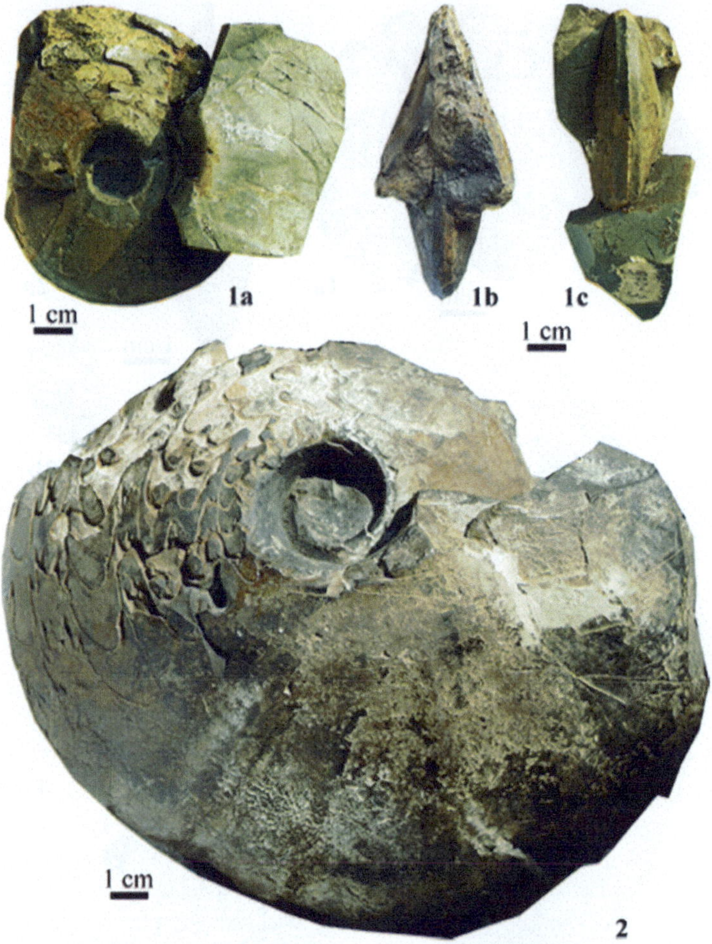

Plate 10.2 (1) *Otoceras concavum* Tozer, FEGI no. 93/803 (field no. 971-3); Nikolkin Klyuch; Nekuchan Formation, submember 3a, *Otoceras concavum* zone. (2) *Otoceras boreale* Spath, morpha B, FEGI no. 31/803 (field no. 351-25); Nikolkin Klyuch; Nekuchan Formation, submember 3b, *Otoceras boreale* zone

responding to the *Tompophiceras morpheous* zone, has been discovered at the Nikolkin Klyuch–Inessin Klyuch watershed (A.M. Popov's coll.).

10.3.3 Levyi Suol and Pravyi Suol-1

The mouth of the Suol Creek is located on the right bank of the Setorym River, 6 km upstream from the mouth of the latter. There are six studied sections recording the P–T boundary transition in the Suol Creek basin: Levyi Suol, Pravyi Suol-1, Pravyi

Suol-2, Pravyi Suol-3 and Nizhny Suol (Fig. 10.1). The most complete and representative sections are Levyi Suol and Pravyi Suol-1 sections.

The Levyi Suol section, recently exposed by a powerful storm event, is located 0.2 km upstream from the Levyi Suol–Pravyi Suol confluence. Pravyi Suol-1 is located just 100 m west of the Levyi Suol section, and so it might be reasonable to consider these as a single section (Fig. 10.5). The Pravyi Suol-1 and possibly Nizhny Suol sections were earlier examined during the excursion organised within the framework of the 27th International Geological Congress in 1984 (Dagys et al. 1984). The majority of the ammonoids described by Dagys and Ermakova (1996) are from the Pravyi Suol-1 section. The Levyi Suol section is undoubtedly the better

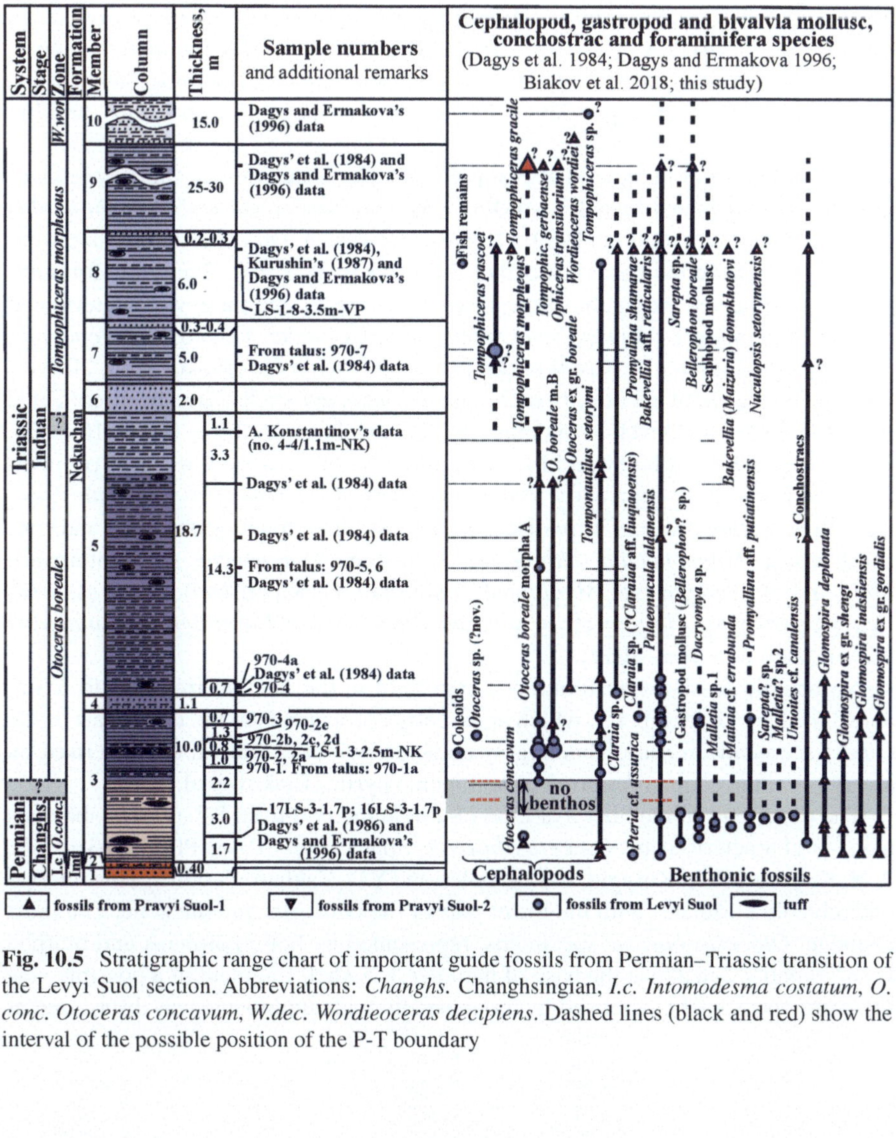

Fig. 10.5 Stratigraphic range chart of important guide fossils from Permian–Triassic transition of the Levyi Suol section. Abbreviations: *Changhs.* Changhsingian, *I.c. Intomodesma costatum*, *O. conc. Otoceras concavum*, *W.dec. Wordieoceras decipiens*. Dashed lines (black and red) show the interval of the possible position of the P-T boundary

exposed of these now, but it was not suitable for detailed research at the time of this earlier excursion.

Attention is drawn to the fact that the lithologic members of the Nekuchan Formation produced in these sections attain a slightly larger thickness than the corresponding members at Nikolkin Klyuch. The lithological description of Pravyi Suol-1 and the stratigraphical distribution of its mollusc and foraminifera fossils, as well as its C-isotope values, have been previously published (Dagys et al. 1984; Kurushin 1987; Dagys and Ermakova 1996; Zakharov et al. 2014, 2015; Biakov et al. 2018).

Member 1 (*Intomodesma postevenicum* subzone of the *Intomodesma costatum* zone) of the Imtachan formation, represented mainly by sandstone, is better exposed in the Pravyi Suol-1 section, 20 m below its top. Sandstone of this subzone yields latest remains of the characteristic typical high-boreal fauna, strongly dominated by *Inoceramus*-like bivalves: *Intomodesma postevenicum* Biakov, *Intomodesma* sp. and *Maitaia* sp. (Biakov et al. 2016, 2018). No fossils have been found in the uppermost 20 m part of member 1, nor in 0.4 m thick member 2 (the basal layer of the Nekuchan Formation represented by sandstone yielding numerous argillite intraclasts).

The lower part of member 3 at Pravyi Suol-1 yields individuals *Otoceras concavum* Tozer (Dagys and Ermakova 1996) which are associated with *Tomponautilus setorymi* Sobolev (Sobolev 1989). Representatives of *Otoceras concavum* were found 1.2 m above the base of member 3 (Dagys et al. 1986). Benthonic foraminifera were collected 0.6 m above the base of member 3, and these were determined by A.V. Yadrenkin as *Ammodiscus septentrionalis* Gerke, *Glomospira deplanata* Kasatkina, *G. indskiensis* Kasatkina and *G.* ex gr. *shengi* Ho (Biakov et al. 2018).

In the Levyi Suol section, two *Otoceras concavum* shells have been collected from the following level of member 3: 1.7 m above its base (Plate 10.3 (2); R.V. Kutygin, E.S. Sobolev and I.L. Vedernikov's coll.). In the lower part of member 3, corresponding apparently to the lower part of the *Otoceras concavum* zone, the following rare bivalves were discovered in association with gastropods and conchostracans: *Palaeonucula aldanensis* Kurushin, *Dacryomya* sp. (dominant), *Malletia?* sp., *Sarepta?* sp., *Myalina* aff. *putiatinensis* (Kiparisova), *Pteria* cf. *ussurica* (Kiparisova), *Maitaia* cf. *errabunda* (Popow) and *Unionites* cf. *canalensis* (Catullo) (Biakov et al. 2018).

No benthic fossils were found at Levyi Suol in the interval from 2.8 to 5.5 m above the base of member 3, which apparently corresponds to the upper part of the *Otoceras concavum* zone. This largely coincides with the interval characterised by a very pronounced manifestation of authigenic pyrite (Biakov et al. 2018).

The upper part of member 3 at Levyi Suol, in interval from 5.7 to 10 m above its base, is characterised by *Otoceras boreale* Spath (Fig. 10.4o; Plate 10.3 (3, 4); A.N. Kilyasov, R.A. Kutygin, E.S. Sobolev and Y.D. Zakharov's coll.). This interval undoubtedly correlates with the lower part of the *Otoceras boreale* zone. The most abundant *Otoceras boreale* specimens, represented by both morpha A and morpha B, are found 7.2 m above the base of member 3. A shell fragment of a possible new *Otoceras* species, characterised by an unusually wide umbilicus, was discovered at

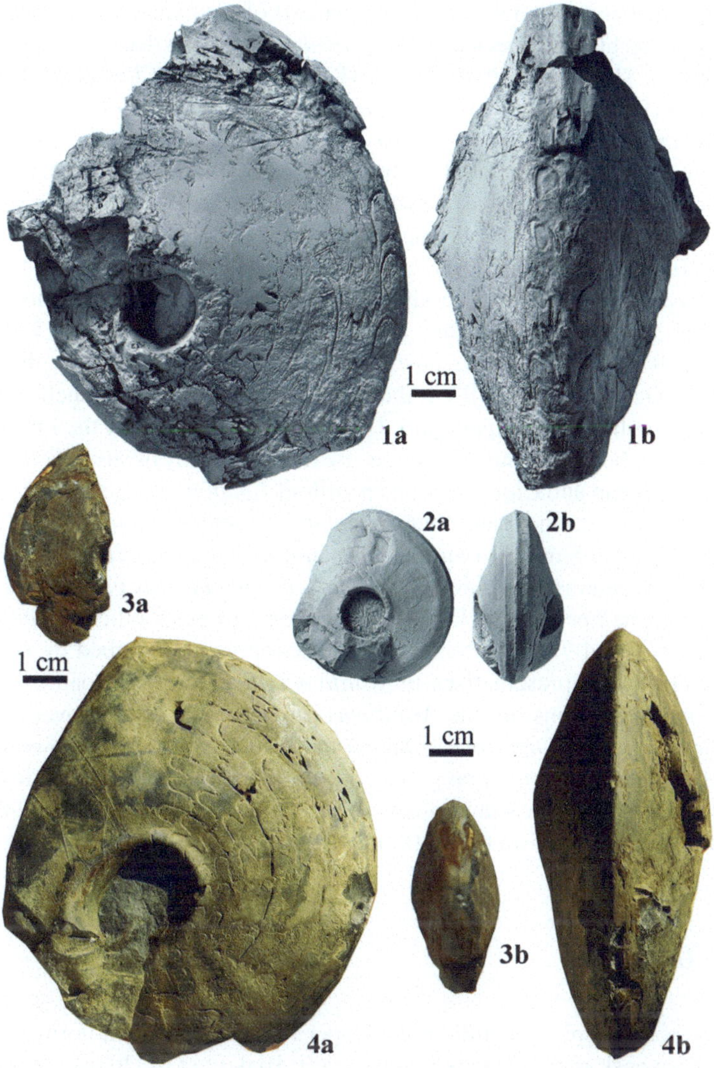

Plate 10.3 (1, 2) *Otoceras concavum* Tozer: (1) DPMGI no. 234/241-1 (field no. 17R1-711.1p); Ustupny; Nekuchan Formation, member 3 (lower part), *Otoceras concavum* zone; (2) DPMGI no. 234/209 (field no. 17LS-3-1.7p+); Levyi Suol; Nekuchan Formation, member 3 (lower part), *Otoceras concavum* zone; (3, 4) *Otoceras boreale* Spath, morpha A: (3) FEGI no. 108/803 (field no. 970-1a); Levyi Suol; Nekuchan Formation, member 3 (upper part), *Otoceras boreale* zone; (4) FEGI no. 105/803 (field no. 970-1); Levyi Suol; Nekuchan Formation, member 3 (upper part), *Otoceras boreale* zone

8.3 m above the base of member 3. The upper part of the member 3 is characterised by the following bivalve species: *Palaeonucula aldanensis* Kurushin (dominant), *Dacryomya* sp., *Myalina* aff. *putiatinensis* (Kiparisova) and ?*Claraia* aff. *linqiaoensis* (He et al.) (Biakov et al. 2018), *Tomponautilus setorymi* Sobolev and more rare coleoids (phragmocone fragment) occur also from this level. In the neighbouring Pravyi Suol-1 section this interval is characterised by the following benthonic fossils: foraminifera *Ammodiscus septentrionalis* Gerke, *Glomospira deplanata* Kasatkina, *G. indskiensis* Kasatkina, *G.* ex gr. *shengi* Ho and *Glomospira* ex gr. *gardialis* (Jones et Parker) (A.V. Yadrenkin's determination; Suol-1 section, Setorym River basin, South Verkhoyansk Biakov et al. 2018) and bivalves *Palaeonucula aldanensis* Kurushin and *Myalina* aff. *putiatinensis* (Kiparisova) (Kurushin 1987). Rare *Otoceras boreale* was collected 0.2 m above the base of member 5 (Y.D. Zakharov's coll.) in the middle part of member 5 (shell fragments; A.M. Popov's coll.) and its upper part. Other fossils, collected in the lower part of the member 5 in the Pravyi Suol-1 section, are represented by bivalves *Palaeonucula aldanensis* Kurushin and *Claraia* sp. and nautiloid *Tomponautilus setorymi* Sobolev, in the Levyi Suol section foraminifera *Ammodiscus septentrionalis* Gerke (Biakov et al. 2018). Member 8 in Levyi Suol yields nautiloid *Tomponautilus setorymi*, associated with fish remains (Sobolev 1989). A few *Tompophiceras pascoei* shells were found in upper members of the Levyi Suol section, but many other *Tompophiceras* from members 8 and 9 of the Pravyi Suol-1 section were described by Dagys and Ermakova (1996). Representatives of *Wordieoceras decipiens* (Spath) have been found in many sections in the *Wordieoceras decipiens* zone, overlaying the *Tompophiceras morpheous* zone (Dagys and Ermakova 1996). However, their stratigraphical position in the Pravyi Suol-1 section has not been strictly determined by these taxa. *Otoceras*-bearing sequences have also been discovered in other sections in the Suol Creek basin (Fig. 10.1).

10.3.4 Ustupny

This is the most complete section of the Upper Permian Imtachan Formation in the South Verkhoyansk region (Domokhotov 1960; Biakov et al. 2016). According to R.V. Kutygin's data, the overlying Nekuchan Formation is also exposed in the watershed part of this section. The lower part of the Ustupny section is similar to the sequences at Pravyi Suol-1 and Levyi Suol. Ammonoid *Otoceras concavum* Tozer (Fig. 10.4c, Plate 10.3 (1); R.V. Kutygin and A.N. Kilyasov's coll.) and nautiloid *Tomponautilus setorymi* Sobolev were found 1.1 m above the base of member 3 of the Nekuchan Formation. *Otoceras boreale* was found at the top of this member, also in association with *Tomponautilus setorymi*. The upper part of the section is characterised by ammonoids of the genus *Tompophiceras*.

10.3.5 Seryogin Creek-1 and -2

The lower part of the member 3 of the Nekuchan Formation, apparently corresponding to the upper Changhsingian *Otoceras concavum* zone, has been identified in locality 339 (Seryogin Creek-1), ca 840 m up from the mouth of Seryogin Creek (Figs. 10.1 and 10.6B). They lie above the dark, poorly sorted fine-grained sandstones that comprise the basal bed (member 2) of the Nekuchan Formation.

The upper part of member 3, which belong to the lower part of the *Otoceras boreale* zone, is exposed at locality 338 (Seryogin Creek-2; Zakharov 1971), situated in the upper part of the southern hillside ca 600 m up from the mouth of Seryogin Creek (Figs. 10.1 and 10.6A). This member is composed of mudstone with numerous calcareous-siliceous concretions that contain *Otoceras boreale* morpha A and *O. boreale* morpha B. Despite the fact that most of the fossils collected from this section were not found in situ, the position of the slope leads us to suggest that the first representatives of *Tompophiceras pascoei* occur from this section in the upper part of member 3, corresponding to the lower part of the *Otoceras boreale* zone. The upper members of the section, composed mainly of mudstone and siltstone, are characterised by *Tompophiceras morpheous* and *Aldanoceras* sp. However, member 10 (interbedding medium-grained and banded fine-grained sandstones), exposed in the lower part of the slope, is conventionally assigned to the *Wordieoceras decipiens* zone.

10.3.6 Seryogin Creek-3 (Shagali) and Dolgachan

In the Seryogin Creek-3 (Shagali) section, situated 460 m up from the mouth of Seryogin Creek, R.V. Kutygin documented the Imtachan/Nekuchan formational contact and found *Otoceras concavum* 2 m above the base of member 3.

Otoceras-bearing deposits of the Dolgachan section, located along the Dolgachan Creek, 720 m up from the mouth of Seryogin Creek, are poorly exposed. *Tompophiceras* was collected by R.V. Kutygin only in the Dolgachan–Seryogin (Shagali) confluence.

10.3.7 Nitrogen Isotopes

Our isotope results, obtained from the Pravyi Suol-1 section are shown in Fig. 10.7. The minimum $\delta^{15}N$ values within the selected N-isotope intervals ("a"–"e") are −0.9‰, and the maximum values do not exceed +3‰ (Zakharov et al. in press); $\delta^{13}C_{org}$ values fluctuate between −30.3 and −26.3‰ (Zakharov et al. 2014, 2015).

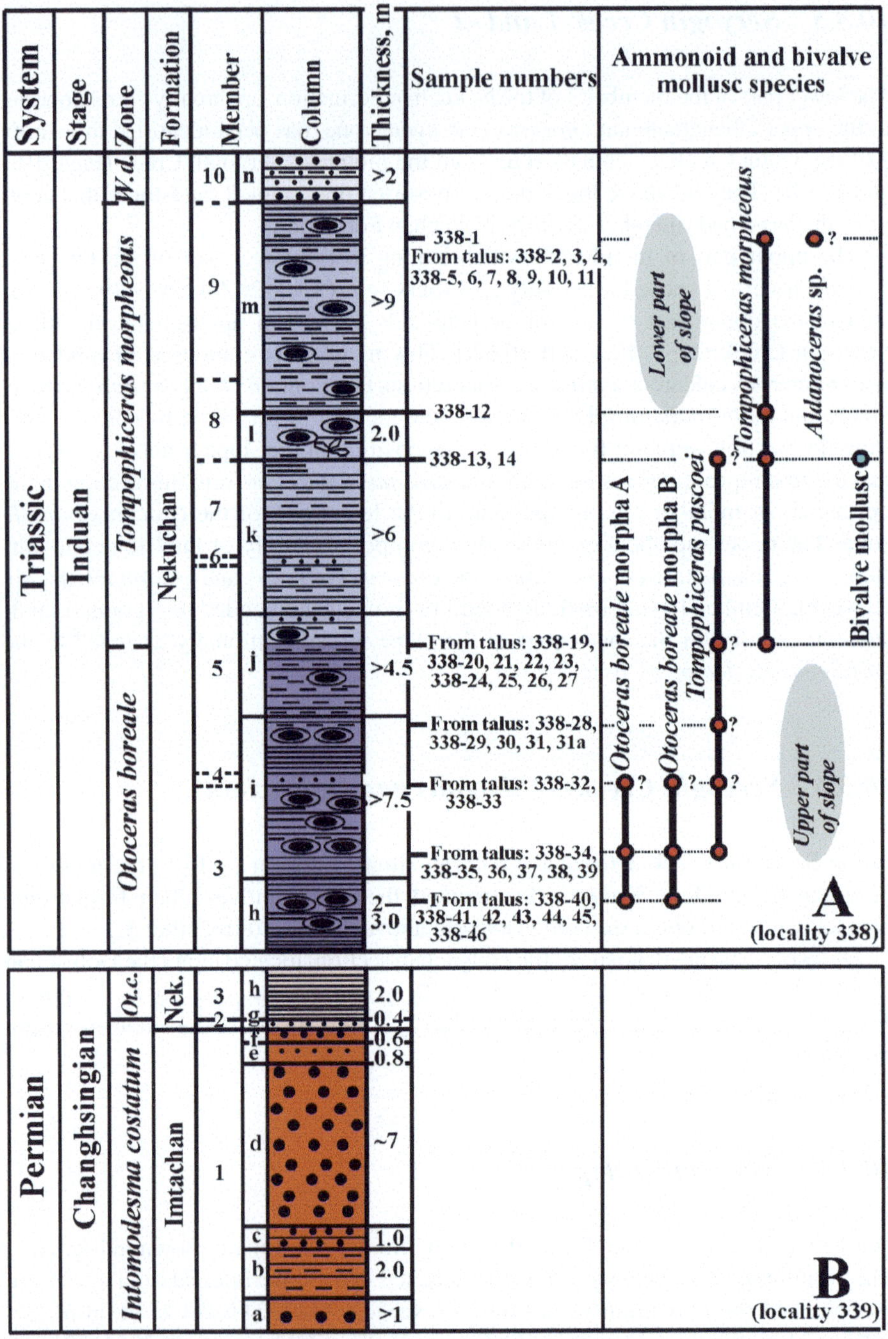

Fig. 10.6 Stratigraphic range chart of important guide fossils from Permian–Triassic transition of the Seryogin Creek-1 and -2 sections. (**B**) Seryogin Creek-1; (**A**) Seryogin Creek-2. Abbreviations: *Ot.c. Otoceras concavum*, *W.d. Wordieoceras decipiens*, Nek. Nekuchan

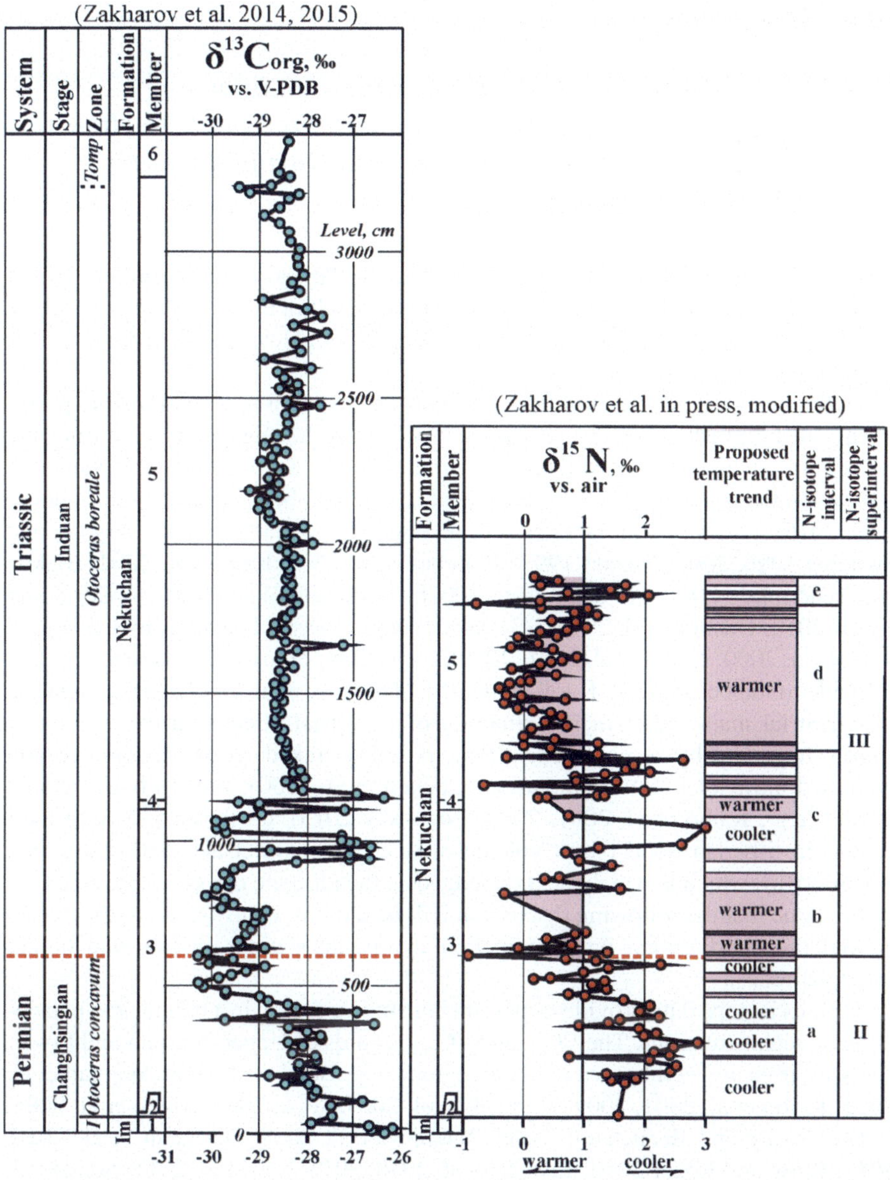

Fig. 10.7 N- and C-isotope composition of Changhsingian and Induan deposits of the Pravyi Suol-1 section, Setorym River basin, South Verkhoyansk area. Abbreviations: *I. Intomodesma costatum*, *Tomp. Tompophiceras morpheous*, Im Imtachan

10.4 Discussion

10.4.1 Reconstruction of Environmental Conditions Influencing Biotic Development in the Middle to High Palaeolatitudes of Eastern Russia During the Late Permian and Early Triassic

It is widely considered that the mass extinction at the end of the Permian was the greatest biotic crisis of the Phanerozoic (fewer than half of the latest Permian marine families survive into the Triassic; Yin and Song 2013). However, the causes of this extinction are still debated.

Of the many triggers proposed for end-Permian extinction, several feature prominently in multiple extinction scenarios: (1) lethally hot temperatures during the latest Permian and into Early Triassic greenhouse (e.g. Knoll et al. 2007; Sun et al. 2012; Goudemand et al. 2013; Romano et al. 2013; Schobben et al. 2014; Grigoryan et al. 2015); (2) anoxia in the oceans (e.g. Wignall and Hallam 1992; Wignall and Twitchet 1996, 2002; Isozaki 1997; Twitchet et al. 2001; Kato et al. 2002; Biakov and Vedernikov 2007; Horacek et al. 2007b; Bond and Wignall 2010; Korte and Kozur 2010; Dustira et al. 2013); (3) catastrophic release of seafloor methane (e.g. Krull et al. 2000; Kaiho et al. 2001, 2009); (4) loading of toxic metals (e.g. Hg) and sulphide to the oceans (e.g. Knoll et al. 2007; Hammer et al. 2019); (5) denudation of terrestrial matter after mass extinction of terrestrial plants and devastation of forests (e.g. Grasby et al. 2013, 2018), possibly linked to ozone damage and increased ultraviolet (UV-B) radiation (Cockell et al. 2000); and (6) bolide/comet impact (e.g., Kaiho et al. 2001, 2009; Lozovsky 2013). Of these, 1–5 have their origins in Siberian flood basalt volcanism (e.g. Hermann et al. 2010; Korte and Kozur 2010), which is widely acknowledged as the ultimate driver of this crisis.

We hope that the following discussion will be a useful contribution to the debate around the end-Permian mass extinction in marine settings in low and higher palaeolatitudes.

Various isotope data provide important constraints for environmental and palaeoclimatic reconstructions. The $^{13}C_{org}$ and $^{13}C_{carb}$ records of Upper Permian and Lower Triassic strata are well known in many regions (e.g. Baud et al. 1989; Wignall et al. 2015; Kaiho et al. 2001, 2009; Horacek et al. 2007a, b, c, 2009; Algeo et al. 2008, 2014; Grasby and Beauchamp 2008; Nakrem et al. 2008; Hermann et al. 2010, 2011; Korte and Kozur 2010; Takashi et al. 2010, 2013; Luo et al. 2011; Song et al. 2013; Zakharov et al. 2014). In Korte and Kozur's (2010) opinion, the Permo–Triassic negative C-isotope excursions, including the major negative $\delta^{13}C$ shift seen globally at the end of the Permian, most likely had their origin in (1) direct and indirect effects of the Siberian Traps and contemporaneous volcanism and (2) the effects of anoxic deepwaters occasionally reaching very shallow marine settings. Short-term events, such as the release of isotopically light methane from the ocean, are questionable as causes for the C-isotope excursions (Korte and Kozur 2010). However, some data, including geochemical (Ce/Ce*) results, argue against global anoxia as a cause for the end-Permian mass extinction (Krull et al. 2000; Kozur 2007; Garbelli et al. 2014).

Despite much progress globally, the O- and N-isotope records for the Upper Permian and the Lower Triassic are still poorly known for many regions, including in Siberia and the Russian Far East. Oxygen isotope values, calculating useful proxy for Late Permian and Early Triassic palaeotemperatures, have so far been calculated from the following few materials: (1) calcite of Wuchiapingian and Changhsingian brachiopods from the Caucasus (Zakharov et al. 1999, 2005), Iran (Schobben et al. 2014) and the Kolyma–Omolon region (Zakharov et al. 2005); (2) apatite of Wuchiapingian and Changhsingian conodonts from South China (Sun et al. 2012; Goudemand et al. 2013; Romano et al. 2013), Iran (Schobben et al. 2014) and Transcaucasia (Grigoryan et al. 2015); and (3) aragonite of Olenekian cephalopods from Arctic Siberia (Zakharov et al. 1975, 1999).

$^{15}N_{org}$ records from the Upper Permian and Lower Triassic are restricted to data from South China (e.g. Luo et al. 2011; Jia et al. 2012; Yin et al. 2012; Algeo et al. 2014; Saitoh et al. 2014), Arctic Canada (Algeo et al. 2012; Knies et al. 2013; Grasby et al. 2015, 2016) and the South Primorye (Zakharov et al. 2018a, b, c) and Kolyma–Omolon (Zakharov et al. 2019) regions of Russia. Other recently published isotope data include sulphur isotope profiles from the P–T boundary transition in Japan and Australia (e.g. Horacek et al. 2010; Takashi et al. 2013).

Stable O-isotope analyses of Late Permian and Early Triassic well-preserved conodont and brachiopod fossils from palaeoequatorial regions and adjacent areas suggest water palaeotemperatures of 20–38 °C (Zakharov et al. 1999, 2005; Schobben et al. 2014; Grigoryan et al. 2015; Joachimsky et al., 2012, 2020) and 28–40 °C and higher (Sun et al. 2012; Goudemand et al. 2013; Romano et al. 2013), respectively. However, there is some difficulty in explaining the low $\delta^{18}O$ values in well-preserved aragonite cephalopod shells from the Boreal Superrealm (Arctic Siberia), which also suggest very high water temperatures assuming that they retained normal salinity (Zakharov et al. 1975, 1999).

Nitrogen is a key nutrient necessary for the existence of living organisms. It is a major component of biomass and is required for photosynthesis (Robinson et al. 2012). The majority of the molecular nitrogen is fixed from air by certain types of cyanobacteria (Bauersachs et al. 2009). As a result, ammonium compounds (NH_4+) and ammonia (NH_3) are usually formed. These products are well assimilated by plants, which in turn are used as food resources by animals.

Of greater interest in the light of our research are data on the N-isotopic composition of marine sediments deposited in both greenhouse and icehouse conditions (e.g. Altabet et al. 1995; Jenkyns et al. 2001; Algeo et al. 2008, 2014). Algeo et al. (2014) indicated that Neoproterozoic and Phanerozoic greenhouse climate modes were characterised by low $\delta^{15}N_{sed}$ values (fluctuating from ~−2 to +2‰), whereas icehouse climate modes showed higher $\delta^{15}N$ values (ranging from ~+4 to +8‰; Fig. 10.8a). A general trend of higher $\delta^{15}N_{sed}$ values during times of cool climate and low $\delta^{15}N_{sed}$values during warm periods is explained apparently by several processes, including (1) the increased microbiological denitrification in cooler conditions, when nitrates (NO_3^-) are reduced to nitrite (NO_2^-) and further to nitrogen gas (N_2)

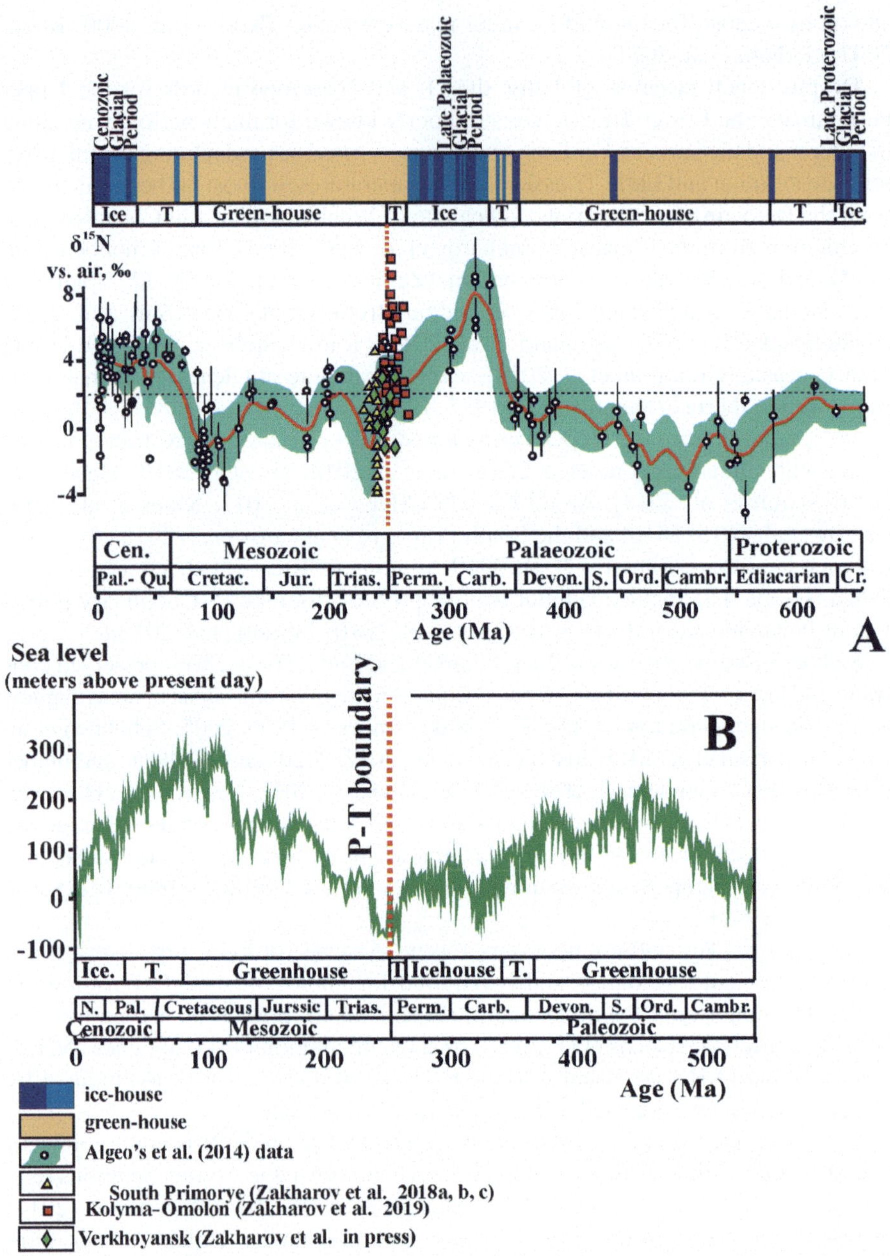

Fig. 10.8 (**a**) Long-term secular variation in marine sediment $\delta^{15}N$ values, based on study units of Proterozoic–Cenozoic age (Algeo et al. 2014) with an addition (this study); (**b**) data on sea-level fluctuation (Boulila et al. 2018). Abbreviations: *Ice* Ice house, *T* Transition, *Pal.–Qu.* Palaeogene–Quaternary, *Cretac.* Cretaceous, *Jur.* Jurassic, *Trias.* Triassic, *Carb.* Carboniferous, *Devon.* Devonian, *S.* Silurian, *Ord.* Ordovician, *Cambr.* Cambrian, *Cr.* Cryogenian

which is returned to the atmosphere, and (2) the increased nitrogen fixation from the air, realised by cyanobacteria in warmer conditions (Luo et al. 2011; Algeo et al. 2014).

The general correlation between $\delta^{18}O$-palaeotemperature and $\delta^{15}N$ curves, as shown for the Kamenushka-1, Kamenushka-2, Pautovaya and Pravyi Suol-1 sections (Zakharov et al. 2018a, b, 2019), supports the idea described in Zakharov et al. (2018c) that intervals with high $\delta^{15}N$ values in the P–T sections of Russia mainly reflect the presence of lower temperature conditions in comparison with intervals with low $\delta^{15}N$ values. This interpretation stimulates our attempts to distinguish both N-isotope superintervals in Permian and Triassic of the eastern part of Russia, as well as their subdivisions (N-isotope intervals), which may indicate climatic and particular temperature variations (Fig. 10.9). An example of the first of these are

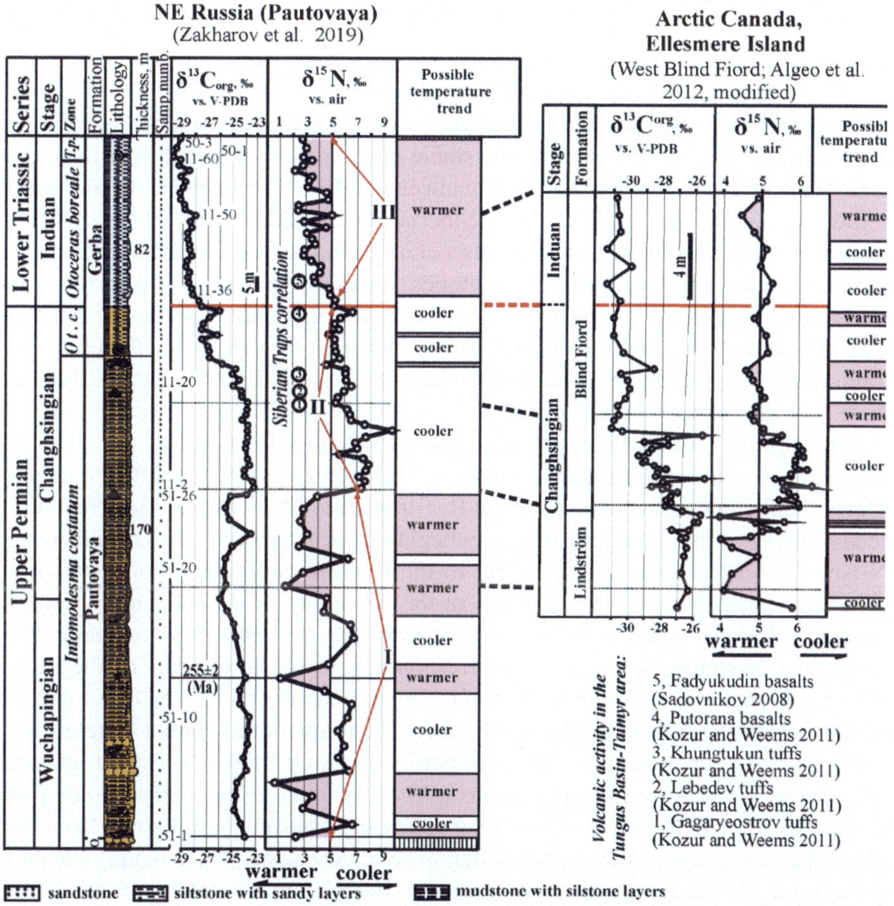

Fig. 10.9 Correlation of C- and N-isotope records from the Permian–Triassic Pautovaya section (Kolyma–Omolon region) with composite record from West Blind Fiord (Arctic Canada). Abbreviations: *Ot.c. Otoceras concavum*, *T. Tompophiceras*, *O* Ovod

N-isotope superintervals I–IX, distinguished in the Wuchiapingian, Changhsingian and Induan of the northern regions of eastern Russia and in the Lower and Middle Triassic of the southern regions (Zakharov et al. 2018a, b, c, 2019). An example of the shorter ones (N-isotope intervals) is "a"–"e", recognised in the P–T boundary transition in the South Verkhoyansk region (Zakharov et al. in press). However, in this discussion we are going to concentrate our attention only on N-isotope superintervals I–III (Fig. 10.9) and on N-isotope intervals "a"–"e" (Fig. 10.7).

By interpreting the N-isotope data to be at least in part a function of temperature, our results from the Kolyma-Omolon region suggest oscillating but generally warm temperature conditions prevailed in the late Wuchiapingian and early Changhsingian (N-isotope superinterval I; Fig. 10.9), as well as in the Sverdrup Basin, Arctic Canada (Algeo et al. 2012). In the middle Changhsingian rather cooler conditions developed (apparently just before the first volcanic activity in the Siberian Traps) before a gradual shift towards warmer and less variable climatic conditions in the late Changhsingian (the upper part of the N-isotope superinterval II; Fig. 10.9). The contemporaneous Gagaryeostrov, Lebedev and Khungtukan tuff accumulations (Kozur and Weems 2011) in northern Siberia might have driven some of this warming trend but apparently did not cause major climatic and environmental instability. This interpretation is supported by the stable development of gymnosperms at the same time in what is now Norway (Hermann et al. 2010; Hochuli et al. 2010) as well as the high taxonomic biodiversity of Siberian flora (Sadovnikov 2008, 2016). A similar climatic development apparently occurred in the Sverdrup Basin, Canada, according to the similarity of the N-isotope data from Pautovaya (Zakharov et al. 2019) and West Blind Fiord (Algeo et al. 2012). A further shift towards warmer and generally quite consistent conditions in Northeast Russia (Kolyma–Omolon and South Verkhoyansk regions) and Arctic Canada is inferred for the earliest Induan (N-isotope superinterval III; Fig. 10.9).

N-isotope data for the Pravyi Suol-1 section (N-isotope intervals "a"–"e", located within the superintervals II (upper part) and III; Fig. 10.7) provide a detailed record of proposed temperature changes in the P–T transition in the Boreal Superrealm. A steady shift towards warmer temperatures beginning near the end of the Changhsigian (N-isotope interval "a", corresponding to the *Otoceras concavum* zone) was followed by markedly warmer conditions at the beginning of the Induan (N-isotope interval "b", corresponding to the lowest part of the *Otoceras boreale* zone). The subsequent N-isotope interval "c", corresponding to the lower part of the same *Otoceras boreale* zone (the upper part of member 3, member 4 and the lower part of member 5) might have been characterised by the frequent alternation of relatively cooler and warmer conditions. This interval is followed by the N-isotope interval "d", characterised by warmer conditions (*Otoceras boreale* zone—the middle part of member 5). The N-isotope data of the following brief interval, known from the middle part of member 5 (N-isotope interval "e"), corresponding to the middle part of the same *Otoceras boreale* zone, seems to once again be characterised by frequent alternations between relatively cooler and warmer conditions.

It is worth noting that comparisons of N-isotope data with data on eustatic sea-level fluctuations in the Phanerozoic time (Fig. 10.8a, b) (e.g. Algeo et al. 2014; Boulila et al. 2018) show a good agreement, this serving as additional evidence for temperature being the dominant driver of both N-isotope variations and sea-level fluctuations..

As mentioned above, previous studies support extreme warmth in equatorial and southern moderate palaeolatitudes for some periods during the Early Triassic (e.g. Schobben et al. 2014). However, the absence of $\delta^{15}N$ values below −0.9‰ in the P–T sections of the Boreal Superrealm (e.g. Verkhoyansk, Kolyma–Omolon and Arctic Canada; Algeo et al. 2012; Zakharov et al. 2019; this study—Fig. 10.8a) in contrast to sections from more southernly palaeolatitudes of Russia (South Primorye—Abrek, Kamenushka-1 and Kamenushka-2; Zakharov et al. 2018a, b) might be explained by the clearly cooler environments of deposition of the former locations at higher palaeolatitudes.

The warming at the very end of the Changhsingian and the beginning of the Induan could have been caused by the latest Changhsingian Putorana basalt event, which occurred during the main phase of Siberian Traps eruptions. It is assumed (e.g. Kozur and Weems 2011) that this activity injected large volumes of greenhouse gases, including CO_2 and CH_4 to the atmosphere, resulting in the Late Permian mass extinction which is documented in many regions of the Tethys (e.g. Baud et al. 1989; Kaiho et al. 2009. This event has been correlated with marked changes in environment in Norway: there is an evidence of extinction of one of the large groups of gymnosperms (Glossopteridales; Hermann et al. 2010) at the P–T boundary there.

Algeo et al. (2012) recognised two P−T marine environmental and biotic crises in Arctic Canada. The contact between the chert of the Lindström formation and the silty shale of the overlying Blind Fiord formation in the Sverdrup Basin records, in their opinion, a regional extinction of siliceous sponges (first Arctic extinction event; Algeo et al. 2012). In the South Verkhoyansk region this event is linked with the disappearance of typical high-boreal fauna (*Inoceramus*-like bivalves), latest remains of which were found 20 m below the top of the Upper Permian Imtachan formation (Biakov et al. 2018).

The second P−T crisis in Arctic Canada is associated with intensified anoxia and a possible change in phytoplankton community composition (Algeo et al. 2012). In South Verkhoyansk, the interval of 2.8–5.5 m above the base of the uppermost Changhsingian–Induan Nekuchan Formation in the Levyi Suol and Pravyi Suol-1 sections is characterised by no benthonic fossils, which were seen in adjacent intervals (Biakov et al. 2018), and the pronounced manifestation of authigenic pyrite, potentially indicating anoxic or euxinic conditions, by analogy with P−T boundary sections in Svalbard (Bond and Wignall 2010). We suggest that this interval corresponds to the second episode of the marine environmental and biotic crisis in Arctic Canada (Algeo et al. 2012) and latest Permian mass extinction event, documented in Tethyan shallow-marine sections (e.g. Yin and Zhang 1996; Burgess et al. 2014; Brosse et al. 2016). In contrast to the first episode of the Late Permian marine environmental crises, the second phase in the Verkhoyansk region did not cause major biotic changes (e.g. Biakov et al. 2018). This assumption, made mainly on the basis of data on benthonic fossils (bivalve molluscs and foraminifera), is consistent with the data on semi-pelagic fossils—ammonoids of the genus *Otoceras* and nautiloid of the species *Tomponautilus setorymi* Sobolev.

10.4.2 The Evolutionary Development and Geographical Differentiation of Otoceratid and Dzhulfitid Ammonoids

Only the following five known ammonoid succession lineages at different levels survived the end-Permian mass extinction (Fig. 10.10): (1) *Paratirolites–Tompophiceras* (at the family level—Dzhulfitidae; Arkhipov 1974; Zakharov and Moussavi Abnavi 2013; Zakharov and Popov 2014); (2) *Hypophiceras triviale* (Spath)–*Hypophiceras martini* (Spath) (at the generic level—*Hypophiceras*; Bjerager et al. 2006); (3) *Otoceras concavum* Tozer–*Otoceras boreale* Spath (at the generic level—*Otoceras*; this study); (4) *Episageceras wynnei* (Waagen)–*Episageceras antiquum* (Popov) (at the generic level—*Episageceras*; Popow 1961; Zakharov 1978; Dagys and Ermakova 1996; Zakharov and Moussavi Abnavi 2013; Zakharov and Popov 2014); and (5) *Episageceras wynnei* (Waagen)–*Episageceras dalailamae* (Diener) (at the generic level—*Episageceras*; Diener 1897). It is notable that all these, with the exception of *Episageceras*, survived into the Triassic within the Boreal Superrealm.

Data relating to the genus *Tompophiceras* (Popow 1961) obtained from northeast Russia (Nikolkin Klyuch and Seryogin Creek sections, South Verkhoyansk region) provide some evidence that its first representatives were coexisting with the earliest Induan *Otoceras boreale* Spath (this study).

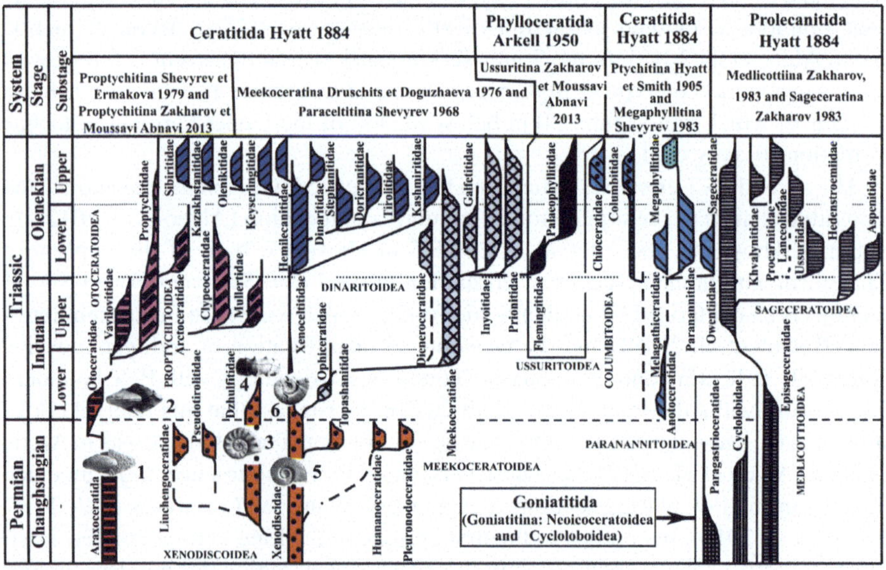

Fig. 10.10 Phylogenetic relationships of the Permian–Triassic goniatitid, prolecanitid, ceratitid and phylloceratid ammonoid superfamilies and families (modified from Zakharov and Moussavi Abnavi 2013). 1, *Avushoceras*; 2, *Otoceras*; 3, *Paratirolites*; 4, *Tompophiceras*; 5, *Xenodiscus*; 6, *Aldanoceras*

Early Induan ammonoids of the genus *Episageceras*, described by Noetling (1904) on the basis of data on the species "*Sageceras*" *wynneri* Waagen from the Upper Permian in the Salt Range, are known in the Boreal Superrealm as *Episageceras antiquum* (Popow) (after Dagys and Ermakova's (1996) revision). *Episageceras* shells were collected by different authors from Kobyume (South Verkhoyansk, *Otoceras boreale* zone; Popow 1961), Burgagandzha (South Verkhoyansk, *Vavilovites turgidus* and *Kingites korostelevi* zones; Zakharov 1978; Dagys and Ermakova 1996) and Okhotsk sea-coast region (*Vavilovites turgidus* zone; Dagys and Ermakova 1996).

Other than *Otoceras concavum* Tozer, *Otoceras gracile* Tozer has also been found from the upper Changhsingian *Otoceras concavum* zone in Arctic Canada (Tozer 1994) and South Verkhoyansk (this study). The earliest Induan *Otoceras boreale* Spath is known from four regions of the Boreal Superrealm: Arctic Canada (Tozer 1994), South Verkhoyansk (Zakharov 1971, 1978, 2002; Arkhipov 1974; Dagys and Ermakova 1996; this study), Svalbard (Petrenko 1963; Korchinskaya 1982; Nakazawa et al. 1987) and Greenland (Spath 1930, 1935).

Among the possible earliest representatives of the genus *Otoceras* from southern higher palaeolatitudes *Otoceras?* sp. may be apparently mentioned. It was recently found in the P–T boundary beds in Western Australia (northern Perth Basin, Redback-2 core; G. Shi unpublished data). The revision of *Otoceras* from the Himalayan province (Chao et al. 2017) shows that a single species of this genus is known there. According to Chao et al. (2017), *Otoceras undatum* (Griesbach), *O. fissisellatum* Diener, *O. parbati* Diener, *O. clivei* Diener, *O. draupadi* Diener, *O. dieneri* Spath and *O. latilobatum* Wang from the Himalayan province are younger synonyms of the *Otoceras woodwardi* Griesbach. If this is true, only a single *Otoceras woodwardi (=Otoceras latisellatum)* zone can be recognised within *Otoceras*-bearing sequences in that region. This corresponds to the *Otoceras boreale* zone of the Boreal Superrealm.

Ceratitid ammonoids of the suborder Otoceratina (Shevyrev and Ermakova (1979)) are represented by the single superfamily Otoceratoidea (Hyatt (1900)) and the three families: (1) the Wuchiapingian Anderssonoceratidae (Ruzhencev (1959)) (*Anderssonoceras*, *Lenticoceltites* and *Pericarinoceras*; Bogoslovskaya et al. 1999); (2) the Wuchiapingian–Changhsingian Araxoceratidae (Ruzhencev (1959)) (*Araxoceras*, *Eoaraxoceras*, *Rotaraxoceras*, *Prototoceras*, *Discotoceras*, *Urartoceras*, *Pseudotoceras*, *Vescotoceras*, *Dzhulfoceras*, *Vedioceras*, *Avushoceras*, *Periptychoceras*, *Anfuceras*, *Kingoceras*, *Abadehceras*, *Julfotoceras*, *Konglingites*, *Jinjiangoceras*, *Kiangsiceras* and *Sanyangites*; Zhou et al. 1999; Bogoslovskaya et al. 1999); and (3) the Changhsingian–Induan Otoceratidae (Hyatt (1900)) (*Anotoceras*, *Otoceras* and *Metotoceras*; Shevyrev 1968).

The suture lines of adult individuals of all Boreal species of the genus *Otoceras* are characterised by the presence of a tie-in at the top of the third lateral saddle, which was apparently inherited directly from Permian ammonoids, some of which (e.g. *Avushoceras*; Zakharov and Pavlov 1986) carry this feature on the third saddle of their suture line.

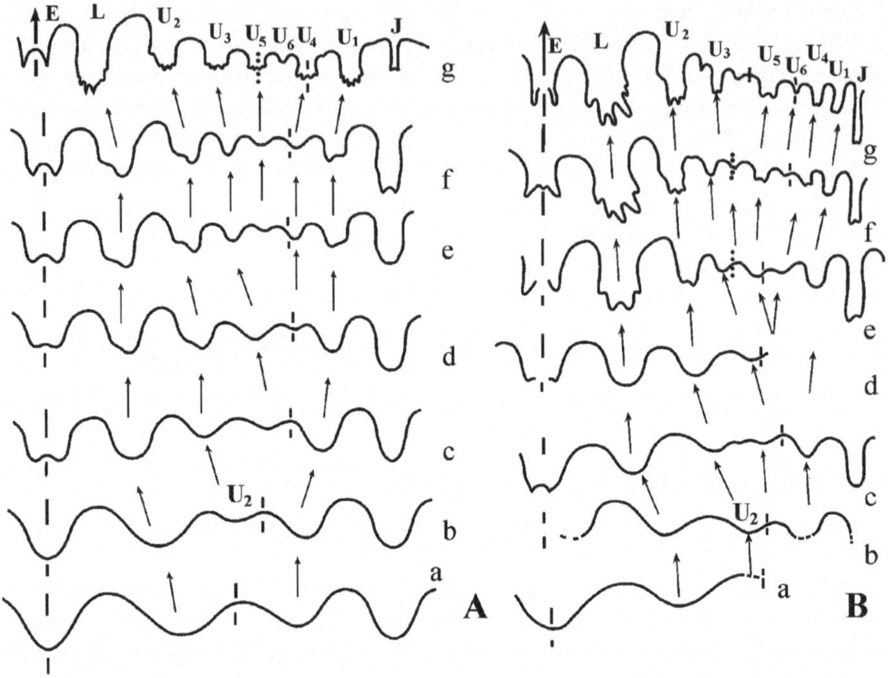

Fig. 10.11 Onthogenetic development of *Otoceras* from the Tethys and the Boreal Superrealm: (**a**) *O. woodwardi* Griesbach suture lines (Schindewolf 1968), with indication of Wedekind's (1916) lobe indexes; (**b**) *O. boreale* Spath suture lines (with indication of Wedekind's (1916) lobe indexes), based on samples 50/803 (stages a–c) and 33/803 (stages e–g) from the Nikolkin Klyuch section (a) at $W = 0.72$ mm; (b) at $H = 1.04$ mm and $W = 1.43$ mm; (c) at $H = 1.25$ mm and $W = 2.1$ mm; (d) at $H = 1.43$ mm and $W = 2.65$ mm; (e) at $H = 2.8$ mm and $W = 4.2$ mm; (f) at $H = 6.2$ mm and $W = 9.5$ mm; (g) at $H = 20.0$ mm and $W = 26.0$ mm

The similarity of onthogenetic development of suture lines in *Otoceras woodwardi* (Schindewolf 1968) and *O. boreale* (Fig. 10.11) which are common in Southern and Northern palaeohemispheres, respectively, gives some grounds for a monophyletic origin of the genus *Otoceras*.

Bando (1973) considered that the late Permian araxoceratid *Avushoceras jakowlevi* Ruzhencev was most intimately related to *Otoceras concavum* Tozer. Our data on the great similarity of suture lines of *Avushoceras jakowlevi* Ruzhencev from the Wuchiapingian–Changhsingian boundary transition in Transcaucasia and representatives of *Otoceras* from the uppermost Changhsingian and lowermost Induan in the Verkhoyansk area (Fig. 10.12) are in agreement with Bando's (1973) suggestion. Furthermore, juvenile *Otoceras* shells closely resemble the adult *Avushoceras* (Zakharov 1978, plate 3), reflecting possible anabolic development in the phylogenetic lineage *Avushoceras–Otoceras*. However, additional data is required to confirm this theory, which should take into account that the inner shell structure for representatives of the suborder Otoceratina is known only from *Otoceras*. The latter is characterised by a very primitive inner shell structure

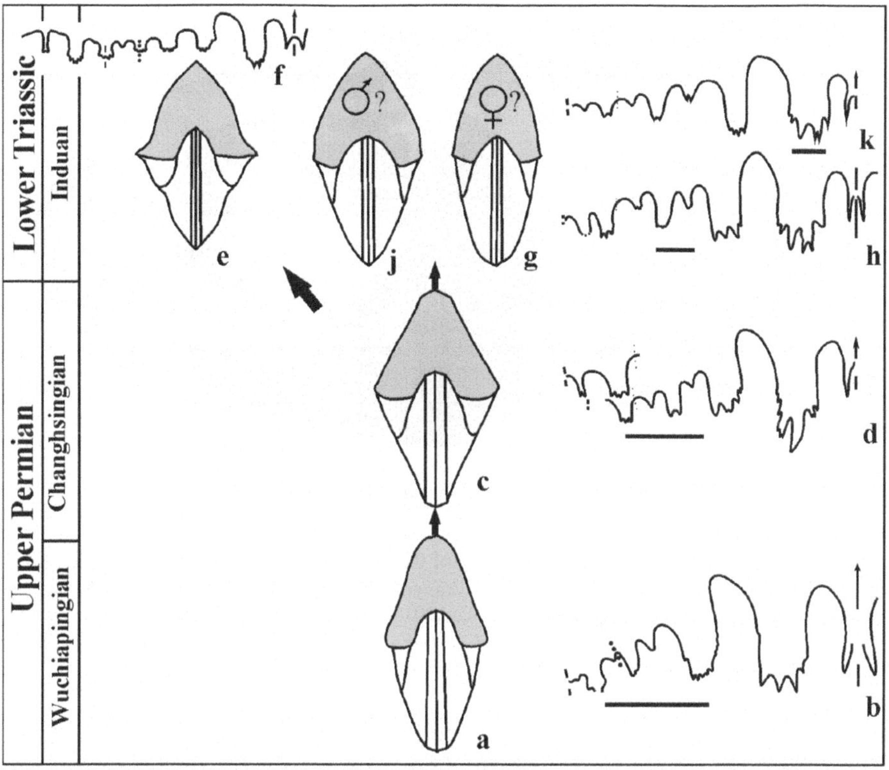

Fig. 10.12 Hypothetical evolutional trend of the Otoceratoidea. (a, b) *Avushoceras jakovlevi* Ruzhencev; (c, d) *Otoceras concavum* Tozer; (e, f) *O. woodwardi* Griesbach; (g, h) *O. boreale* Spath (morpha A); (j, k) *O. boreale* Spath (morpha B)

including retrochoanitic septal necks in both early and late ontogenetic stages and ventral position of the siphuncle in early ontogenetic stages (Zakharov 1978, Plate 3).

Late Permian Otoceratoidea (Araxoceratidae and Anderssonoceratidae) are known only from equatorial and neighbouring palaeoregions, such as South China (Zhao et al. 1978), Iran (Bando 1973; Zakharov et al. 2010), Transcaucasia (e.g. Kotlyar et al. 1983) and Mexico (Spinosa et al. 1970). However, representatives of the genus *Otoceras* are characterised by their bipolar distribution (Zakharov et al. 2008): in the Boreal Superrealm, Himalayan province of the Tethys (and possibly Gondwana Superrealm) (Fig. 10.13). We associate the migration of ancestral forms of *Otoceras* and some associated ammonoid genera to higher palaeolatitude areas mainly with climatical changes at the very end of the Permian as inferred from O- and N-isotope data. As shown above, the coolest conditions for the Late Permian occurred during the middle Changhsingian. A subsequent gradual increase in temperature in the late Changhsingian, as suggested by N-isotopic evidence (this study) and by O-isotope thermometry data (the palaeotemperature of the habitat of conodonts in the North-

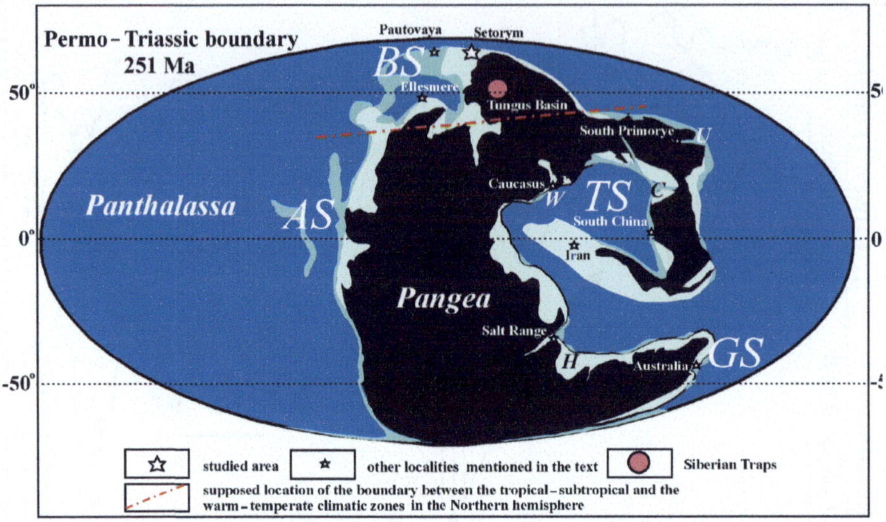

Fig. 10.13 Simplified palaeogeographical map of the Permian–Triassic boundary interval showing the relative location of the studied locality (Pravyi Suol, Setorym River basin) and other localities mentioned in the text (modified from Scotese 2014). Superrealms: *BS* Boreal. *TS* Tethyan. *AS* American, *GS* Gondwanan (Zakharov et al. 2008). Provinces: *C* Cathasian, *U* Ussurian, *W* Western-Tethyan (Zakharov et al. 2008)

Western Iran during the late Changhsingian reached ~38 °C; Schobben et al. 2014), caused the migration of ancestral forms of the genera *Otoceras*, *Tompophiceras* and others from the Iran–Transcaucasia area to cooler settings in the Boreal Superrealm. The general absence of *Otoceras* in the lowest Induan of the palaeoequatorial and palaeosubequatorial areas (South China, Transcaucasia and Iran) farther supports this proposed migration. Some groups of goniatitid, prolecanitid and ceratitid ammonoids that did not have any ability to migrate to higher (cooler) palaeolatitude areas became extinct presumably due to the extremely warm and possibly anoxic conditions that were most pronounced in equatorial and near-equatorial regions. This is consistent with the observation that almost all the known ammonoid survivors of the end-Permian mass extinction did so only within the Boreal Superrealm. In light of this, the Boreal Superrealm can be considered as a key refugium for the survival of otoceratid and some other ammonoids through the end-Permian mass extinction.

10.5 Conclusions

1. The ammonoid assemblages of the upper Changhsingian *Otoceras concavum* zone in the South Verkhoyansk region, as in Arctic Canada, are represented by two species belonging to the family Otoceratidae: *Otoceras concavum* Tozer and *Otoceras gracile* Tozer. The first representatives of *Tompophiceras* (Dzhulfitidae)

as well as *Episageceras* (Sageceratidae) in the South Verkhoyansk region have been found in the lowest division of the Indian (*Otoceras boreale* zone).
2. The absence of very low $\delta^{15}N_{sed}$ values (below −0.9‰ values that are commonly seen in the middle and low palaeolatitude areas) in Permian and Triassic strata of the Boreal Superrealm suggests accumulation under a relatively cooler climatic regime. However, the global trend towards warming across the P–T boundary transition is also seen in the Boreal regions.
3. According to biostratigraphical and palaeogeographical data, almost all the known ammonoid phylogenetic lineages that survived the mass extinction at the end of the Permian were able to accomplish this survival of only within the Boreal Superrealm. This suggests that the region was a key refugium for the survival of at least some otoceratid and dzhulfitid ammonoids during the development of lethally hot temperatures and possibly anoxic conditions in the Tethys. Since there is no evidence for Permian–Triassic glaciation in the Boreal region it seems likely that the conditions in this region were more favourable for the development of ammonoids even during the period of the end-Permian mass extinction, which was most pronounced in the Tethys.
4. A comparative morphological analysis of the species belonging to the genus *Otoceras*, which has bipolar geographic distribution, leads us to assume its monophyletic origin from araxoceratid *Avushoceras*, known from the Wuchiapingian–Changhsingian boundary transition in Transcaucasia.

Acknowledgements N-isotope analyses by E. Riegler (BLT Wiselbburg Research Center HBLFA Francisco-Josephinum, Austria) are gratefully acknowledged. We are also grateful to I.V. Brynko, I.V. Budnikov, S.S. Burnatyj, I.L Vedernikov, A.N. Kilyasov, V.I. Makoshin, A.N. Naumov and A.M. Popov from Magadan, Vladivostok and Yakutsk for their organising and taking part in our expeditions in the South Verkhoyansk region. Our special thanks to Tretyakovs, Felix F. and Maxim F., for the opportunity to use the Yakutsk Federal University field base on the left bank of the Vostochnaya Khandyga River (the Tompo Training Ground). This research was funded by the grants RFBR no. 18-05-00023, 18-05-00191, and 20-05-00604, in part by the Program for Improvement of the Competitiveness of Kazan (Volga River) Federal University among the world's leading research and educational centers, and also in part (monographic study of new invertebrate collections from the South Verkhoyansk region) by the Russian Scientific Foundation, grant no. 19-17-00178. A monographic study of P-T cephalopods is done on state assignments for a number institutions of the Russian Academy of sciences (FEGI, DPMGI, and IPGG). D.P.G. Bond acknowledges funding from the Natural Environment Research Council (Grant NE/JO1799X/1) and the Royal Society (International Exchanges scheme-project "Volcanic and climatic impacts on Permian biota across Russian ecological zones").

References

Algeo TJ, Rowe H, Hower JC, Schwark L, Hermann A, Heckel PH (2008) Oceanic denitrification during Late Carboniferous glacial-interglacial cycles. Nat Geosci 1:709–714

Algeo T, Henderson CM, Ellwood B, Rowe H, Elswick E, Bates S, Lyons T, Hower JC, Smith C, Maynard B, Nays LE, Summons RE, Fulton J, Freeman KN (2012) Evidence for a

diachronous Late Permian marine crisis from Canadian Arctic region. Geol Soc Am Bull 124(9/10):1424–1448
Algeo TJ, Meyers PA, Robinson RS, Rowe H, Jiang GQ (2014) Icehouse-greenhouse variations in marine denitrification. Biogeosciences 11:1273–1295
Altabet MA, Francois R, Murray DW, Ptell WL (1995) Climate-related variations in denitrification in the Arabian Sea from sediment $^{15}N/^{14}N$ ratios. Nature 373:506–509
Arkhipov YV (1974) Stratigrafiya triasaovykh otlozhenij vostochnoj Yakutii (Triassic stratigraphy of Eastern Yakutia). Yakutskoye knizhnoye izdatelstvo, Yakutsk, 273 p (in Russian)
Bando Y (1973) On the Otoceratidae and Ophiceratidae. Sci Rep Tohoku Univ 6:337–351
Baud AM, Margaritz M, Holser WT (1989) Permian–Triassic of the Tethys: carbon isotope stratigraphy. Geol Rundsch 78:649–677
Bauersachs T, Schouten S, Compaore J, Wollenzien U, Stal LJ, Damsté JSS (2009) Nitrogen isotopic fractionation associated with growth on dinitrogen gas and nitrate by cyanobacteria. Limnol Oceanogr 54(4):1403–1411
Biakov AS, Vedernikov IL (2007) Evidence for anoxia in deep environments of Northeast Asia. Dokl Earth Sci 417A(9):1325–1327
Biakov AS, Zakharov YD, Horacek M, Richoz S, Kutygin RV, Ivanov YY, Kolesov EV, Konstantinov AG, Tuchkova MI, Mikhalitsyna TI (2016) New data on the structure and age of the terminal Permian strata in the South Verkhoyansk region (northeastern Asia). Russ Geol Geophys 57:282–293
Biakov AS, Kutygin RV, Goryachev NA, Burnatny SS, Naumov AN, Yadrenkin AV, Vedernikov IL, Tretyakov AV, Brynko IV (2018) Discovery of the late Changhsingian bivalve complex and two fauna extinction episodes in northeastern Asia at the end of the Permian. Dokl Biol Sci 480:78–81
Bjerager M, Seidler L, Stemmerik L, Surlyk F (2006) Ammonoid stratigraphy and sedimentary evolution across the Permian–Triassic boundary in East Greenland. Geol Mag 143(5):635–656
Bogoslovskaya MF, Kuzina LF, Leonova TB (1999) Classification and distribution of Late Palaeozoic ammonoids. In: Rozanov AY, Shevyrev AA (eds) Iskopaemye tsefalopody noveishiye dostizhenuya v ikh izuchenii (Fossil cephalopods: recent advances in their study). PIN RAN, Moscow, pp 89–124, (in Russian)
Bond DG, Wignall PB (2010) Pyrite framboid study of marine Permian–Triassic boundary sections: a complex anoxic event and its relationship to contemporaneous mass extinction. Geol Soc Am Bull 122(7/8):1265–1279
Boulila S, Laskar J, Haq BU, Galbrun B, Hara N (2018) Long-term cyclicities in Phanerozoic sea-level sedimentary record and their potential drivers. Supplementary material (SM1 & SM2). https://arxiv.org/pdf/1803.05623
Brosse M, Bucher H, Goudemand N (2016) Quantitative biochronology of the Permian-Triassic boundary in South China based on conodont unitary associations. Earth Sci Rev 155:153–171
Burgess SD, Bowring S, Shen S (2014) High-precision timeline for Earth's most severe extinction. Proc Natl Acad Sci 111(9):3316–3321
Chao Z, Bucher H, Shen S-Z (2017) Griesbachian and Dienerian (Early Triassic) ammonoids from Qubu in the Mt. Everest area, southern Tibet. Palaeoword 26:650–662
Cockell CS, Cathing DC, Davis WL, Snook K, Kepner RL, Lee P, McKay CP (2000) The ultraviolet environment of Mars: biological implications past, present and feature. Icarus 146(2):343–359
Dagys A, Ermakova S (1996) Induan (Triassic) ammonoids from north-eastern Asia. Rev Paléobiol 15(2):401–447
Dagys AS, Arkhipov YV, Bytchkov YM (1979) Stratigrafiya triasovoj sistemy Severo-Vostoka Azii (Triassic stratigraphy of north-eastern Asia). Nauka, Moscow, 244 p (in Russian)
Dagys AS, Arkhipov YV, Truschelyev AM (1984) Excursion 054. Permian and Triassic deposits in Yakutiya. Svodnyj putevoditel ekskursij 052, 053, 054. 055 (27 Mezhdunarodnyi Geologicheskij Kongress) Yakutskaya ASSR, Sibirskaya Platforma (Summary Field Guide Book of excursions 052, 053, 05 and 055 (27th International Geological Congress)). Nauka, Novosibirsk, p 68–89. (in Russian)

Dagys AS, Dagis AA, Kazakov AM, Konstantinov AG, Kurushin NI (1986) Lower Induan biostratigraphy of eastern Verkhoyansk. In: Yanshin AL, Dagys AS (eds) Biostratigrafiya mezozoya Sibiri i Dalnrgo Vostoka (Mesozoic biostratigraphy of Siberia and Far East). Nauka, Novosibirsk, pp 21–31, (in Russian)

Diener C (1897) The Cephalopoda of the Lower Trias. Palaeontologica Indica 15(2/1):1–181

Domokhotov SV (1960) Induan stage and *Otoceras* zone of the eastern Verkhoyansk area. In: Kobelyatskij IA (ed) Materialy po geologg i poleznym iskopaemym Yakutskoj ASSR. 1 (Materials on geology and minerals of the Yakut ASSR. 1). Yakutskoye knizhnoye izdatelstvo, Yakutsk, pp 111–120, (in Russian)

Dustira AM, Wignall PB, Joachimski M, Blomeier D, Hartkopf-Fröder C, Bond DPG (2013) Gradual onset of anoxia across the Permian–Triassic boundary in Svalbard, Norway. Palaeogeogr Palaeoclimatol Palaeoecol 374:303–313

Garbelli C, Angiolini L, Brand U, Shen S, Jadoul F, Azmy K, Posenato R, Cao C (2014) The paradox of the Permian global oceanic anoxia. Permophiles 61:26–28

Goudemand N, Romano C, Brayard A, Hochuli PA, Bucher H (2013) Comment on "Lethally hot temperatures during the Early Triassic greenhouse". Science 339:1033a–1033c

Grasby SE, Beauchamp B (2008) Intrabasin variability of the carbon-isotope record across the Permian–Triassic transition, Sverdrup Basin, Arctic Canada. Chem Geol 253:141–150

Grasby SE, Sanei H, Beachamp B, Chen Z (2013) Mercury deposition through the Permo-Triassic biotic crisis. Chem Geol 351:209–216

Grasby SE, Beauchamp B, Bond DPG, Wignall P, Talavera C, Galloway JM, Piepjohn K, Reinhardt L, Blomeier D (2015) Progressive environmental deterioration in northwestern Pangea leading to the latest Permian extinction. Geol Soc Am Bull 127:1331–1347

Grasby SE, Beauchamp B, Knies J (2016) Early Triassic productivity crises delayed recovery from world's worst mass extinction. Geology 44:779. https://doi.org/10.1130/G38141.1

Grasby SE, Beauchamp B, Bond DPG, Wignall P, Sanei H (2018) Mercury anomalies associated with three extinction events (Capitanian crises, latest Permian extinction and the Smithian/Spathian extinction in NW Pangea). Geol Mag 153:285–297

Grigoryan AG, Alekseev AS, Joachimski MM, Gatovsky YA (2015) Permian–Triassic biotic crisis: a multidisciplinary study of Armenian sections. In: Nurgaliev DK (ed) XVIII International Congress on the Carboniferous and Permian (August 11–15, 2015, Kazan, Russia), Kazan, Kazan University Press, abstract volume (p 74)

Hammer D, Jones MT, Schneebeli-Hermann E, Hansen BB, Bucher H (2019) Are Early Triassic extinction events associate with mercury anomalies? A reassessment of the Smithian/Spathian boundary extinction. Earth Sci Rev 195:179. https://doi.org/10.1016/j.earscirev.2019.04.016

Hermann E, Hochuli PA, Bucher H, Vigran JO, Wessert H, Bernasconi SM (2010) A close-up view of the Permian-Triassic boundary based on expanded organic carbon isotope records from Norway (Trøndelag and Finnmark platform). Global Planet Change 74:156–167

Hermann E, Hochuli PA, Méhay S, Bucher H, Brühwiler T, Ware D, Hautmann M, Roohi G, ur-Rehman K, Yaseen A (2011) Organic matter and palaeoenvironmental signals during the early Triassic biotic recovery: the salt range and Surghar range records. Sediment Geol 234:19–41

Hochuli PA, Vigran JO, Hermann E, Bucher H (2010) Multiple climatic changes around the Permian–Triassic boundary event revealed by an expanded palynological record from mid-Norway. GSA Bull 122(5/6):884–896

Horacek M, Brander R, Abart R (2007a) Carbon isotope record of the P/T boundary and the Lower Triassic in the Southern Alps: evidence for rapid changes in storage of organic carbon. Palaeogeogr Palaeoclimatol Palaeoecol 252:347–354

Horacek M, Richoz S, Brandner R, Krystyn L, Spötl C (2007b) Evidence for recurrent changes in Lower Triassic oceanic circulation if the Tethys: the Delta ^{13}C record from marine sections in Iran. Palaeogeogr Palaleoclimatol Palaeoecol 252:355–369

Horacek M, Wang X-D, Grossman EL, Richoz S, Cao Z (2007c) The carbon-isotope curve from the Chaohu section, China: different trends at the Induan-Olenekian boundary or diagenesis? Albertiana 35:41–45

Horacek M, Koike T, Richoz S (2009) Lower Triassic $\delta^{13}C$ isotope curve from shallow-marine carbonates in Japan, Panthalassa realm: confirmation of the Tethys $\delta^{13}C$ curve. J Asian Earth Sci 3(6):481–490. https://doi.org/10.1016/j.jseaes.2008.05.005
Horacek M, Brandner R, Richoz R, Povoden E (2010) Lower Triassic sulphur isotope curve of marine sulphates from the Dolomites, N-Italy. Palaeogeogr Palaeoclimatol Palaeoecol 290(1–4):65–70. https://doi.org/10.1016/j.palaeo.2010.02.016
Hyatt A (1900) Ammonoidea. In: Zittel KA (ed) Textbook of palaeontology. C.R. Eastman, London, pp 502–592, 1st English ed
Isozaki Y (1997) Permo-Triassic boundary superanoxia and stratified superocean: records from lost deep sea. Science 276:235–238
Jenkyns HC, Gröcke DR, Hesselbo SP (2001) Nitrogen isotope evidence for mass denitrification during the early Toarcian (Jurassic) oceanic anoxic event. Paleoceanography 16:593–603
Jia C, Huang J, Kershaw S, Luo G (2012) Microbial response to limited nutrients in shallow water immediately after the end-Permian mass extinction. Geobiology 10:60–71
Joachimsky MM, Lai X, Shen S, Jiang H, Luo G, Chen B, Chen J, Sun Y (2012) Climate warming in the latest Permian and the Permian-Triassic mass extinction. Geology 40:195–198
Joachimski MM, Alekseev AS, Grigoryan A, Gatovsky YA (2020) Siberian Trap volcanism, global warming and the Permian-Triassic mass extinction: New insights from Armenian Permian-Triassic sections. The Geological Society of America Bulletin 132: 427–443. https://doi.org/10.1130/B35108.1
Kaiho K, Kajiwara Y, Nakano Y, Miura Y, Chen ZQ, Shi GR (2001) End-Permian catastrophe by a bolide impact: evidence of a gigantic release of sulfur from the mantle. Geology 29:815–818
Kaiho K, Chen ZQ, Sawda K (2009) Possible causes for a negative shift in the stable carbon isotope ratio before, during and after the end-Permian mass extinction in Meishan, South China. Aust J Earth Sci 56:799–808
Kato Y, Nakao K, Isozaki Y (2002) Geochemistry of Late Permian to Early Triassic pelagic cherts from southwest Japan: implications for an oceanic redox change. Chem Geol 182:15–34
Knies J, Grasby SE, Beuchamp B, Schubert CJ (2013) Water mass denitrification during the latest Permian extinction in the Sverdrup Basin, Arctic Canada. Geology 41(2):167–170
Knoll AH, Bambach RK, Payne JL, Pruss S, Fischer WW (2007) Palaeophysiology and end-Permian mass extinction. Earth Planet Sci Lett 256:295–313
Korchinskaya MV (1982) Obyasnitelnaya zapiska k stratngraficheskoj scheme mezozoya (triasa) Svalbarda (Explanatory note on the biostratigraphic scheme of the Mesozoic (Trias) of Spitsbergen). Sevmorgeologiya, Leningrad, 99 p (in Russian)
Korostelev VI (1972) Stratigrafia triasovykh otlozhenij Vostochnogo Verkhoyanya (Triassic stratigraphy of the Eastern Verkhoyansk area). Yakutskoye knizhnoye izdatelstvo, Yakutsk, 174 p (in Russian)
Korte C, Kozur HW (2010) Carbon-isotope stratigraphy across the Permian-Triassic boundary: A review. J Asian Earth Sci 39:215–235
Kotlyar GV, Zakharov YD, Koczyrkevicz BV, Kropacheva GS, Rostovcev KO, Chedija IO, Vuks GP, Guseva EA (1983) Pozdnepermskij etap evolyutsii organicheskogo mira (Dzhulfinskij i Dorashamskij yarusy SSSR) (Evolution of the latest Permian biota (Dzhulfian and Dorashamian regional stages in the USSR)). Nauka, Leningrad, 199 p (in Russian)
Kozur HW (2007) Biostratigraphy and event stratigraphy in Iran around the Permian-Triassic boundary (PTB): implications for the causes of the PTB biotic crisis. Global Planet Change 55:155–176
Kozur HW, Weems RE (2011) Detailed correlation and age of continental late Changhsingian and earliest Triassic beds: implications for the role of the Siberian trap in the Permian–Triassic biotic crisis. Palaeogeogr Palaeoclimatol Palaeoecol 308:22–40
Krull ES, Retallack GJ, Campbrll HJ, Lyon GL (2000) $\delta^{13}C_{org}$ chemostratigraphy of the Permian–Triassic boundary in the Maitai Group, New Zealand: evidence for high-latitudinal methane release. N Z J Geol Geophys 43:21–32

Kurushin NI (1987) Drevnejshie triasovye dvustvorchatye mollyuski Yakutii (The oldest Triassic bivalves from Yakutiya). In: Dagys AS (ed) Boreal'nyj trias (Boreal Triassic). Nauka, Moscow, pp 99–110, (in Russian)

Lozovsky VR (2013) Permian-Triassic crisis and its possible reason. Byulyuten Moskovskogo Obschestva Ispytatelej Prirody. Otdeleniye Geologicheskoye 88(1):49–58. (in Russian)

Luo G, Wang Y, Kump LR, Algeo TJ, Yang H, Xie S (2011) Enhanced nitrogen fixation in the immediate aftermath of the latest Permian marine mass extinction. Geology 39(7):647–650

Nakazawa K, Nakamura K, Kimura G (1987) Discovery of *Otoceras boreale* Spath from West Spitsbergen. Proc Japan Acad Ser B 63(6):171–174

Nakrem HA, Orchard M, Weitschart W, Hounslow MW, Beaity TW, Mørk A (2008) Triassic conodonts from Svalbard and the Boreal correlations. Polar Res 27(3):523–537

Noetling F (1904) Ueber *Medlicottia* Waag. und *Episageceras* n. g. aus den permischen und triadischen Schichten indiens. Neues Jahrbuch für Mineralogy, Geologie und Paläontologie 19:334–376

Parfenov LM, Kuzmin MI (2001) Tectonics, geodynamics and metallogeny of the Sakha Republic (Yakutia). MAIK Nauka Interperiodica, Moscow, 571 p

Petrenko VM (1963) Some important finds of Early Triassic fauna on Spitsbergen. Uchyonye Zapiski NIIGA 3:50–54. (in Russian)

Popow YN (1956) *Otoceras* in the lower Triassic of eastern Verkhoyansk. In: Kobelyatsky IA (ed) Materialy po geologii i poleznym iskopaemym Severo-Vostoka SSSR, vol 10. Magadan, Magadanskoye Knizhnoye Izdatelstvo, pp 64–81. (in Russian)

Popow YN (1958) Find of *Otoceras* in the lower Triassic of eastern Verkhoyansk. Izvestiya Akademii Nauk SSSR Ser Geol 12:105–109. (in Russian)

Popow YN (1961) Triasovye ammonoidei Severo-Vostoka SSSR (Triassic ammonoids of the North-Eastern USSR). Tudy NIIGA 79:1–180. (in Russian)

Robinson RS, Kienast M, Albuquerque AL, Altalet M, Contreras S, Holz R, Dubots N, Francois R, Galbraith E, Hsu T, Ivanochko T, Jaccard S, Kao S, Kiefer T, Kienast S, Lehmann M, Martinez P, McCarthy M, Möbius J, Pederson T, Quan TM, Ryabenko E, Schmittner A, Schneider R, Schneider-Mor A, Thunell R, Yang J (2012) A review of nitrogen isotope alteration in marine sediments. Paleoceanography 27:1–13

Romano C, Goudemand N, Vennemann TW, Ware D, Schneebeli-Hermann E, Hochuli PA, Brühwiler T, Brinkmann W, Bucher H (2013) Climatic and biotic upheavals following the end-Permian mass extinction. Nat Geosci 6:57–60

Ruzhencev VE (1959) Classification of the superfamily Otocerataceae. Paleontologocheskij Zhurnal 2:57–67. (in Russian)

Sadovnikov GN (2008) On the position of the "point of the lower Triassic boundary in global stratotype". Stratigrafiya Geologicheskaya Correlyatsiya 16(1):34–50

Sadovnikov GN (2016) Evolution of the biome of the trap plateau of Central Siberia. Paleontologicheskij Zhurnal 5:87–99. (in Russian)

Saitoh M, Ueno Y, Nishizawa M, Isozaki Y, Takai K, Yao J, Ji Z (2014) Nitrogen isotope chemostratigraphy across the Permian-Triassic boundary at Chaotian, Sichuan, South China. J Asian Earth Sci 93:113–128

Schindewolf OH (1968) Studien zur Stammesges chichte der Ammoniten, Lieferung 7. Abhandlungen der mathematisch-naturwissenschaftlichen Klasse. Akademie der Wissenschaften und der Literatur in Mainz 3:1–171

Schobben M, Joachimski MM, Korn D, Leda L, Korte C (2014) Palaeotethys seawater temperature rise and an intensified hydrological cycle following the end-Permian mass extinction. Gondw Res 26:675–683

Scotese CR (2014) Atlas of Permo–Triassic paleogeographic maps (Mollweide projection), maps 43—52, Volumes 3 & 4 of the PALEOMAP Atlas for ArcGIS, PALEOMAP Project, Evanston, IL. doi: https://doi.org/10.13140/2.1.2609.9209

Shevyrev AA (1968) Triassic ammonoids of the South USSR. Trudy PIN 110:1–272. (in Russian)

Shevyrev AA, Ermakova SP (1979) Systematics of ceratites. Paleontologicheskij Zhurnal 1:52–58. (in Russian)

Sobolev ES (1989) Triasovye nautilidy Severo-Vostochnoy Azii (Triassic nautilids of northeastern Asia). Nauka, Siberian Branch, Novosibirsk, 192 p (in Russian)

Song H, Tong J, Algeo TJ, Horacek M, Qiu H, Tian L, Chen Z-Q (2013) Large vertical $\delta^{12}C_{DIC}$ gradients in Early Triassic seas of the South China craton: implications for oceanographic changes related to Siberian Traps volcanism. Global Planet Change 105:7–20

Spath LF (1930) The Eo-Triassic invertebrate fauna of East Greenland. Medd Grønland 83(1):1–90

Spath LF (1935) Additions to the Eo-Triassic invertebrate fauna of East Greenland. Medd Grønland 98(2):1–115

Spinosa C, Furnish WM, Glenister BF (1970) Araxoceratidae, Upper Permian ammonoids, from the Western Hemisphere. J Paleo 44(4):730–736

Sun Y, Joachimski MM, Wignall PB, Yan C, Chen Y, Jiang H, Wang L, Lai X (2012) Lethally hot temperatures during the Early Triassic Greenhouse. Science 338:366–370

Takashi S, Kaiho K, Oba M, Kakegawa T (2010) A smooth negative shift of organic carbon isotope ratios at an end-Permian mass extinction horizon in central pelagic Panthalassa. Palaeogeogr Palaeoclimatol Palaeoecol 292:532–539

Takashi S, Kaiho K, Kori RS, Hori RS, Gorjan P, Watanabe T, Yamakita S, Aita Y, Takemura A, Spörli KB, Kakegawa T, Oba M (2013) Sulfur isotope profiles in the pelagic Panthalassic deep sea during the Permian-Triassic transition. Global Planet Change 105:68. https://doi.org/10.1016/J.gloplacha.2012.12.006

Tozer ET (1994) Canadian Triassic ammonoid faunas. Geol Surv Can Bull 467:1–663

Twitchet RJ, Looy CV, Morant ER, Vissehes H, Wignall PB (2001) Rapid and synchronous collapse of marine and terrestrial ecosystems during the end-Permian biotic crisis. Geology 29(4):351–354

Wedekind R (1916) Über Lobus, Suturallobus und Inzision. Zentralblatt fü Geologie und Paläontologie 8:185–195

Wignall PB, Hallam A (1992) Anoxia as a cause of the Permian-Triassic mass extinction: facies evidence from northern Italy and the western United States. Palaeogeogr Palaeoclimatol Palaeoecol 102:215–237

Wignall PB, Twitchet RJ (1996) Oceanic anoxia and the end-Permian mass extinction. Science 272:1155–1158

Wignall PB, Twitchet RJ (2002) Extent, duration, and nature of the Permian-Triassic superanoxic event. Geol Soc Am Spec Paper 356:395–413

Wignall PB, Bond DPG, Sun Y, Grasby SE, Beauchamp B, Joachimski MM, Blomeier DPG (2015) Ultra-shallow-marine anoxia in an Early Triassic shallow-marine clastic ramp (Spitsbergen) and the suppression of benthic radiation. Geol Mag 153:316–331

Yin H, Song FK (2013) Mass extinction and Pangea integration during the Paleozoic-Mesozoic transition. Sci China Earth Sci 56(1):1–13

Yin H, Zhang K (1996) Eventostratigraphy of the Permian-Triassic boundary at Meishan section, South China. In: Yyn H (ed) The Palaeozoic-Mesozoic boundary (candidates of global stratotype section and point of the Permian–Triassic boundary). China University of Geosciences Press, Wuhan, pp 84–96

Yin H, Xie S, Luo G, Algeo T, Zhang K (2012) Two episodes of environmental change at the Permian–Triassic boundary of the GSSP section Meishan. Earth Sci Rev 115:162–172

Zakharov YD (1971) *Otoceras* of the boreal realm. Paleontologicheskij Zhurnal 3:50–59. (in Russian)

Zakharov YD (1978) Rannetriasovye ammonoidei vostoka SSSR (Early Triassic ammonoids of the east USSR). Nauka, Moscow, 224 p (in Russian)

Zakharov YD (1995) The Induan–Olenekian boundary in the Tethys and Boreal realm. Annali dei Musei Civici di Rovereto Sezione Archeologia, Storia e Scienze Naturali 11:133–156

Zakharov YD (2002) Ammonoid succession of Setorym River (Verkhoyansk area) and problem of Permian–Triassic boundary in Boreal realm. J China Univ Geosci 13(2):107–123

Zakharov YD (2003) Ammonoid succession in the Lower Triassic of the Verkhoyansk area and problem of the Otocerataceae, Xenodiscaceae and Proptychitaceae phylogeny. In: Kryukov AP, Yakimenko LV (eds) Problemy evolyutsii (Problems of evolution), vol 5. Dalnauka, Vladivostok, pp 244–262. (in Russian)

Zakharov YD, Moussavi Abnavi N (2013) The ammonoid recovery after the end-Permian mass extinction: evidence from the Iran-Transcaucasia area, Siberia, Primorye and Kazakhstan. Acta Palaeontol Pol 58(1):127–147

Zakharov YD, Pavlov AM (1986) The first find of an araxoceratid ammonoid in the Permian of the eastern part of the USSR. In: Zakharov YD, Onoprienko YI (eds) Permo-triasovye sobytiya v razvitii organicheskogo mira severo-vostochnoj Asii (Permian-Triassic events during the biotic evolution of North-East Asia). Academy of Sciences of the USSR, Far-Eastern Scientific Centre, Vladivostok, pp 74–85. (in Russian)

Zakharov YD, Popov AM (2014) Recovery of brachiopod and ammonoid faunas following the end-Permian crisis: additional evidence from the lower Triassic of the Russian Far East and Kazakhstan. Aust J Earth Sci 25(1):1–44

Zakharov YD, Naidin DP, Teis RV (1975) Oxygen isotope composition Early Triassic cephalopod shells of Arctic Siberia and salinity of the Boreal basins at the beginning of the Mesozoic. Izvestiya Akademii Nauk SSSR Ser Geol 4:101–113. (in Russian)

Zakharov YD, Boriskina NG, Cherbadzhi AK, Popov AM, Kotlyar GV (1999) Main trends in Permo–Triassic shallow-water temperature changes: evidence from oxygen isotope and Ca-Mg ratio data. Albertiana 23:11–22

Zakharov YD, Biakov AS, Baud A, Kozur H (2005) Significance of Caucasian sections for working out carbon-isotope standard for Upper Permian and Lower Triassic (Induan) and their correlation with the Permian of north-eastern Russia. J China Univ Geosci 16(2):141–151

Zakharov YD, Popov AM, Biakov AS (2008) Late Permian to Middle Triassic palaeogeographic differentiation of key ammonoid groups: evidence from the former USS. Polar Res 27(3):441–468

Zakharov YD, Mousavi Abnavi N, Ghaedi M (2010) New species of Dzhulfian (Late Permian) ammonoids from the Hambast Formation of Central Iran. Paleontol J 44(6):614–631

Zakharov YD, Biakov AS, Horacek M (2014) Global correlation of basal Triassic layers in the light of the first carbon isotope data on the Permian–Triassic boundary in Northeast Asia. Russ J Pac Geol 8(1):3–19

Zakharov YD, Biakov AS, Richoz S, Horacek M (2015) Importance of carbon isotopic data of the Permian-Triassic boundary layers in the Verkhoyansk region for the global correlation of the basal Triassic layer. Dokl Earth Sci 460(1):1–5

Zakharov YD, Horacek M, Popov AM, Bondarenko LG (2018a) Nitrogen and carbon isotope data of Olenekian to Anisian deposits from Kamenushka/south Primorye, far-eastern Russia and their palaeoenvironmental significance. Aust J Earth Sci 29(4):837–853

Zakharov YD, Horacek M, Shigeta Y, Popov AM, Bondarenko LG (2018b) N and C isotopic composition of the lower Triassic of southern Primorye and reconstruction of the habitat conditions of marine organisms. Stratigr Geol Correl 26(5):534–551

Zakharov YD, Horacek M, Shigeta Y, Popov AM, Maekawa T (2018c) N- and C isotopic composition of the lower Triassic of south Primorye and reconstruction of habitat conditions of marine organisms after end Permian mass extinction at the end of the Permian. Dokl Earth Sci 478(2):161–165

Zakharov YD, Biakov AS, Horacek M, Goryachev NA, Vedernikov IL (2019) The first data on the N isotopic composition of the Permian and Triassic of northeastern Russia and their significance for paleotemperature reconstructions. Dokl Earth Sci 484(1):21–24

Zakharov YD, Horacek M, Biakov AS (in press) First data on N-isotope composition of the Permian-Triassic in the Verkhoyansk region and their significance for reconstruction of marine environments. Stratigrafiya. Geologicheskaya Korrelyatsiya (in Russian)

Zhao J, Liang X, Zheng Z (1978) Late Permian cephalopods of South China. Palaeontol Sin New Ser B 154(12):1–194

Zhou Z, Glenister BF, Furnish WM, Spinosa C (1999) Multi-episodal extinction and ecological and differentiation of Permian ammonoids. In: Rozanov AY, Shevyrev AA (eds) Fossil cephalopods: recent advances in their study. Paleontological Institute RAS, Moscow, pp 195–212

Chapter 11
Stress, Development, and Evolution in Coral Reef Communities

Neil W. Blackstone and Austin P. Parrin

Abstract Attention to coral reef communities has increased as concerns about climate change have mounted. Integral to these reefs are colonial corals and their cnidarian relatives. Reef building depends on symbiosis with photosynthetic dinoflagellates. When environmental stress becomes extreme, these dinoflagellates are lost, and corals bleach. Symbiotic cnidarians thus return to their primitive, non-symbiotic state. This symbiosis impacts all aspects of coral biology. Colony development begins with the formation of a primary polyp, which corresponds to development in unitary animals. The subsequent development of a colony, however, is without parallel in unitary animals and is particularly sensitive to input from environmental factors. These factors include water currents and patterns of food availability. For symbiotic cnidarians, availability of light has major impacts as well. Additionally, light functions in mechanisms of host-symbiont conflict mediation. Environmental signaling pathways are thus central to cnidarian development and result in considerable phenotypic plasticity. Plasticity poses challenges for the classification of corals. Modern molecular studies have revealed that well-known coral species are in some cases not distinct but rather either polymorphic or frequent hybridizers. At the same time, cryptic lineages—morphologically similar but genetically distinct—have been identified in other species. High levels of genetic polymorphism or reticulate evolution, or both, could have dramatic impacts on the availability of heritable variation in corals and hence their responsiveness to selection caused by climate change. These recent data suggest that much remains to be learned about the genetic and phenotypic plasticity of corals, which together may be the key to alleviating the scourge of coral bleaching. Nonetheless, evolutionary theory suggests that bleaching may be a consequence of mechanisms of conflict mediation, which are essential to maintaining the coral-dinoflagellate symbiosis.

Keywords Clonal animals · Colonial animals · Environmental stress · Octocorals

N. W. Blackstone (✉)
Department of Biological Sciences, Northern Illinois University, DeKalb, IL, USA
e-mail: neilb@niu.edu

A. P. Parrin
Department of Biological Sciences, Georgia State University, Atlanta, GA, USA

J. Guex et al. (eds.), *Morphogenesis, Environmental Stress and Reverse Evolution*, https://doi.org/10.1007/978-3-030-47279-5_11

11.1 Introduction

Concerns about the effects of climate change on coral reefs have brought increasing attention to these reef communities and to the corals that are integral to their existence (e.g., Hoegh-Guldberg 1999; Cornwall 2019). While scleractinian hexacorals ("stony" corals) produce much of the calcium carbonate that actually builds the reef, octocorals ("soft" corals) make up much of what covers the reef (Fig. 11.1). These corals and many of their cnidarian relatives are colonial, that is, structural individuals that are made up of iterated modules. The component modules ("polyps" in cnidarians) typically can acquire and process resources, and these resources can be shared via connections to other modules. In such animals, development begins with the formation of a primary polyp, which in many ways corresponds to development in unitary (i.e., non-colonial) animals (e.g., Bode 2003). Subsequently, colonies develop by processes that have no parallel in unitary animals. During colony development, corals and other cnidarians typically extend connective tissues, termed either coenosarc (in hexacorals and hydroids) or coenenchyme (in octocorals), and iterate genetically identical polyps on these tissues.

Fig. 11.1 A coral reef in microcosm. A large colony of *Phenganax parrini*, an octocoral, dominates the lower center of the reef with colonies of scleractinian hexacorals *Pocillopora damicornis*, *Seriatopora hystrix*, and *Blastomussa merleti* above

Colonial animals thus comprise a subset of clonal animals, i.e., animals that reproduce asexually by cloning. Clonal and colonial organisms differ dramatically from unitary ones in their evolutionary ecology (Jackson et al. 1985). For instance, a clone's size and how it occupies space may have much greater relevance to its fitness than post-zygotic age. Further, clonal animals provide useful exemplars of the highly faceted concept of individuality and the value of a multilevel theory of evolution (Buss 1987; Radzvilavicius and Blackstone 2018).

The divergent biology of clonal animals is particularly apparent when they are compared to the model systems of modern biology (e.g., worms, flies, and mice). Such "assembly-line animals" undergo rapid development, and subsequently their morphology remains relatively unaffected by environmental inputs (Blackstone 2007). Rather, the environment principally determines post-zygotic rates of survival and sexual reproduction. Colonial animals, on the other hand, may continue to develop for essentially their entire life span, and they can be extremely long lived. The effects of the environment on colony morphology can thus be profound, affecting both the geometry of connective tissues and the disposition of polyps on these tissues. Environmental impacts may also result in the mechanical fragmentation of a colony, leading to asexual reproduction, or partial colony death, or both. In colonies with feeding and reproductive polyps or other polymorphisms (Cartwright et al. 1999), the environment largely dictates the proportions of these polyp types.

Clonal and colonial animals are particularly likely to form symbiotic associations (Wulff 1985; Burgess et al. 2017). Indeed, reef building may require symbiosis with photosynthetic algae, and all modern reef-building cnidarians are symbiotic with *Symbiodinium* dinoflagellates (Hoegh-Guldberg 1999). When environmental stress becomes extreme, these dinoflagellates are lost, and corals bleach. Symbiotic cnidarians thus return to their primitive, nonsymbiotic state. While this nonsymbiotic state is not likely to be stable for these modern corals, it can be considered an example of reverse evolution when subject to extreme stress. As elaborated below, the evolutionary dynamics of symbiosis are likely central to both the process and the prognosis of coral bleaching.

This chapter outlines how the environment impacts colony development and life history in corals and other cnidarians. Since reef-building corals are symbiotic with photosynthetic dinoflagellates, the role of light in host-symbiont cross talk will also be clarified. The considerable phenotypic plasticity that results from environmental inputs makes corals difficult to classify but may also provide resilience in the face of environmental stress. Phenotypic variation may obscure underlying genetic variation that in turn may be due to hybridization or polymorphism. Such genetic variation may provide the raw material for selection to craft corals that are resistant to environmental stress (Cornwall 2019). Further, reticulate evolution can allow introgression of alleles from resistant forms of one taxon into another, less resistant taxon. Clarifying these issues has important implications for understanding the ability of corals to adapt to a rapidly changing environment. Nevertheless, the constraints associated with mediating evolutionary conflict in the coral-dinoflagellate symbiosis may hinder adaptation to environmental stress.

11.2 Environmental Signaling in Colony Development

Food supply and other environmental factors directly affect colony development in cnidarians. For instance, heterotrophic hydroid colonies will cluster feeding polyps more or less closely depending on the rate of feeding (Blackstone 2001, 2003). Feeding triggers contractions of mitochondrion-rich myoepithelial cells at the base of polyps, which in turn affect both the rate of gastrovascular circulation and emissions of reactive oxygen species (Fig. 11.2, Harmata et al. 2013, 2015). Both likely serve as developmental signals (Blackstone and Bridge 2005). A colony that finds itself in a food-rich environment may thus elaborate a system of closely spaced polyps, while one growing in a food-poor environment may form long threads of connecting tissues and widely spaced polyps.

These phenotypically plastic growth forms correspond to extensively studied between-taxon differences, characterized as "sheets" and "runners" (Buss and Blackstone 1991). Runners and sheets differ not only in morphology, but in life history traits as well. Runners typically exhibit widely spaced polyps. They are fast growing, early reproducing, and notoriously poor in between-species competition.

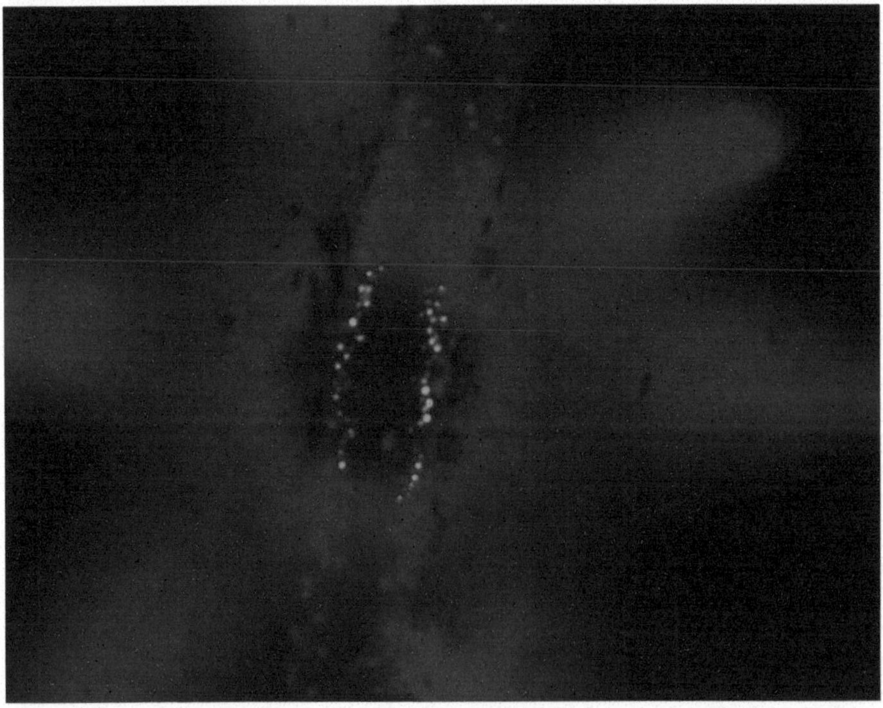

Fig. 11.2 Fluorescent micrograph of a section of a living polyp of *Podocoryna carnea*, visualized with an inverted microscope. Cell nuclei appear blue, while reactive oxygen species appear green, emitted by mitochondrion-rich myoepithelial cells. The latter are concentrated at polyp-stolon junctions (see Harmata et al. 2015)

On the other hand, sheets exhibit closely spaced polyps. They are slow-growing, late-reproducing, and effective between-species competitors. In this context, the role of genetic assimilation may be considered for the origin of these distinct forms (Blackstone 1999). For instance, representatives of a taxon that often found themselves in a food-poor environment would exhibit a runner-like form via environmental signaling. Recruitment of genes to the relevant developmental processes could stabilize this form so that it becomes the default developmental program. At the same time, metamorphosis of the planula larva might only be triggered by microorganisms that inhabit the target environment favoring runner-like growth. A parallel process can be envisioned for the evolution of sheet-like forms.

In contrast to most hydroids, corals are typically at least partly autotrophic, containing photosynthetic dinoflagellate symbionts, which have been classified as *Symbiodinium* spp. (see LaJeunesse et al. 2018, for recent updates to the classification). Colonial corals, which include the major reef builders as well as virtually all octocorals, are particularly likely to be symbiotic (Burgess et al. 2017). Photosynthesis is extremely sensitive to environmental inputs because, as a chemiosmotic mechanism, it connects external sources and sinks of electrons (Allen 1993). Colonial corals thus must coordinate and integrate signals from these symbionts as well as food-related and other signals from the environment. Unlike some hydroids, corals do not seem to use myoepithelial contractions to circulate gastrovascular fluid within the colony. Rather, ciliary currents are used (Harmata et al. 2013). In part, this may reflect the difficulty involved in distinguishing the signals of mitochondrion-rich myoepithelial cells (Fig. 11.2) from those of symbionts (Fig. 11.3), since both involve reactive oxygen species (see below for further discussion).

The centrality of the gastrovascular system to environmental signaling has drawn considerable attention to this system (Schierwater et al. 1992; Buss 2001; Blackstone and Bridge 2005). A cnidarian colony exhibits a continuous gastrovascular cavity connecting polyps via the coenosarc or coenenchyme. The gastrovascular system circulates food and other nutrients and removes waste from the colony. Patterns of pressure, shear stress, and surface tension can vary throughout the system depending on various environmental inputs. In this way, the gastrovascular system integrates information from the environment with the metabolic state of the colony, providing an effective mediator of environmental signaling (Buss 2001; Blackstone et al. 2005).

While some hydroids employ myoepithelial cells to drive gastrovascular flow, to the extent that is known corals do not do this and instead use cilia. Muscle-driven flow differs dramatically from cilia-driven flow, perhaps most notably in that the former is sequentially bidirectional, while the latter is simultaneously bidirectional (Fig. 11.4). Of particular interest in this regard is the colonial hydroid *Myrionema* (Fig. 11.5), which has photosynthetic *Symbiodinium* symbionts in the manner of many corals (Fitt 2000). Whether and how this impacts the functioning of the gastrovascular system and signaling with reactive oxygen species is not currently known. Further examination of *Myrionema* might shed light on these and other questions.

Fig. 11.3 Polyp of an octocoral, *Sympodium* sp., showing the symbionts on the surface of the polyp stalk and particularly in the tentacles

11.3 Water Currents

Currents can have a major impact on colonial cnidarians (Patterson 1992; Burgess et al. 2017). Colonies that extend into the water column must develop appropriate structural support to resist drag forces due to water movement. Further, currents carry the zooplankton that some corals feed on. Spacing and arrangement of modules must be responsive to provisioning by these resources (Jackson et al. 1985). Water currents, prey items, and light levels can interact in a variety of ways to affect colony development at a macroscale (e.g., colony geometry) and at a microscale (e.g., the distribution and density of polyps).

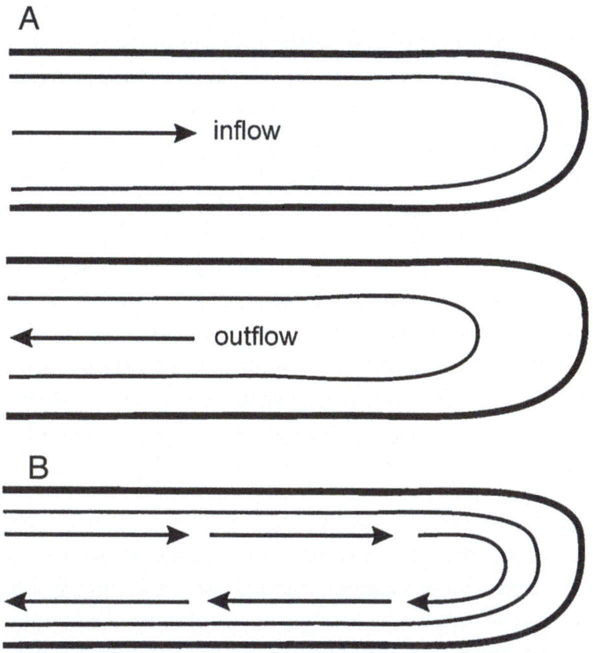

Fig. 11.4 Schemata of the cnidarian stolons: (**a**) a hydroid, in which myoepithelial contractions result in sequentially bidirectional flow, and (**b**) an octocoral, in which ciliary action results in simultaneously bidirectional flow

Fig. 11.5 The colonial hydroid *Myrionema*, which is symbiotic with *Symbiodinium*. *Xenia*, an octocoral, appears on the left and in the background

11.4 Coral-Symbiont Cross Talk

As with many animals, cnidarians may host an array of microorganisms (e.g., Bang et al. 2018). Particularly for reef-building corals, foremost among these symbionts are *Symbiodinium* dinoflagellates. This symbiosis presents both challenges (evolutionary conflict) and opportunities (photosynthesis). In terms of development, the growth of the colony must be coordinated with a growing symbiont population. This can be achieved by rapid multiplication of symbionts in developing tissue, or movement of symbionts from older to newer parts of the colony, or uptake of new symbionts from the environment, or some combination of these mechanisms. The challenge of supplying symbionts to new tissue is complicated by evolutionary conflict, which is inherent to all symbioses (Blackstone and Golladay 2018). In a symbiosis, cooperation is not an automatic outcome. Hosts and symbionts may respond to divergent selective forces. A "defector" symbiont can be selected to sequester resources from the host and symbiont community and to use these resources for its own replication. Such defectors may gain a replicatory advantage compared to cooperative symbionts that at least in part forgo reproduction and share resources with the larger community. On the other hand, by sharing resources with the host, the cooperative symbiont community establishes a durable environment for their long-term persistence. Despite these long-term advantages, the host and the larger symbiont community (the higher level unit) remain vulnerable to exploitation by defectors. Cooperation can emerge only if mechanisms of conflict mediation evolve to suppress defectors. While often conceptualized in bilateral terms (e.g., mutualism or parasitism), mechanistically these evolutionary interactions are multilateral and multilevel. In other words, even when a host-symbiont community appears to be dominated by mutualistic interactions, defecting symbionts can still arise and flourish unless they are controlled by mechanisms of conflict mediation.

Environmental input can be the key to conflict mediation. Replication of symbionts is typically limited by the host, although the mechanistic basis for this remains unclear (Davy et al. 2012). Combined with limited replication of symbionts, the biophysics of photosynthesis dictate that excess photosynthate must be exported, i.e., shared. If the export mechanisms are inactivated by mutation (Fig. 11.6), the redox state of the now-defecting symbiont will shift in the direction of reduction. High levels of reactive oxygen species (partially reduced forms of oxygen) will

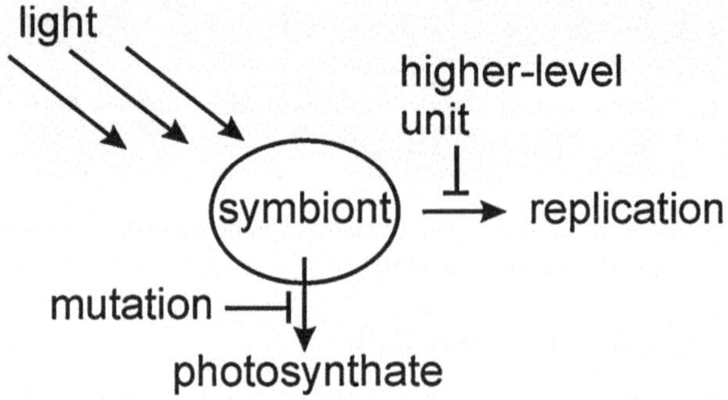

Fig. 11.6 Conflict and conflict mediation in a cnidarian-*Symbiodinium* system. With symbionts housed inside cells, replication can be constrained by the higher level unit, the host, and the cooperative symbiont community, forcing symbionts to export photosynthate. In the presence of light, a symbiont that ceases to export photosynthate can be identified by high levels of reactive oxygen species, and their host cells can undergo programmed cell death, thus eliminating the defecting symbiont

form, in turn triggering programmed cell death. This hypothesis can explain features of the coral-dinoflagellate symbiosis that otherwise appear mysterious or even contradictory. Symbiotic corals in sunny, shallow water often release half of the photosynthate that is produced daily (Burgess et al. 2017). Given such overproduction of substrate, exposing symbionts to high light levels seems of little value. When considered in terms of conflict mediation, however, exposing symbionts to high light levels, while at the same time limiting their replication, ensures that defectors essentially self-identify and self-eliminate by overproducing reactive oxygen species (Blackstone and Golladay 2018).

11.5 Cryptic Species, Polymorphism, and Reticulate Evolution

Corals are proving very difficult to classify as species, the essential unit of biology. Modern molecular studies have revealed that well-known coral "species" are in some cases not distinct but rather either polymorphic or frequent hybridizers (Forsman et al. 2017). At the same time, some taxa that have heretofore been considered single species have been revealed to be comprised of a number of cryptic lineages, raising questions of how these species maintain distinct ecological niches (McFadden et al. 2017). As with phenotypic plasticity, the lack of correspondence between morphological and genetic boundaries presents complications for the study of corals. Nevertheless, these results imply that some coral populations may incorporate considerable amounts of genetic variation.

11.6 Stress, Plasticity, and Evolution

In coping with environmental stress, organisms may alter their physiology, morphology, or other aspects of their biology (e.g., Gates and Edmunds 1999). Such plasticity can mitigate the effects of stress. Given their biology as colonial organisms, representatives of coral reef communities are expected to exhibit considerable plasticity. Nevertheless, the increasing frequency and severity of coral bleaching events (e.g., Hughes et al. 2018) suggest that this plasticity is insufficient to cope with increasingly severe environmental stress.

At a populational level, if there is sufficient heritable variation, environmental stress can lead to differential survival and reproduction. Evolutionary adaptation to the environmental stressors can thus occur. Coral reef communities may have limited potential for evolutionary adaptation. In part, this reflects their underlying biology. Colonial animals tend to be extremely large and long lived. A healthy reef may require very little recruitment of sexually produced corals to maintain itself for centuries. At the same time, a major selective event that eliminates many adult corals cannot be immediately remedied by new recruits. These recruits may need to grow for centuries before the fabric of the reef was restored. Meanwhile, the reef community can change rapidly on ecological time scales, and non-coral species (e.g., macroalgae) can quickly become ascendant in habitats once dominated by reefs (e.g., Hughes 1994).

Corals, however, may harbor considerable heritable variation and thus potential for adaptation. Either polymorphism or hybridization, or both, can provide significant amounts of genetic variation (Forsman et al. 2017). Further, reticulate evolution via hybridization can allow introgression of resistant alleles from one lineage to another (cf., Oziolor et al. 2019). Whether corals can evolve rapidly when confronted with increasing stress remains an open question (Cornwall 2019).

The crux of bleaching nevertheless seems to be the evolutionary conflict between hosts and symbionts (Blackstone and Golladay 2018). A robust symbiosis requires robust mechanisms of conflict mediation. A population structure of many small groups can often lead to such an outcome. No matter how strongly defectors are selected for at the individual level, with many small groups, purely by chance (i.e., genetic drift) some groups will comprise only cooperators. These groups of cooperators will then be strongly selected for at the group level and outcompete groups with more defectors. This scenario likely explains the evolution of the eukaryotic symbiosis and the secondary symbioses that gave rise to dinoflagellates among others (Radzvilavicius and Blackstone 2018). Corals, however, have a population structure that is entirely unfavorable in this respect, perhaps best characterized as relatively few very large and very long-lived groups. In other words, corals are large, long lived, and relatively scarce, as compared to, say, insects, and a single colony contains many trillions of symbionts. Under these conditions defecting symbionts are strongly selected for, and additional mechanisms of conflict mediation are a necessity. Housing symbionts within host cells is a way to create many small groups within a single colony. Linking these within-cell groups to photosynthate export effectively mediates evolutionary conflict (Fig. 11.6).

The catch, however, is that under high levels of environmental stress in which photosynthesis itself fails, many symbionts may emit high levels of reactive oxygen species. It is under these circumstances that bleaching occurs (Weis 2008; Parrin et al. 2017). Bleaching is a by-product of failed conflict mediation or, put another way, of a failed higher level unit. In a sense, when subject to extreme stress, the higher level unit breaks down, and the coral reverts to a nonsymbiotic state. It is perhaps instructive that under the same circumstances, *Symbiodinium* themselves do not bleach. Dinoflagellates are a product of a secondary symbiosis between two eukaryotic cells, one of which contained chloroplasts. This symbiosis does not break down during bleaching, arguably because it exhibits more effective mechanisms of conflict mediation (e.g., genome loss) that entirely prevent this. Coral reef communities, for all their grandeur, are built on a rather fragile symbiosis with weak mechanisms of conflict mediation that break down when subject to too much environmental stress. Resistant symbionts, or additional mechanisms of conflict mediation, or both, may be necessary for corals to adapt to high levels of environmental stress. Additional focus on the evolution of conflict mediation may be useful in this context.

Acknowledgments The workshop, "Evolution of Hierarchical Organization in Adaptive Systems," at Wissenschaftskolleg zu Berlin in January 2019, was particularly helpful in the development of these ideas.

References

Allen JF (1993) Control of gene expression by redox potential and the requirement for chloroplast and mitochondrial genomes. J Theor Biol 165:609–631

Bang C, Dagan T, Deines P, Dubilier N, Duschl WJ, Fraune S, Hentschel U, Hirt H, Hülter N, Lachnit T, Picazo D, Pita L, Pogoreutz C, Rädecker N, Saad MM, Schmitz RA, Schulenburg H, Voolstra CR, Weiland-Bräuer N, Ziegler M, Bosch TCG (2018) Metaorganisms in extreme environments: do microbes play a role in organismal adaptation. Zoology 127:1–19

Blackstone NW (1999) Redox control in development and evolution: evidence from colonial hydroids. J Exp Biol 202:3541–3553

Blackstone NW (2001) Redox state, reactive oxygen species and adaptive growth in colonial hydroids. J Exp Biol 204:1845–1853

Blackstone NW (2003) Redox signaling in the growth and development of colonial hydroids. J Exp Biol 206:651–658

Blackstone NW (2007) A food's-eye view of the transition from basal metazoans to bilaterians. Integr Comp Biol 47:724–733

Blackstone NW, Bridge DM (2005) Model systems for environmental signaling. Integr Comp Biol 45:605–614

Blackstone NW, Golladay JM (2018) Why do corals bleach? Conflict and conflict mediation in a host/symbiont community. Bioessays 40:1800021

Blackstone NW, Kelly MM, Haridas V, Gutterman JU (2005) Mitochondria as integrators of information in an early-evolving animal: insights from a triterpenoid metabolite. Proc R Soc B 272:527–531

Bode HR (2003) Head regeneration in hydra. Dev Dyn 226:225–236

Burgess SC, Ryan WH, Blackstone NW, Edmunds PJ, Hoogenboom MO, Levitan DR, Wulff JL (2017) Metabolic scaling in modular animals. Invert Biol 136(4):456–472

Buss LW (1987) The evolution of individuality. Princeton University Press, Princeton, NJ

Buss LW (2001) Growth by intussusception in hydractiniid hydroids. In: Jackson JBC, Lidgard S, McKinney FK (eds) Evolutionary patterns. University of Chicago Press, Chicago, IL, pp 3–26

Buss LW, Blackstone NW (1991) An experimental exploration of Waddington's epigenetic landscape. Philos Trans R Soc B 332:49–58

Cartwright P, Bowsher J, Buss LW (1999) Expression of a Hox gene, *Cnox-2*, and the division of labor in a colonial hydroid. Proc Natl Acad Sci U S A 96:2183–2186

Cornwall W (2019) The reef builders. Science 363:1264–1269

Davy SK, Allemand D, Weis VM (2012) Cell biology of cnidarian-dinoflagellate symbiosis. Microbiol Mol Biol Rev 76:229–261

Fitt WK (2000) Cellular growth of host and symbiont in a cnidarian-zooxanthellar symbiosis. Biol Bull 198:110–120

Forsman ZH, Knapp ISS, Tisthammer K, Eaton DAR, Belcaid M, Toonen RJ (2017) Coral hybridization or phenotypic variation? Genomic data reveal gene flow between *Porites lobata* and *P. compressa*. Mol Phylogenet Evol 111:132–148

Gates RD, Edmunds PJ (1999) The physiological mechanisms of acclimatization in tropical reef corals. Am Zool 39:30–43

Harmata KL, Parrin AP, Morrison PR, McConnell KK, Bross LS, Blackstone NW (2013) Quantitative measures of gastrovascular flow in octocorals and hydroids: toward a comparative biology of transport systems in cnidarians. Invert Biol 132:291–304

Harmata KL, Somova EL, Parrin AP, Bross LS, Glockling SL, Blackstone NW (2015) Structure and signaling at hydroid polyp-stolon junctions, revisited. Biol Open 4:1087–1093

Hoegh-Guldberg O (1999) Climate change, coral bleaching and the future of the world's coral reefs. Mar Freshw Res 50:839–866

Hughes TP (1994) Catastrophes, phase shifts, and large-scale degradation of a Caribbean coral reef. Science 265:1547–1551

Hughes TP, Anderson KD, Connolly SR, Heron SF, Kerry JT, Lough JM, Baird AH, Baum JK, Berumen ML, Bridge TC, Claar DC, Eakin M, Gilmour JP, Graham NAJ, Harrison H, Hobbs J-PA, Hoey AS, Hoogenboom M, Lowe RJ, McCulloch MT, Pandolfi JM, Pratchett M, Schoepf V, Torda G, Wilson SK (2018) Spatial and temporal patterns of mass bleaching of corals in the Anthropocene. Science 359:80–83

Jackson JBC, Buss LW, Cook RE (eds) (1985) Population biology and evolution of clonal organisms. Yale University Press, New Haven, CT

LaJeunesse TC, Parkinson JE, Gabrielson PW, Jeong HJ, Reimer JD, Voolstra CR, Santos SR (2018) Systematic revision of Symbiodiniaceae highlights the antiquity and diversity of coral endosymbionts. Curr Biol 28:2570–2580

McFadden CS, Haverkort-Yeh R, Reynolds AM, Halàsz A, Quattrini AM, Forsman ZH, Benayahu Y, Toonen RJ (2017) Species boundaries in the absence of morphological, ecological or geographical differentiation in the Red Sea octocoral genus *Ovabunda* (Alcyonacea: Xeniidae). Mol Phylogenet Evol 112:174–184

Oziolor EM, Reid NM, Yair S, Lee KM, Guberman VerPloeg S, Bruns PC, Shaw JR, Whitehead A, Matson CW (2019) Adaptive introgression enables evolutionary rescue from extreme environmental pollution. Science 364:455–457

Parrin AP, Somova EL, Kern PM, Millet TA, Bross LS, Blackstone NW (2017) The use of in vivo microscopy to image the cnidarian stress response. Invert Biol 136:330–344

Patterson MR (1992) A mass transfer explanation of metabolic scaling relations in some aquatic invertebrates and algae. Science 225:1421–1423

Radzvilavicius AL, Blackstone NW (2018) The evolution of individuality revisited. Biol Rev 93:1620–1633

Schierwater B, Piekos B, Buss LW (1992) Hydroid stolonal contractions mediated by contractile vacuoles. J Exp Biol 162:1–21

Weis VM (2008) Cellular mechanisms of cnidarian bleaching: stress causes the collapse of symbiosis. J Exp Biol 211:3059–3066

Wulff JL (1985) Clonal organisms and the evolution of mutualism. In: Jackson JBC, Buss LW, Cook RE (eds) Population biology and evolution of clonal organisms. Yale University Press, New Haven, CT, pp 437–466

Chapter 12
Fossil Benthic Foraminifera Morphologic Adaptation (Kleptoplastidy) Within Low-Oxygen-Bottom Water Environments, Coupled with Geochemical Insights from the Late Cretaceous in the Levant Basin

Aaron Meilijson, Sarit Ashckenazi-Polivoda, and Peter Illner

Abstract Following a multi-proxy analysis of the Upper Cretaceous high-productivity sequence from proximal and distal basins in Israel, Meilijson et al. (Paleobiology 42:77–97, 2016) provided evidence indicating that different benthic foraminifera species could survive and sustain large populations under long-term anoxic to dysoxic bottom water conditions. They proposed that massive blooms of triserial (buliminid) benthic foraminifera with distinct apertural and test morphologies during the Campanian managed to survive anoxic conditions by their capability to sequester diatom chloroplasts (kleptoplastidy) and associate with bacteria, in a similar manner as their modern analogues. This advantageous capability as well as other adaptations such as using nitrate instead of oxygen for their respiratory pathways, or changes in food type arriving to the seafloor, were all affected by the substantial shift in the depositional environment following the Campanian/Maastrichtian boundary. However, several of the hypothesis and assumptions presented in this chapter called for a continued study of the Upper Cretaceous deposits in the Levant, to better constrain the oceanographic and bottom water process in which these organisms lived.

A. Meilijson (✉)
The Dr. Moses Strauss Department of Marine Geosciences, Charney School of Marine Sciences, University of Haifa, Haifa, Israel
e-mail: ameilijs@campus.haifa.ac.il

S. Ashckenazi-Polivoda
Dead Sea and Arava Science Center, Masada National Park, Mount Masada Dead-Sea, Israel
e-mail: sarit@adssc.org

P. Illner
Institute for Mineralogy and Geochemistry, Karlsruhe University, Karlsruhe, Germany

J. Guex et al. (eds.), *Morphogenesis, Environmental Stress and Reverse Evolution*, https://doi.org/10.1007/978-3-030-47279-5_12

Here we report on a high-resolution investigation focused on the inorganic geochemical properties of two sections within the high-productivity setting of the Late Cretaceous in the Levant. Benthic foraminiferal assemblages were compared with the trace metal enrichment, bottom water renewal and water column oxygen levels, on a high-productivity seafloor. Our work focused on the occurrence and distribution of redox-sensitive/sulphide-forming trace metals obtained by analytical approaches (bulk sediment composition, ED- and WD-XRF), in the organic-rich sediments. On basis of the bulk sediment geochemistry, a principal component analysis distinguished between two factors for both sections: Factor 1 mirrors the degree of bottom-water oxygenation (Cu, S, Ni, Zn, Cr and Corg) and includes elements representing enhanced phosphorite deposition (P_2O_5, U, As, Mo, Y). Factor 2 reflects the interplay between the input of biogenic carbonate (Ca) and terrigenous material (TiO_2, Rb, SiO_2, Fe_2O_3(t), Ce, Ga, V and Al_2O_3). An additional factor was used in the distal and deeper of the two sections to distinguish between times in which dominance of siliceous or calcareous biogenic sedimentation occurred. We observe that the lowest part of the Maastrichtian contained the strongest reducing conditions, whereas the upper part was affected by a lesser degree of oxygen deficiency. Geochemical results of the molybdenum-to-organic carbon ratio reveal a change in the water mass circulation to more restrictive condition within the lower Maastrichtian, which coincides with reported sea-level rise. Based on factor analysis of the elemental distribution we demonstrate a clear connection between diatom abundance and peaks in the abundance of foraminifera species thought to have used kleptoplastidy as a morphological adaptation to cope with environmental instability, advocating previous assumptions and hypothesis. Additionally, it is evident that along the section in which fluctuations in the relative abundance of primary producers occurred, substantial shifts also transpired in the relative contribution of terrigenic material to the accumulating sediments. The synchronous occurrence of abnormally high numbers of low-diversity benthic foraminifera demonstrates the existence and success of functional adaptations. Our identification of morphological adaptations in *Praebulimina prolixa*, which are identical to those recognized in modern diatom-based kleptoplastidy of benthic foraminifera, acts as the missing link for understanding this complex system. It does so by tying between productivity, oceanography, continental-marine interactions and remarkable biochemical reciprocity and adaptiveness of present and deep-time ecosystems.

Keywords Anoxia · Foraminifera · Late Cretaceous · Environmental instability · Phytoplankton · Kleptoplastidy · Trace elements · Multivariate analysis

12.1 Introduction

The majority of benthic foraminifera colonize oxygenated pore waters and surface sediments, yet a considerable number of living species belonging to this group are known to tolerate short- or long-term anoxic-dysoxic and even sulphidic conditions (Bernhard 1993; Bernhard and Reimers 1991; Geslin et al. 2011, 2014; Moodley et al. 1998) by means of diverse physiological and respiratory adaptations such as complete or bacterial mediated denitrification, kleptoplastidy (sequestration of algae plastids by an organism) or other symbiotic associations with bacteria or archaea (Bernhard 2003; Bernhard et al. 2003, 2006, 2010, 2012a, b; Gooday et al. 2000; Hogslund et al. 2008; Koho and Piña-Ochoa 2012; Leiter and Altenbach 2010; Piña-Ochoa et al. 2010; Pucci et al. 2009; Risgaard-Petersen et al. 2006). Meilijson et al. (2016) presented a multi-proxy micropalaeontological and geochemical set of evidence from the Upper Cretaceous high productivity organic-rich sediments from the Levant Basin, demonstrating that fossil foraminifera were able to successfully colonize anoxic-dysoxic bottom waters, by using adaptations similar to those found in living species. A regional shift from buliminid to diverse trochospiral dominated assemblages was recorded in an interval with a distinct anoxic geochemical signature coinciding with a regional change in lithology. This change was triggered by an alteration in the type of primary producers from diatoms to calcareous nannoplankton, possibly causing modifications in benthic foraminiferal morphological and physiological adaptations to life in the absence of oxygen. Massive blooms of triserial (buliminid) benthic foraminifera with distinct apertural and test morphologies during the Campanian were supposedly possible due to their ability to sequester diatom chloroplasts and associate with bacteria, in a similar manner as their modern analogues. Following the shift from diatom-dominated to calcareous nannoplankton-dominated primary production at the Campanian/Maastrichtian boundary, diverse trochospiral forms existed during the Maastrichtian by using nitrate instead of oxygen for their respiratory pathways in a denitrifying environment. Species belonging to the Stilostomellidae and Nodosariidae families might have been affected by the change in food type arriving to the seafloor after the phytoplankton turnover at the Campanian/Maastrichtian boundary, in a similar manner as their mid-Pleistocene descendants prior to their extinction. Here, we shortly review these results and then critically compare them with further work performed on the Upper Cretaceous section, related to environmental shifts as deduced by elemental analysis.

Calcareous benthic foraminifera are important components of modern and ancient marine ecosystems and a key tool in palaeoceanography (Sen Gupta 1999). Recent studies have also shown that these foraminifera are able to calcify under anoxic conditions, at various depths in the sediment, with or without nitrates (Geslin et al. 2014). These adaptations allow them to survive and even be active in the most extreme oxygen minimum zones, such as those prevailing in modern upwelling systems (Høgslund et al. 2008).

The discovery of these versatile adaptations among phylogenetically diverse and geographically widespread appearances of foraminifers (e.g. denitrification in Piña-Ochoa et al. 2010) presents a new and potentially groundbreaking challenge to the field of palaeoceanography. The geologic record bears many episodes of widespread bottom-water anoxia and benthic foraminifera were reported in some of these strata (Friedrich 2010). Meilijson et al. (2016) provided the first reported evidence to validate the notion that certain fossil benthic foraminifera were able to survive anoxia by utilizing similar physiological adaptations as their modern descendants.

A major obstacle for such interpretation is providing direct evidence for the co-occurrence of benthic foraminifera and anoxic environments. Friedrich (2010) termed the difficulty to prove this co-occurrence within mid-Cretaceous oceanic anoxic events (OAE) as the 'anoxic benthic foraminifera' paradox, likely resulting from an artefact of sample spacing, i.e. the merging of hundreds to thousands of years in a single sample. Several authors have demonstrated extremely low bottom-water oxygen levels during the deposition of the Upper Cretaceous organic-rich sediments of the Levant margins while hosting a (mostly) diverse and flourishing community of benthic foraminifera (Almogi-Labin et al. 2012; Alsenz et al. 2015; Ashckenazi-Polivoda et al. 2011, 2018; Meilijson et al. 2014, 2016, 2018; Schneider-Mor et al. 2012).

In this chapter we discuss the evidence for morphological adaptations of benthic foraminiferal species to life in extreme-low-oxygen environments, as presented in Meilijson et al. (2016), while including new elemental data regarding the depositional environment of the studied sequence, promoting a discussion of other potential environmental conditions influencing the shifts observed in foraminiferal populations. Our work focuses on the Upper Cretaceous Levantine high-productivity sequence (upper Menuha, Mishash and lower Ghareb formations in Israel), which shows that different benthic foraminifera species with diverse morphologies were able to tolerate and successfully colonize anoxic-dysoxic (i.e. anoxic—no dissolved oxygen, dysoxic—0.1–1 ml/l; Altenbach et al. 2011; Sen Gupta 1999) bottom-water environments. The combination of foraminiferal records and geochemical proxies allowed us to recognize faunal shifts and their environmental settings and to take a further step in inferring specific biological evolutionary adaptations that enabled benthic foraminifera to successfully colonize anoxic-dysoxic environments, in a similar manner as their modern successors.

12.1.1 The Late Cretaceous Southern Tethyan Upwelling System

The Upper Cretaceous Levantine high-productivity sequence is a product of an extensive upwelling system that prevailed in the Southern Tethys for ~19 Ma from late Coniacian to mid-Maastrichtian (Meilijson et al. 2014; Soudry et al. 2006; Fig. 12.1). The upwelling induced a high-nutrient regime with extremely high primary productivity in the upper part of the water column and oxygen depletion at the seafloor (Almogi-Labin et al. 1993; Ashckenazi-Polivoda et al. 2011).

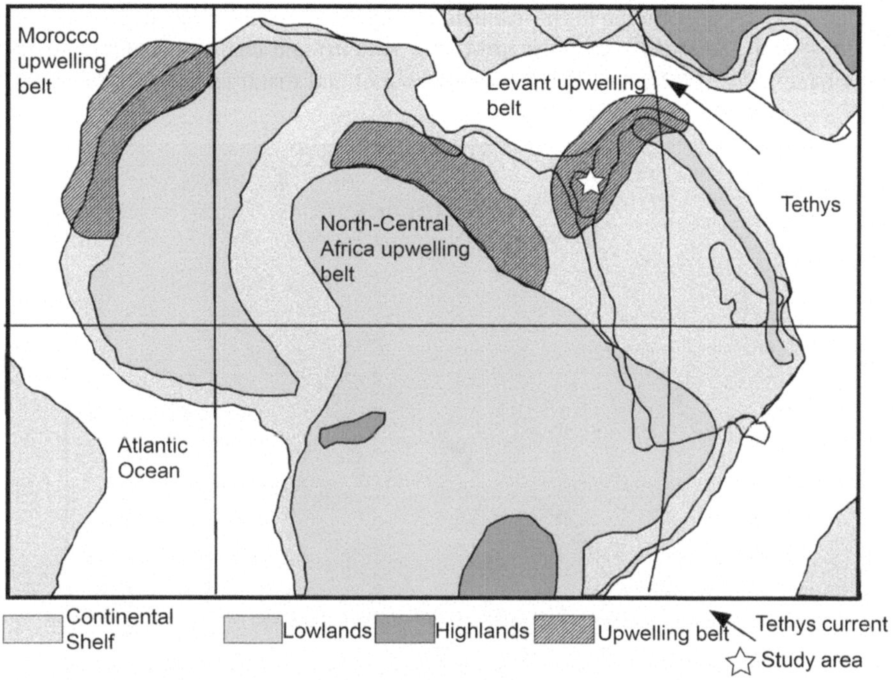

Fig. 12.1 Palaeogeographic reconstruction showing the upwelling belts that developed along the southern Tethys margin during the late Coniacian-Maastrichtian. The star marks the location of the study area (from Meilijson et al. 2016)

The diverse upper Coniacian, Santonian and Campanian sedimentary sequence (upper Menuha and Mishash formations) is enriched in silica (e.g. porcellanite beds, chert nodules and massive and brecciated chert beds) and phosphate (carbonate-fluorapatite). These lithologies disappear above the Campanian/Maastrichtian boundary (~72.1 Ma) when organic-rich carbonates (locally referred to as oil shale) become the dominant deposit throughout the entire region, accompanied by significantly elevated organic carbon levels, during the lower part of the Ghareb Formation. This lithological shift was triggered by a change in the dominant forms of primary producers in the upper water column: from diatoms along the siliceous Campanian to calcareous nannoplankton (e.g. coccolithophorids) in the Maastrichtian, as also evident by thiophenic biomarkers (Meilijson et al. 2016; Sinninghe Damste et al. 1990). Additional indirect evidence for the dominance of diatoms in the water column during this period is based on the high dinoflagellate P/G ratio of peridinoids (heterotroph diatom consumers) over gonyaulacoid (autotrophs thriving in oligotrophic stratified conditions) recorded by Eshet et al. (1994). A constant supply of diatoms from the upper water column is also indicated by the silica (porcellanite and chert)-enriched intervals in both the chert and the phosphate members of the Campanian Mishash Formation (Bein et al. 1990).

Our study focuses on the high-productivity deposits from two basins in southern (Negev–Saraf and PAMA sections in Mishor Rotem) and central (Shefela–Aderet core) Israel (Fig. 12.2), which represent proximal and distal (relative to the shore

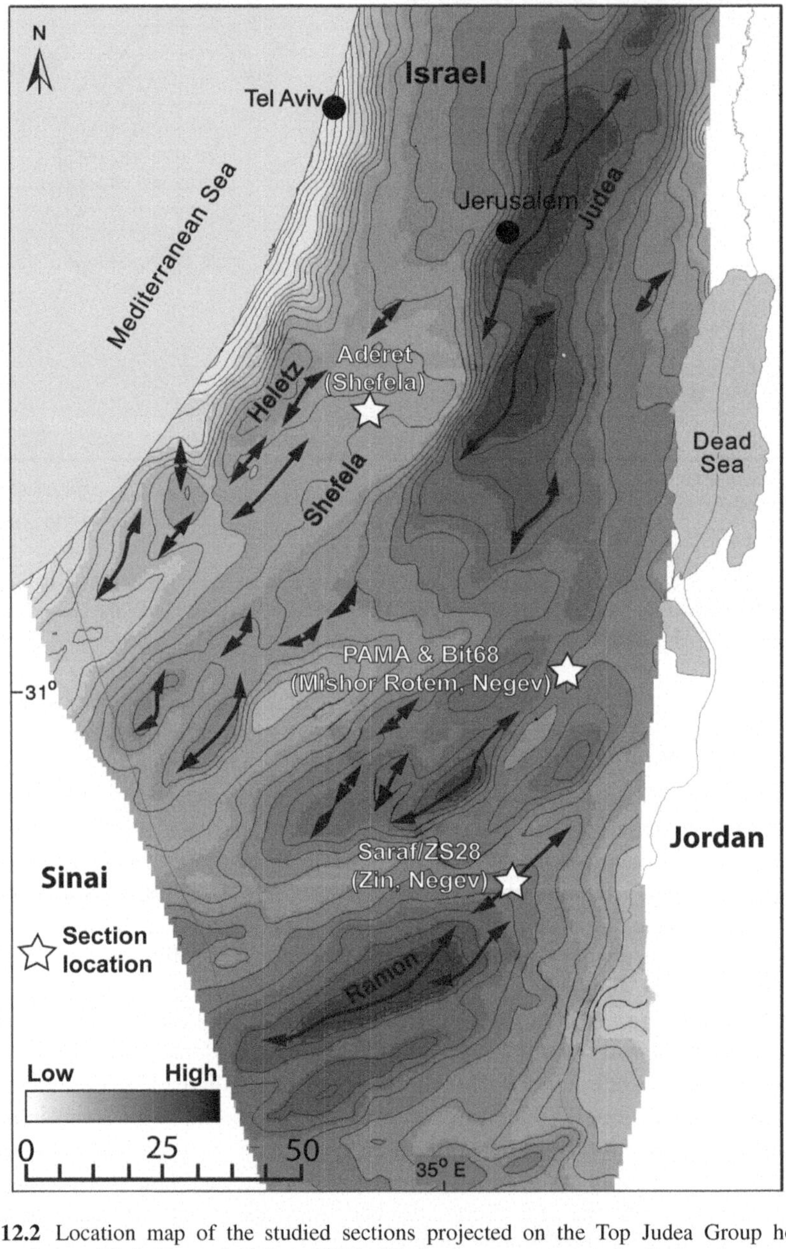

Fig. 12.2 Location map of the studied sections projected on the Top Judea Group horizon structural map (Fleischer and Gafsou 2003). Black arrows represent the Syrian Arc structural pattern (from Meilijson et al. 2016)

line) locations, respectively, within the upwelling belt. These deposits contain highly abundant and diverse benthic and planktic foraminiferal assemblages, and therefore provide an ideal test case to study the association of benthic foraminifera and anoxic environments.

12.1.2 Physiological Adaptations, Kleptoplastidy and Bacterial Symbionts of Benthic Foraminifera in the Studied Section

The unique sedimentary sequence of the Late Cretaceous upwelling regime provides clear evidence for a widespread and long-lived (~19 Myr; Meilijson et al. 2014, 2016) colonization of anoxic-to-dysoxic bottom-water environments by diverse benthic foraminiferal communities. The micropaleontological records suggest that the prominent change from buliminid dominated to diverse trochospiral and uniserial dominated assemblages, following the Campanian/Maastrichtian boundary and coinciding with a distinct regional change in lithology, was triggered by a shift in the type of primary producers in the upper water column. This shift enforced a change in the life strategies used by the benthic foraminifera to survive these conditions. Meilijson et al. (2016) discussed possible ways in which these environmental changes affected the benthic foraminiferal assemblages along the Cretaceous section of the Levant. These included kleptoplastidy, denitrification and food type dependency. Following is a review on inferred adaptations related to kleptoplastidy and bacterial symbionts which may have facilitated the survival of these diverse species under the Late Cretaceous anoxic-dysoxic conditions of the Levant.

Kleptoplastidy is the ability of heterotrophic organisms, including foraminifera, to preserve chloroplasts of algal prey they eat (e.g. Pillet et al. 2011). In this unusual 'symbiotic' association, the photosynthetic organelle is retained by the hosting foraminifera (Bernhard and Bowser 1999). This chloroplast husbandry plays an important role in surviving deepwater aphotic dysoxic environments (Bernhard 2003; Bernhard and Bowser 1999).

Living foraminifera sequestering plastids include both coiled and bi-triserial forms. All of these species share morphologic features that enable extraction of chloroplasts from their algal prey. These include 'teeth' rimming the apertures, toothed fossettes, a serrated tooth plate, elaborate test ornamentation, double-folded lip apertures and/or tubercles (Bernhard and Bowser 1999 and references therein; Fig. 12.3). Contact against the sharp tubercles or teeth rasps the moving prey, ultimately disarticulating the frustules and tearing the cell wall (Bernhard and Bowser 1999; Austin et al. 2005; Fig. 12.3d). Bernhard and Bowser (1999) sug-

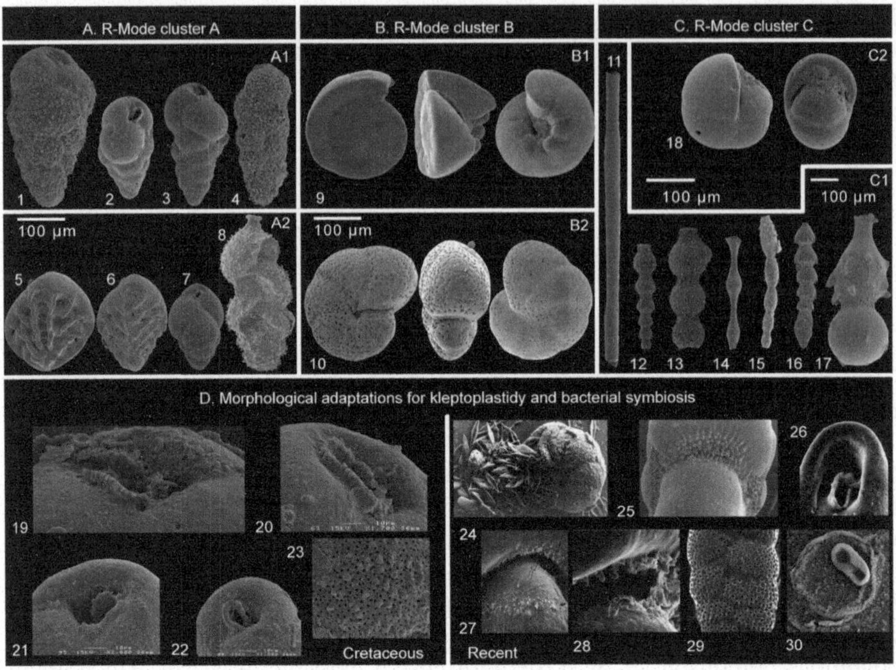

Fig. 12.3 (**a–c**) SEM micrographs of the foraminiferal assemblages demonstrating strong morphologic homogeneity. (**a**) Biserial and triserial. (**b**) Trochospiral; (**b1**) smooth tests; (**b2**) perforated by macropores. (**c**) Mainly uniserial. (**d**) Morphological adaptations for kleptoplastidy and bacterial symbiosis; Cretaceous. (19–20) Loop-shaped line of teeth surrounding the aperture and a tooth plate (*Praebulimina prolixa*). (21) Serrated tooth plate (*Praebulimina* sp.). (22) Tooth plate and double-folded lip aperture (*Praebulimina prolixa*Neobulimina canadensis). (23) Surface covered with tinypustules and pores (P. *prolixa*). Recent (24–25 from Austin et al. 2005). (24) (*Haynesina germanica* with a large diatom feeding bundle. (25) Teeth-like tubercles (H. *germanica*). (26–28) From Bernhard and Bowser (1999). (26) Double-lip aperture and tooth plate (*Bulimina elegantissima*). (27) Teeth lining the entire aperture (*Nonionella stella*). (28) Empty diatom frustule and apertural teeth (N. *stella*). (29–30) From Bernhard et al. (2010). (29) Surface pores (*Bolivina pacifica*). (30) Close-up of two pores of B. *pacifica* showing bacterial ectobionts (figure from Meilijson et al. 2016)

gested that these ornamentations can be used to identify fossil species likely to have sequestered chloroplasts.

Several of these features are found among the species that bloomed during the Campanian in the Levant, an interval also marked by a high abundance of diatoms (Meilijson et al. 2016; Fig. 12.3). These species and morphological attributes include (Fig. 12.3d, 19–23) the following: *Praebulimina prolixa*, the most dominant species in this part of the sequence, has a surface covered with tiny pustules, a loop-shaped line of 'teeth' surrounding the aperture (not apparent in literature descriptions of this species elsewhere) and a tooth plate; *Praebulimina* sp. 2 has a serrated tooth plate; and *N. canadensis* has an extensive tooth plate and a double-folded lip aperture. This raises the possibility that the buliminid blooms might have been a result of their advantageous adaptation to low-oxygen settings by using diatom chloroplast sequestration. While these buliminids are generally considered endobenthic (Corliss and Chen 1988; Corliss 1991; Thomas 1990; Widmark and Malmgren 1992), according to the TROX model (Jorissen et al. 1995, 2007) in food-rich oxygen-poor environments these taxa would migrate to the uppermost part of the sediment and on top of it, acquiring an epibenthic mode. This would have facilitated them in obtaining fresh diatom chloroplasts.

Another supporting indication for kleptoplastidy during the diatom-dominated Campanian is the distribution of several *Nonionella* species identified throughout the region, representing one of the very few spiral species found in high abundances in the Campanian interval. Species of *Nonionella* have been reported to perform kleptoplastidy in modern upwelling belts (Bernhard and Bowser 1999; Bernhard et al. 2003).

Several foraminifera species were found to host bacteria that are known to aid aerobic inhabitants of sulphidic environments (Bernhard 2003; Bernhard et al. 2006, 2010; Kuhnt et al. 2013), some of which have also been found to sequester chloroplasts. These studies show that ectobionts (Bernhard et al. 2010) are directly associated with the pores of the foraminifera test (Kuhnt et al. 2013), indicating that perforated tests may provide an adaptive advantage for associations with ectobionts. Consequently, the highly perforated test of *P. prolixa* (macropores) and *N. canadensis* (micropores) might also indicate an adaptation to a sulphidic bottom-water environment.

Subsequently, Meilijson et al. (2016) proposed that the dominant species of the bulminid assemblages bloomed during the Campanian (Fig. 12.4), specifically in association with the siliceous deposits, due to their adaptation of diatom-related kleptoplastidy. In the relatively proximal deposits of the Negev (Fig. 12.2), where siliceous lithologies are most apparent, the buliminid dominance reaches 98% of the assemblage and is uninterrupted (Fig. 12.4). However, in the more distal area of the Shefela basin (Figs. 12.2 and 12.4), where siliceous lithologies are much less apparent, the buliminid dominance is interrupted by several discreet intervals in which different trochospiral species become dominant (Fig. 12.4; see full assemblage description in Meilijson et al. 2016). The regional primary producer turnover from diatoms to calcareous nannoplankton (Meilijson et al. 2016, 2018) which occurred around the Campanian/Maastrichtian boundary marks the practical disappearance from the entire region of these potentially kleptoplastidic species.

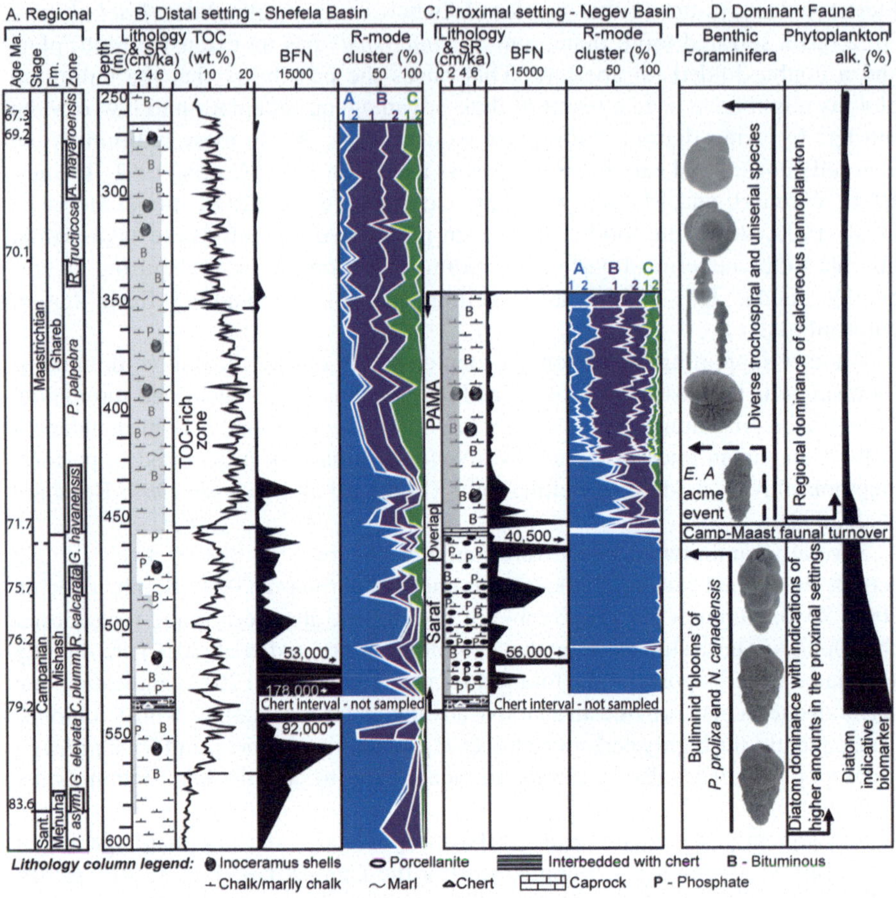

Fig 12.4. Faunal and environmental variations in the Shefela and Negev basins. (**a**) Regional data after Meilijson et al. (2014): Santonian (Sant.), *Dicarinella asymetrica* (D. asym.), *Contusotruncana plummerae* (C. plumm). (**b**) Data from the Shefela basin, Aderet core; sedimentation rates (SR), total organic carbon (TOC), benthic foraminifera specimens/gram dry sediment (BFN), relative abundance of the statistical analysis R mode clusters (see Fig. 12.3 caption for cluster description). (**c**) Negev area (Saraf core—Almogi-Labin et al. 1993; Mishor Rotem section—Ashckenazi-Polivoda 2011) depth scale is linearly accommodated according to the regional ages. (**d**) Faunal shift from biserial/triserial to trochospiral and uniserial dominated foraminiferal assemblages in the TOC-rich zone following the Campanian/Maastrichtian boundary. Phytoplankton dominance is evaluated based on published biomarker records in Jordan (Sinninghe Damsté et al. 1990). *Elhasaella alanwoodi* (E. A.), alkylthiophene (alk.) (figure from Meilijson et al. 2016)

12.2 Geochemical Proxies for Depositional and Oceanographic Reconstruction of the Late Cretaceous in the Levant

Concentrations of trace elements (V, Cr, Ni, Cu, Zn, Ga, As, Rb, Sr, Y, Zr, Nb, Mo, Cd, Ba, La, Ce, Pb, Th, U) and some major elements (SiO_2, TiO_2, Al_2O_3, Fe_2O_{3tot}, CaO, MnO, K_2O, SO_2, P_2O_5) were determined for 351 sediment samples from the PAMA (Negev Basin) and Aderet core (Shefela Basin) (Fig. 12.2) by means of energy-dispersive X-ray fluorescence (ED-XRF), using an Epsilon 5 spectrometer (PANalytical). To complement the element spectrum (Mg, Na), a selected set of samples were analysed with wavelength-dispersive (WD)-XRF, using an S4 Explorer WD spectrometer (BrukerAXS, Germany). The bulk mineral composition of the samples was analysed by means of X-ray diffraction (XRD) analysis (Kristalloflex D500, Siemens, Germany) at 40 kV and 25 mA in a 0–2θ geometry. $CuK\alpha_1$ radiation was used at angles between 3° and 63° with an increment of 0.02°, resulting in a continuous scanning of 0.5° per minute.

Statistical analyses were employed on the results of the geochemical analyses obtained by energy-dispersive X-ray fluorescence, using the STATISTICA 6.0 software package. To verify the geochemical results a correlation matrix was generated. To assess inter-element relationships, a factor analysis was performed. This method groups related variables, and thus geochemically coherent groups, into a limited number of factors that account for a substantial proportion of the variance of the data. We use the routine function for a principal component analysis (PCA). After extraction of principal components, factor axes were rotated by normalized Varimax method to facilitate the interpretation loadings. This method is used for determining depositional factors and additional factors defining the diagenesis in the sediments. In terms of a better statistical significance, only factor loadings showing a high statistical significance (>0.7) were used. Obtained results show factor loadings with an opposite algebraic sign relative to each other. Factor loadings are numerical values between +1 and −1 and represent how each factor is associated with the extracted variable from a dataset, which comprised originally more than 20 variables. Significant factor magnitude is labelled in blue for positive and in red for negative factor loadings. A negative sign of factor loadings indicates an inverse relationship, and vice versa. The length of the bar is proportional to the amount of the factor loadings and indicates the strength of the correlation between the corresponding elements and each of the factors. Based on these categories, it is possible to group results from the factor analysis into two related factors. Due to the high number of samples, samples from the Maastrichtian Oil Shale Member (OSM) were best suited for a multivariate statistical analysis. The number of samples from the top Marl Member (MM) and from the base of the profile Phosphate Member (PM) was too low, and so the statistical requirements were not met (Fig. 12.5).

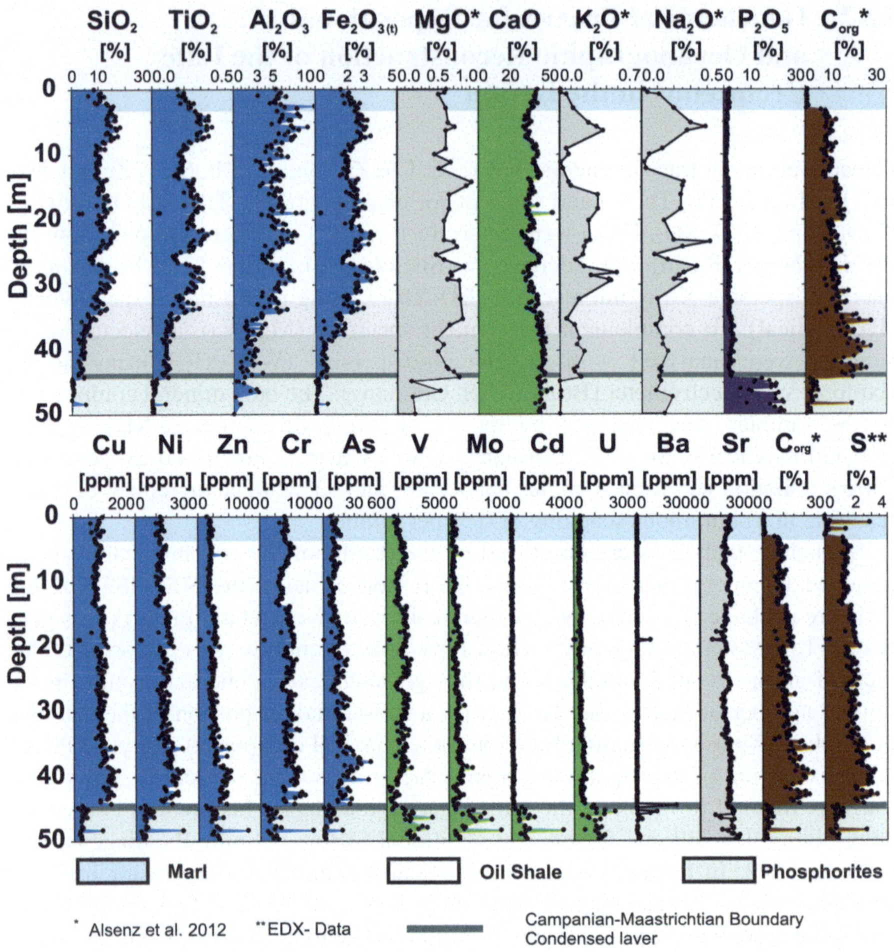

Fig. 12.5 Distribution of major and minor elements throughout the PAMA section (Mishor Rotem, Negev) profile. The 50 m profile comprises the uppermost part of the Phosphorite Member (PM), the Oil Shale Member (OSM) and the lowermost part of the Marl Member (MM). The dark green-coloured bar marks a condensed layer that signifies the Campanian/Maastrichtian boundary. This interval also indicates a sharp transition horizon between the different chemical compositions of the units. (**a**) Blue columns show the concentration of elements that are commonly strongly related to the terrigenic detrital input. Green-coloured columns (CaO) exhibit no remarkable changes along the profile. White columns are elements that are linked to light detrital sedimentation (MgO, Na_2O, K_2O). (**b**) Blue columns display the similar behaviour of chalcophile and oxy-anions with low mobility under reducing conditions with total organic carbon (C_{org}) and total sulphur (S) content, and their highest concentration at the base of the lower oil shales (LOS) 44–33 mbsf, respectively. Green-coloured columns exhibit oxyanion elements (V, Mo, U) and Cd (chalcophile) that show a favoured enrichment in the PM. Barium and Sr display no remarkable pattern along the profile

12.2.1 The Proximal Mishor Rotem Basin from Southern Israel (PAMA; Negev Sub-basin)

General Elemental Distribution in the Mishor Rotem Sediments

Major Elements

SiO_2, CaO, Al_2O_3, P_2O_5, Fe_2O_3, SO_3 and organic carbon (C_{org}) are major constituents in all of the lithological units, whereas MgO, TiO_2, Na_2O and K_2O appear in much lower concentrations (Fig. 12.5). However, a distinct variance is recorded in the major elemental composition along the section.

Phosphorite Member (PM) 49.9–45.2 m

The PM is clearly dominated by CaO with a range from 27.9 to 48.7 wt. %, and by P_2O_5 with concentrations ranging from 10.2 to 15.4 wt. %. Average MgO contents range from 0.2 to 0.5 wt. %. The concentrations of sodium, Fe_2O_3, Al_2O_3, SiO_2, MnO, TiO_2 and K_2O (10–100 ppm) are close to the detection limit. With the exception of CaO, Na_2O and Sr, all analysed elements show a clear shift in concentrations at what is known as the 'condensed layer' (45–44.4 m) which appears at the boundary between the PM and OSM (Ashckenazi-Polivoda 2011; Meilijson et al. 2014).

Oil Shale Member (OSM) 44.2–3.2 m

Calcium oxide is the major constituent in the OSM and ranges from 23.6 to 49 wt. %. Silicon ranges from 4.9 to 19.4 wt. % and shows, together with Fe_2O_3, a concomitantly increasing trend in concentrations from the base to the top of this member. Titanium concentrations range in the same magnitude (0.1–0.4 wt. %), compared to those of the PM, while Al_2O_3 ranges between 1 and 9 wt. %. Magnesium oxide values are comparable to the PM and the MM ranging from 0.5 to 0.9 wt. %. Potassium and sodium concentrations are similar throughout the OSM. The K_2O and Na_2O concentrations range from 0 to 0.4 wt. %. P_2O_5 shows a large decreasing trend in the OSM compared to that of the PM, dropping from 4.0 at the base to 0.8 wt. % at the top of the OSM. Manganese concentrations are close the detection limit (10 ppm).

Marl Member (MM) 3–0

In the MM, the average value of CaO is 41 wt. %, higher than that of the OSM. Silicon oxide concentrations are comparable to the OSM, ranging from 10.8 to 13.3 wt. %. The Al_2O_3 values show slightly lower values in comparison to the OSM, averaging at 5.4 wt. % within the MM. The average Fe_2O_3 concentration is 2.1 wt. %. Phosphorus content is very similar to the OSM with an average concentration of

1.9 wt. %. The average concentration of MgO is 0.6 wt. %. Titanium oxide (average 0.2 wt. %) shows an exceptionally low concentration within a distinct part of the profile. Sodium shows a concentration of 0.1 wt. %, while MnO and K_2O are not detectable.

Trace Elements

In the PM, trace metals such as Cu, Ni, Zn and Cr show their lowest values along the section (Fig. 12.5). P, V, Mo, U and Cd include their highest concentrations at the top of the member. All elements sharply decline in concentrations at 45 m within the condensed layer, which separates the PM from the overlying OSM.

A general decreasing trend in the concentrations of micronutrients such as Ni, Cu, Zn and Cr occurs within the OSM, positively correlating with the content of organic carbon. Oxyanion building elements with low mobility under reducing conditions, such as Mo, V, U and As, display their highest concentrations in the basal part of the OSM, and sharply decrease from 35 m and up-section. Moreover, elements like Rb, Ga and Ce, which are in conjunction to the detrital component, mirror the same increasing up-section trend, as also reported for the main elements. There are no significant deviations in terms of trace element concentrations throughout the transition from the OSM to the overlying MM.

Factor Analysis

Factor 1 accounts for 44% of the total variance. The first factor mirrors the degree of bottom-water oxygenation (Cu, S, Ni, Zn, Cr, Mo, C_{org}; Fig. 12.6), while also including elements which stand for conditions that promote phosphorite deposition (P_2O_5, U, As, Y). Rare earth elements, which are typical for conditions that favour the deposition of Ca-phosphate (phosphorite/apatite), are in close relation to the constant (low) values of P throughout the remaining profile.

The bipolar factor 2 summarizes 40% of the total variance, and is characterized by elements with high positive factor loadings for TiO_2, Rb, SiO_2, $Fe_2O_{3(t)}$, Ce, Ga, V and Al_2O_3 (Fig. 12.6). Additionally, CaO comprises a negative algebraic sign with an oppositely high negative loading.

On the basis of these results, factor 1 is used for demonstrating shifts in bottom-water oxygen content, whereas factor 2 is regarded as the combined interplay between the input of bioclastic, mainly carbonatic, material and terrigenic detrital input. The factor analysis from the bulk rock analysis in Mishor Rotem explains 84% of the total variance (Fig. 12.6).

Factor analysis and depth ranking allowed us to identify different parts within the OSM. These are (1) lower oil shale I (LOSI) in the depth between 44 and 38.2 m; (2) lower oil shale II (LOS II) in the depth between 38 and 33.6 m; and (3) upper oil shale (UOS) in the depth between 33.4 and 3 m.

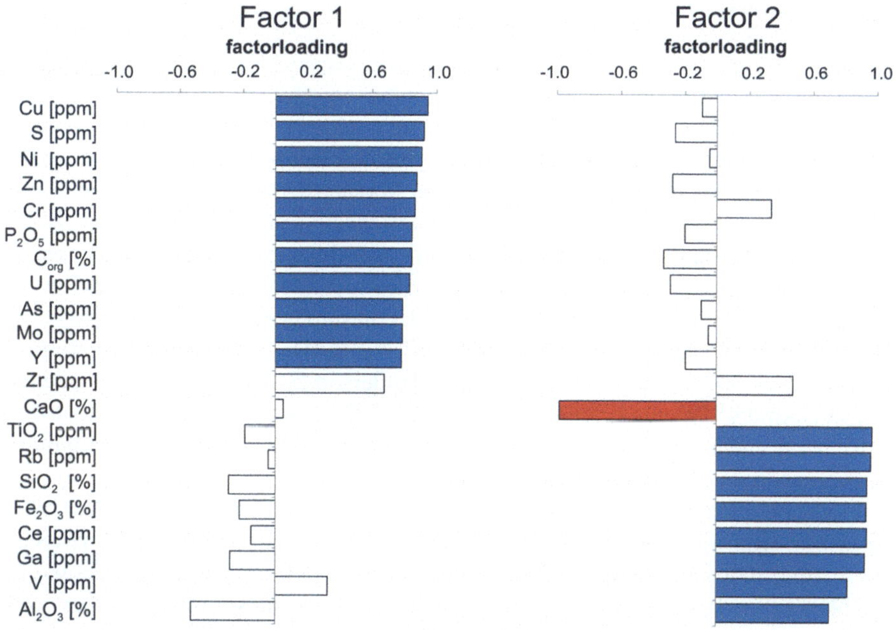

Fig. 12.6 Bar plots display the extracted factors from the bulk data set that comprises originally more than 20 variables. Factor 1 is interpreted in terms of the accumulation of redox-sensitive (chalcophile) and oxyanion-forming elements. Additionally, this factor also includes elements that favour a phosphorous component (P_2O_5, U, As, Y). Factor 2 includes variables that group elements, which are preferably interpreted in terms of terrigenic/detrital input. Interestingly, factor 2 indicates a negative correlation between CaO and the terrigenic component. This is interpreted in terms of the combined interplay between terrigenic/clastic input and calcareous/bioclastic sedimentation (foraminifera). Blue-labelled bars signify positive factor loadings, whereas red-labelled bars display negative, opposite factor loadings. Only loadings with a high statistical significance of >0.7 are shown

Factor 1: Degree of Bottom-Water Oxygenation

The elements clustered in the first factor (F1) typically accumulate in organic-rich sediments, deposited under poorly aerated and oxygen-depleted bottom waters. In addition to total organic carbon (C_{org}) and sulphur, this factor includes sulphide-forming chalcophile trace elements (Ni, Cu, Zn), and oxyanion building elements with low mobility under reducing conditions (Mo, U). Moreover, the combined occurrence of P_2O_5, As and Y, within this factor, is interpreted in terms of an occurrence of a phosphatic component. Following is a discussion of the elemental contribution to each of the factors.

Sulphide-Forming Elements (Ni, Cu, Zn)

Nickel

There are at least three possible sources for an enrichment of nickel in marine sediments. Inorganic (detrital) sources of nickel could originate from ultrabasic and mafic rocks, weathered and mobilized, strongly contributing to the elevated concentration of this element in seawater (Algeo and Maynard 2004; Calvert and Pedersen 1993). Organic sources include marine organisms that live in anaerobic bottom waters (Lewan and Maynard 1982). Another possibility for this metal enrichment can occur by downward diffusion from interstitial water, by metalation of tetrapyrrole (Lewan and Maynard 1982; Piper and Perkins 2004). Upon sulphate-reducing conditions, Ni can be incorporated in pyrite and pentlandite $(FeNi)_9S_8$. The average Ni concentration in seawater is ~0.005 ppm (Henderson and Henderson 2010), whereas a concentration in coastal sediments is ~39 ppm (Turekian 1977). Nickel shows a nutrient-like distribution and is commonly observed in high concentrations in anoxic/sulphidic areas compared to oxidizing environments.

In samples from the OSM, the average Ni content is 147 ± 29 ppm ($N = 197$; Fig. 12.5). Highest values (reaching 226 ppm) are found in the LOS I and II parts of the section, while in the upper OSM a decreasing trend in Ni values is observed, reaching 138 ± 19 ppm ($N = 134$). In the factor analysis, nickel plots together with Cr, C_{org} and sulphur (Fig. 12.7). In the UOS, Ni shows a positive correlation with C_{org} ($r = 0.67$, $N = 134$, $P < 0.05$; Fig. 12.7), and in the LOS II ($r = 0.55$, $N = 19$, $P < 0.05$) and the LOS I ($r = 0.87$, $N = 19$, $P < 0.05$).

A preferred correlation of Ni with V is observed in the LOS I ($r = 0.93$, $N = 24$, $P < 0.05$), but not in the LOS II, only to correlate once again in the UOS ($r = 0.57$, $N = 134$, $P < 0.05$) (Fig. 12.8). A weak negative correlation ($r = -0.48$, $N = 123$, $P < 0.05$) with Al_2O_3 might suggest that nickel is inadequately bonded to the clay fraction and instead forms a preferred association with the sulphurized (Alsenz et al. 2015; Meilijson et al. 2018) organic matter. The Ni/V proportion is determined by the environmental conditions at the source locality where they were deposited. Diffusion from the overlying water column and formation of metallo-organic com-

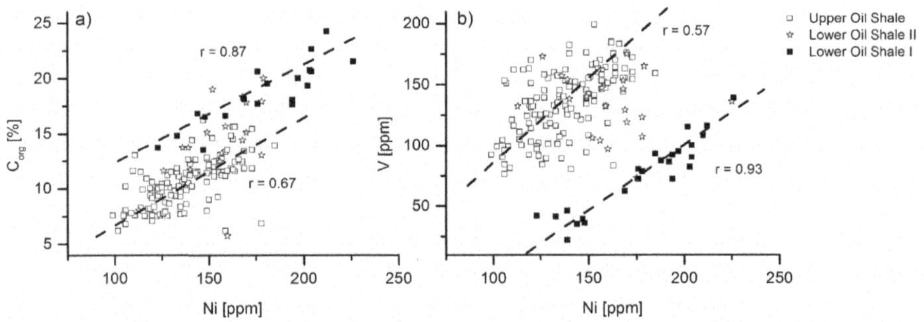

Fig. 12.7 Scatterplots of Ni vs. total Corg (**a**) and vs. V (**b**) within the OSM. All correlation coefficients are significant at $P < 0.05$

plexes and their incorporation in organic-rich sediments lead to an enrichment of V and Ni in the sediment. Normally, reservoir alteration (migration, maturation) could change this proportion.

For the research area, the organic matter (kerogen) has been determined as immature, and no migration of hydrocarbons has been shown to occur (Meilijson et al. 2018). This has been repeatedly confirmed for the Upper Cretaceous deposits in the Levant and nearby deposits based on biomarker analysis and rock-eval pyrolysis, demonstrating an immature kerogen of type I (Bein et al. 1990; Bein and Amit 1982), or type IIs (Meilijson et al. 2018; Spiro et al. 1983), 's' signifying that it is rich in sulphur.

Based on the high stability of the Ni/V proportion in different environmental condition, it is repeatedly used as a proxy related to the interpretations of source rocks and depositional environments (e.g. Barwise 1990). Barwise (1990) has shown that an Ni/V ratio <0.5 is explained by a preferred incorporation of vanadium into the porphyrin, instead of nickel (Fig. 12.8).

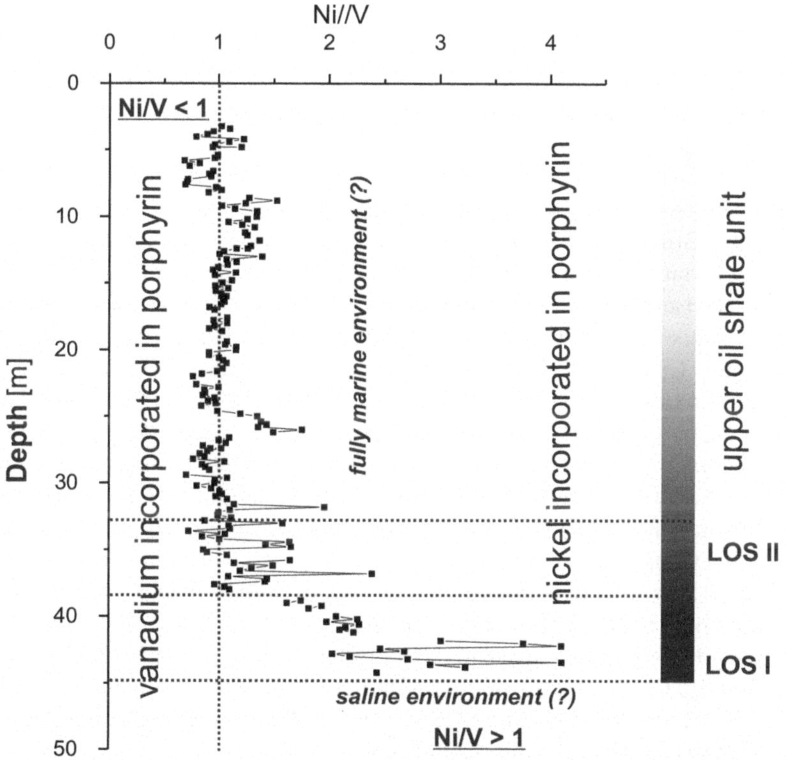

Fig. 12.8 Ni-to-V relationship, used as an empirical tool for identifying variations in the depositional history and distinct sources of organic matter. Results show that at the base of the OSM high values of Ni/V >2 occur, whereas the UOS is characterized by Ni/V values of ~1. Following the transition from the LOS to UOS, a significant change in the type of the organic matter is deduced. A transition from a saline to a more normal marine environment is interpreted following the transition between the lower and the upper OSM

An Ni/V ratio higher than one is explained by a low terrigenic input and preservation of porphyrin into an environment that changes from a marine carbonate (non-siliciclastic) to a marine siliciclastic and/or lacustrine/non-marine environment. Based on the work of Barwise (1990) and Lewan (1984), the Ni/V ratio (>2) measured at Mishor Rotem (Fig. 12.8) for the LOS II suggests a low preservation of porphyrin and incorporation of nickel instead of vanadium into the organic matter. Up-section slightly oscillating Ni/V values in the upper OSM ranging around one might indicate a steady input of marine algae (porphyrin). This differs from the LOS which is marked by Ni/V values lower than two and a trend to a higher Ni/V ratio (Fig. 12.8). Based on this premise, a tentative conclusion might be that the source of organic matter has changed significantly. According to Barwise (1990), a preferred higher Ni/V ratio, at the base of the OSM, might be interpreted in terms of a more saline bottom-water environment. This is in strong contrast to the overlying part of the OSM. A sharp contrast between these units might indicate the beginning of a new episode in primary surface water production and so a change in the type of organic matter.

Copper

One of the trace elements that are most commonly associated with organic-rich sediments is copper, also considered as a micronutrient together with Ni and Zn. All three elements Ni, Zn and Cu display a nutrient-type oceanic distribution in plankton (mostly diatoms) where they appear in high concentrations (12, 80 and 10 ppm, respectively (Bruland 1980, 1983)). However, the average concentration for Cu in shale sedimentation is 45 ppm (Schwartz 1976). Copper occurs in two valence states—Cu (II) and (I)—as Cu^{2+} can replace Fe^{2+} and Mg^{2+} in mineral structures and it mainly occurs in sulphide minerals (Okrusch and Matthes 2005). In oxic marine environments, Cu is mainly bonded to organometallic ligands (Algeo and Maynard 2004). In anoxic marine environments, Cu^{2+} is reduced to Cu^{1+} and can be incorporated in sulphide phases, mainly pyrite and chalcopyrite ($CuFeS_2$). This mostly occurs in environments that are highly affected by bacterial sulphate reduction (Huerta-Diaz and Morse 1990). In pelagic sediments Cu is coupled to a low sedimentation rate, and can possibly be diagenetically fixed by clay minerals, e.g. smectite. Copper can also be scavenged on the surface of Mn/Fe-oxyhydroxides and released by reductive dissolution, or by decaying organic matter, to the surrounding pore water (Tribovillard et al. 2006).

Cr, Ni and Zn distribution positively correlate with C_{org} in the PAMA section. Concomitantly, Cu displays a high correlation coefficient to sulphur in the UOS ($r = 0.83$, $P < 0.05$, $N = 134$), the LOS II ($r = 0.9$, $P < 0.05$, $N = 22$) and the LOS I ($r = 0.81$, $P < 0.05$, $N = 28$) (Fig. 12.9a).

A positive strong correlation is observed between copper and total sulphur content (Fig. 12.9a). This might be due to the high degree in organic matter sulphurization, which was shown to occur within the Upper Cretaceous iron-limited environment of the Levant (Alsenz et al. 2015; Meilijson et al. 2018). A high correlation is also shown between copper and rubidium in the LOS relative to the UOS (Fig. 12.9b).

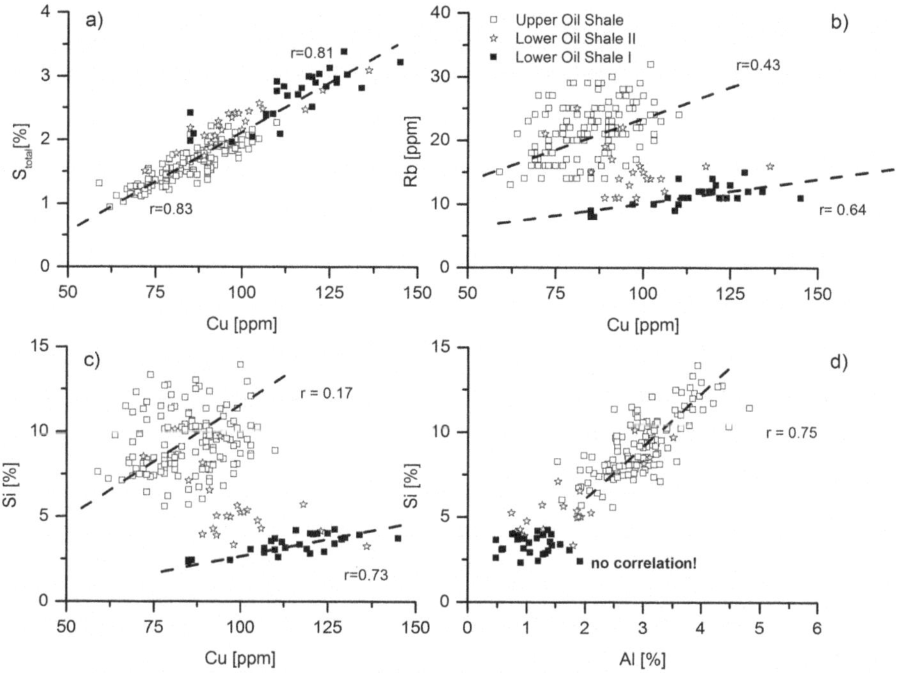

Fig. 12.9 Scatterplots of elemental ratios. All correlation coefficients are significant at $P < 0.05$. (**a**) A positive strong correlation between copper and total sulphur content. (**b**) A high correlation between copper and rubidium in the lower OSM relative to the upper OSM. (**c**) A similar pattern as the latter occurs between copper and silica. (**d**) The silica-to-aluminium plot indicates a strong correlation within the UOS, but a non-correlation in the LOS

This might be caused by a higher detrital input that induced a coupling between Cu and clay minerals (Rb). This correlation is positively affected by the low sedimentation rates in the OSM (Meilijson et al. 2014; Ashckenazi-Polivoda et al. 2011) coupled with deposition of fine detrital material with Cu, forming a bond with the high amounts of organic matter. The lower correlation coefficient in the UOS is due to the decrease in organic matter production. A strong positive correlation of Cu with Si is recorded in the LOS I ($r = 0.73$, $P < 0.05$, $N = 28$), and a weak negative correlation in LOS II ($r = -0.57$, $P < 0.05$, $N = 22$) and a very weak but statistically significant positive correlation are found in the UOS ($r = 0.17$, $P < 0.05$, $N = 134$) (Fig. 12.9c). Copper is known to be incorporated by diatoms. The strong Cu/Si correlation at the base of the OSM could be due to the relatively high diatom abundance in the underlying Campanian deposits, still influencing the content of sedimentation. Another contribution to the strong correlation might be the low sedimentation rates calculated for this interval (1–2.4 cm/kyr^{-1}; Ashckenazi-Polivoda et al. 2011; Meilijson et al. 2014), coupled with a strong degradation of organic matter, that leads to a release of Cu. The lower correlation coefficient in the upper OSM can be explained by a decrease in the primary production (diatoms). A non-correlation of Si with Al is measured in the LOS, while the UOS displays a very strong positive correlation of Si and Al ($r = 0.75$, $P < 0.05$, $N = 134$) (Fig. 12.9d). This can be explained by the higher occurrence of opal in the LOS and the fixation of Si in clay minerals or feldspar in the UOS. To conclude, our results support the notion of a biogenic source (diatoms) for the release of trapped trace elements by decaying of phytoplankton and their later fixation in a reducing environment onto the surface of sulphurized organic matter.

Oxyanions (U, Cr, As)

Uranium

Statistical analysis plots the uranium concentration within the PAMA samples with high factor loadings (0.83) within factor 1 (Fig. 12.5). There are two main processes through which uranium accumulates in marine settings. Figure 12.10(a, b) shows possible scenarios in which uranium may be deposited and fixed. The different environments in which an incorporation or adsorption of uranium on apatite takes place are shown in Fig. 12.10a. Sulphate-reducing bacteria (SRB) create an environment in which HS^-, a reducing agent for U^{6+}, is produced (Langmuir 1978). In an anoxic environment U^{4+} is immobile and is adsorbed on apatite, under the same conditions in which H_2S, HS^- and S^{2-} are stable and occur in pore water (Dybek 1963). In contrast, under oxic environments in the presence of SO_4^{2-} and HSO_4^-, most uranium is mobile in the form of U^{6+} (uranyl carbonate). The behaviour of uranium in different Eh/pH conditions is described according to the following equation (Eq. 12.1). The reduction of U^{6+} to U^{4+} takes place due to the activity of sulphate-reducing bacteria (Klinkhammer and Palmer 1991; Langmuir 1978):

$$4UO_2\left(CO_3\right)_3^{4-} + HS^- + 15H^+ \rightarrow 4UO_2 + SO_4^{2-} + 12CO_2 + 8H_2O \qquad (12.1)$$

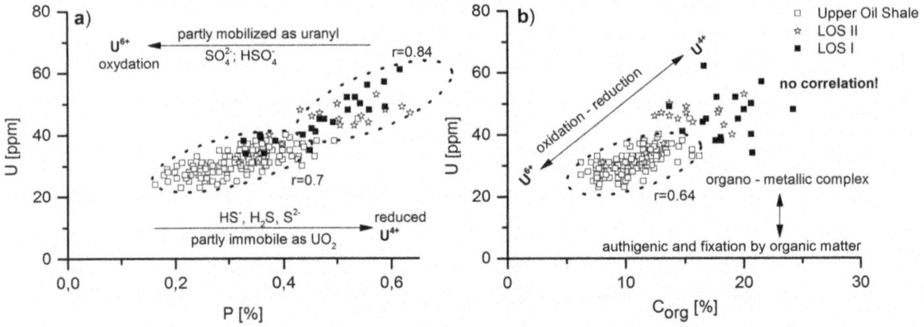

Fig. 12.10 Scatterplots of P vs. U (**a**), and Corg with U (**b**). Notice the shift to more aerated conditions in the UOS relative to the LOS

The relationship between organic matter and uranium is shown in Fig. 12.10b. Many authors have shown that uranium is fixed by organic matter, specifically where the highest organic matter flux occurs (Böning et al. 2009; Kniewald and Branica 1988; Morford et al. 2009). Our results demonstrate a low U concentration in the LOS relative to the overlying UOS. The correlation between U and C_{org} ($r = 0.64$, $P < 0.05$) may confirm that under anoxic circumstances, a part of the dissolved uranium can be reduced inorganically neither by iron and vanadium, nor by metal-reducing bacteria (Lovley 1995). This may lead to the formation of organometallic ligands which are adsorbed onto the surface of organic matter (McManus et al. 2005). In times of a short-term oxygen supply, the U^{4+} can be oxidized to U^{6+}, which is then fixed by organometallic complexes. This might explain the higher U/C_{org} correlation in the upper part of the OSM (Fig. 12.10b). The oxidation of organic matter is accompanied by a reduction of U^{6+} to U^{4+} by bacteria as shown in Eq. (12.2) (Barnes and Cochran 1993):

$$CH_2O + 2UO_2\left(CO_3^{2-}\right)_3^{4-} + H_2O + H^+ \rightarrow 2UO_2 + 6HCO_3^- + CO_2 \quad (12.2)$$

We speculate that the incorporation of U^{4+} in the PAMA section takes place under anoxic/sulphidic conditions below the sediment-water interface. In such an environment, U only occurs as U^{4+} which is stable in a pH range of 3–4 (Langmuir 1978). In contrast to the base of the OSM, possible up-section oscillating oxygenation might have facilitated the fixation of uranyl ions as U^{6+}, onto the surface of the deposited organic matter, or onto the surface of apatite. The fixation of uranyl ions signifies a fluctuating sedimentation regime with (short-term) injections of oxygen into the pore water. Consequently, it is possible to use the uranium content as an indicator for the oxygenation state in the bulk sediment (Fig. 12.10). Thus, according to the uranium distribution, and in terms of differences in redox potential, it appears that slightly higher bottom-water oxygen levels occurred in the UOS relative to the LOS (Fig. 12.10).

Chromium

Chromium also plots in factor 1 (chalcophile component), which implies a strong affinity to sulphur and organic carbon. The average seawater content of chromium in the present is 0.002 ppm at 3.5‰ salinity (Turekian 2010), compared to ~90 ppm measured in shale sediment samples (Schwartz 1976). This is in strong contrast to the organic-rich carbonates of the Ghareb Formation, where the content of chromium ranges between 250 and 700 ppm. Chromium is mostly supplied by the detrital fraction as hexavalent Cr and, under anoxic condition, stabilizes as trivalent Cr (Cranston 1983; Tribovillard et al. 2006, 2008). In an anoxic environment, the Cr(VI) is reduced to a Cr(III) by forming aqua hydroxyl-cations ($Cr(OH)_3$, $(Fe,Cr)(OH)_3$), and can be adsorbed onto the surface of Fe/Mn oxyhydroxides. A further possibility is given by the complexation by humic acids (Algeo and Maynard 2004).

Because of the large ion radii, Cr^{3+} is not known to form sulphides or to be taken up by Fe-sulphides. Cranston (1983) claimed that Cr is mainly incorporated up to 6.6 ppm into the hard parts of siliceous organisms (e.g. valves in diatoms). Tribovillard et al. (2008) show that Cr in the Orca Basin has either an authigenic or an organic origin. Borchers et al. (2005) support the observation made by Cranston (1983) which observed a correlation of Cr with TOC ($r = 0.63$), possibly indicating a connection between Cr content and productivity intensity. Borchers et al. (2005) also report an average content of 83 ppm Cr in the diatom ooze in the Namibian upwelling belt. In the Ghareb Formation, a correlation to the C_{org} content ($r = 0.67$, $n = 197$; Fig. 12.11a) was observed, possibly supporting an affinity to productivity. In anoxic environments, a reduction of Cr(VI) to Cr(III) takes place (Arias and Tebo 2003; Smillie et al. 1981). In these environments, the $Cr(OH)_3$ diffuses from the overlying water column downwards, along a redox gradient, onto the sediment/water interface where it is complexed by humic acids or is adsorbed on the surface of clay minerals (mostly illite).

Correlation Between As and P_2O_5 (Apatite)

Many mono (Na^+)-, di (Sr^{2+}, Pb^{2+}, Ba^{2+}, Mn^{2+})-, tri (REE^{3+})-, tetra (Th^{4+}, U^{4+}) and hexa (U^{6+})-valent cations substitute for Ca in apatite minerals (Pan and Fleet 2019). The same holds true for the substitution of anions $(CO_3)^{2-}$, $(SO_4)^{2-}$ and $(AsO_4)^{3-}$ for the tetrahedral $(PO_4)^{3-}$. Factor 2, with high loadings for P_2O_5 and rare earth elements (La, Y—Xenotime group), relates to conditions which favour the deposition of Ca phosphate (apatite). Arsenate is found in rivers and lakes, transported to the open sea, where it is bonded to or adsorbed on Fe/Mn oxyhydroxides, sulphates and carbonates (Ruokolainen et al. 2000). Therefore, it can be a source for biological uptake by phytoplankton. Phytoplankton can take As up in the upper water column and transport it down to the sediment as biogenic debris (i.e. faecal pellets), but not to the intermediate water where it is regenerated like phosphor (Cutter and Cutter 1995). Arsenic is thermodynamically stable as As^{5+} in oxic environments, whereas

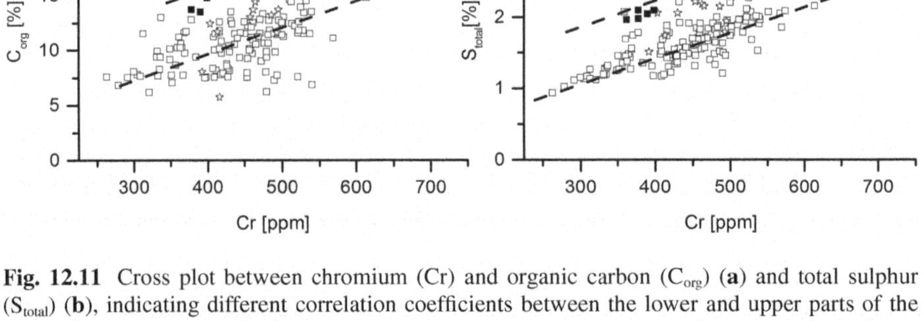

Fig. 12.11 Cross plot between chromium (Cr) and organic carbon (C_{org}) (**a**) and total sulphur (S_{total}) (**b**), indicating different correlation coefficients between the lower and upper parts of the OSM. The degree of correlation depends on the type or organic matter and the amount of sulphur

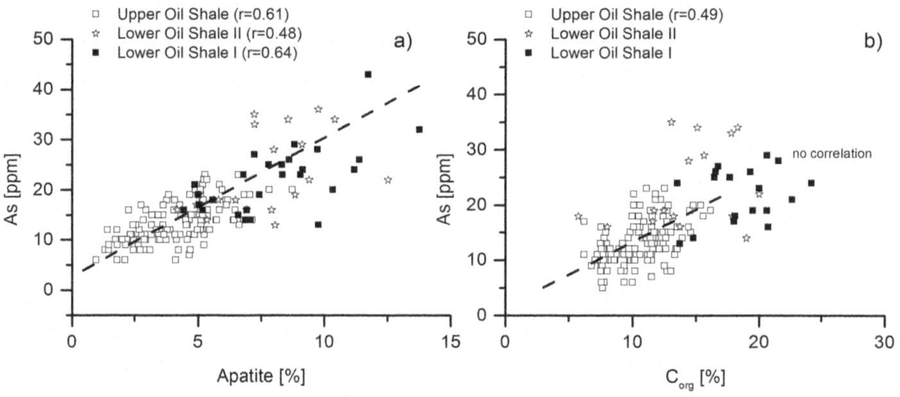

Fig. 12.12 Cross correlation between apatite and arsenic (As) (**a**) and C_{org} with As (**b**) within the OSM. Formula for calculation of apatite content was taken from American Mineralogist database

under anoxic conditions it occurs as As^{3+}. The specification of As depends on the intensity of the primary production, activity of bacteria and (de)composition of organic matter (Sohrin et al. 1997). There are at least two explanations for the correlation between arsenic and apatite observed throughout the OSM (Fig. 12.12). The diagenetic formation of arsenates ideally leads to the precipitation of johnbaumite [($Ca_5(AsO_4)_3(OH)$)] or svabite [$Ca_5(AsO_4)_3F$]. Johnbaumite is stable in a range of circumneutral conditions (Yokoyama et al. 2012). However, the observed lack of correlation with CaO does not support the existence of these minerals in the section. An increasing plausibility of the diadochy substitution of AsO_4 for the tetrahedral $(PO_4)^{3-}$ comes from an observed correlation between arsenic and apatite (Fig. 12.12a). The correlation coefficient decreases upwards along the profile, whereas a weak correlation, although statistically significant, is observed between As and C_{org} (Fig. 12.12b). This might be due to the decreasing apatite content in the upper OSM (Fig. 12.12a). Another possible explanation might be related to the preferred adsorption of arsenic onto organic matter. Here we propose that a release of trace elements by re-oxidative weathering of organic matter and subsequent dissolution/reprecipitation process led to a secondary enrichment of arsenic in the sediment. Moreover, a precipitation of pyrite together with arsenic as arsenopyrite is conceivable, which might be supported by its precipitation in a suboxic/euxinic environment (Huerta-Diaz and Morse 1990).

Deepwater Renewal Time (τ_{DW}) and Bottom-Water Oxygenation

During the Campanian/Maastrichtian transition, micropalaeontological results display a sharp decline in benthic and planktic foraminiferal assemblage compositions within the hard interval known as the 'cap rock' in the Negev sub-basins (Ashckenazi-Polivoda 2011). This event was interpreted in terms of a complete reorganization of the water mass circulation (Ashckenazi-Polivoda 2011;

Eshet and Almogi-Labin 1996). Additionally, the Upper Cretaceous sediments of the Levant were deposited in a succession of numerous silled sub-basins (Garfunkel et al. 1981; Meilijson et al. 2014). Towards the inner shelf area, the deepwater mass circulation should become more or less restricted due to the decrease in inflow of freshwater from the outer shelf and possibly due to sills created as part of the Syrian Arc deformational belt (Krenkel 1924). While sea level was believed to be higher than the edges of the sub-basin (Meilijson et al. 2014), the silled basins still acted as barriers, promoting sluggish circulation within their depocenters. The basin's deepwater renewal time (τ_{DW}) should be generally controlling the short-term (ca. million-year) drawdown of the trace metal inventory of seawater, a phenomena reported to occur repeatedly in conjunction with oceanic anoxic events, which results in elevated rates of trace metal removal within the widespread anoxic facies (Algeo and Rowe 2012). The molybdenum concentration in the sediment depends on the concentration of dissolved Mo in seawater and the amount of organic carbon in the sediment (Algeo and Lyons 2006). In times of highly reducing/euxinic conditions, molybdenum correlates with the organic carbon concentration in the sediment. A calculation of the rsMo/C_{org} ratio therefore leads directly to the Mo concentration in seawater. Deviations in the regression slope (rs) can be used to reveal information about the τ_{DW} (Fig. 12.13). The application of rsMo/C_{org} as a proxy for severe water mass restriction was originally shown by Algeo and Lyons (2006) and Algeo and Rowe (2012). In modern marine sediments the concentration of Mo is ~105 nM, which corresponds to an rsMo/Mo of ~45. This value reflects

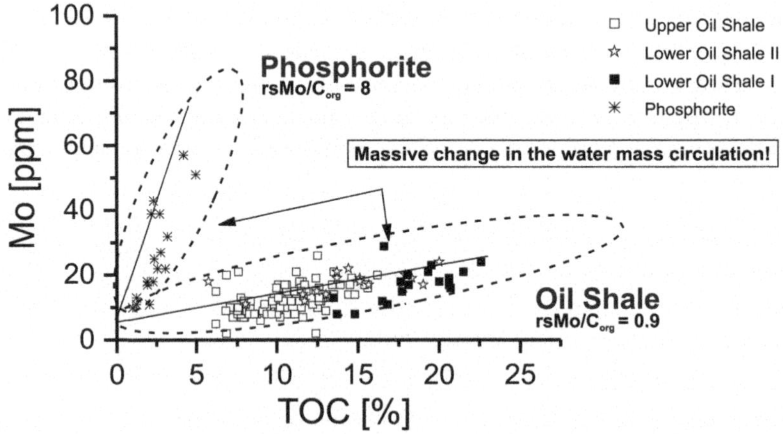

Fig. 12.13 Mo-to-TOC plot for the Upper Cretaceous section from Mishor Rotem in the southern Negev Basin. Notice the clear distinction between the regression line (rs) of the Phosphate Member (PH; Campanian) samples and the overlying Oil Shale Member (OSM; Maastrichtian) samples, indicating a much lower τDW in the latter

conditions of open water mass exchange. A decreasing trend in rsMo/C_{org} is interpreted in terms of a more restricted water mass with a lower τ_{DW}.

Our results suggest that from the underlying PM to the OSM in the Negev, the rsMo/C_{org} decreases rapidly from ~8 to ~1 (Fig. 12.13), with a regression slope of 8 (PM) and 0.9 (OSM) and an intercept on the Mo axis of 7 ppm and 5.5 ppm, respectively. For the overlying Marl Member, no correlation between the Mo/C_{org} is observed. Values for the PM imply a water mass circulation comparable to the Framvaren Fjord in Norway, a permanently super-anoxic fjord in southern Norway with very restricted water circulation (Algeo and Lyons 2006; Yao and Millero 1995). The overlying OSM shows a rather puzzling picture. The slope of the regression line (rsMo/C_{org}; Fig. 12.13) is exceptionally low in this unit and could be interpreted in terms of a (even further) reduction in bottom-water circulation, which might be much lower (fivefold) than in the modern Black Sea (rsMo/C_{org} = 4.5). An ancient example for severe water mass restriction was demonstrated in the Mississippi Barnett Formation, which yielded an rsMo/ C_{org} of ~2 (Algeo and Rowe 2012). Also, the Lower Toarcian Black Shales (Cleveland Basin) exhibit an extreme example for low rsMo/C_{org} values of ~0.5; this extraordinary low value is explained by the combination of warm and wet climate and the installation of a deep pycnocline, where local sills barred an exchange with the open ocean, despite higher sea levels (McArthur et al. 2008). The non-correlation between Mo and organic carbon observed in the overlying Marl Member potentially indicates a very abrupt return to open normal marine conditions in terms of τ_{DW} and bottom-water oxygenation.

To conclude, the Campanian/Maastrichtian boundary marks a substantial change in the deepwater circulation from low τ_{DW} and water circulation at the upper PM to a massive reduction in the deepwater circulation in the OSM. This also coincides with a substantial shift in foraminiferal assemblages, from dominance of bulminid to mainly trochospiral benthic foraminiferal morphologies in the PAMA section (Ashckenazi-Polivoda et al. 2011) and in the Shefela Basin (Meilijson et al. 2015). And indeed, after passing the Campanian/Maastrichtian boundary, a regional rapid in the sea level is reported (Meilijson et al. 2014), which could have triggered a deepening of the bottom-water line, consequently leading to the installation of a pycnocline. In combination with the sills that hamper the deepwater circulation between different basins, these causes would promote black shale accumulation and preservation in highly stagnate bottom-water mass conditions, specifically in the lower organic-rich carbonate unit.

Factor 2: Bioclastic vs. Terrigenic Input

The elements clustered in the second factor (F2) typically delineate the interplay between carbonate and terrigenous input. Following is a description of these attributes.

Bioclastic Input

In factor 2, calcium shows high factor loadings (–0.98) in an opposite trend relative to the element suite indicative of detrital deposition (Fig. 12.5). X-ray diffraction estimates of the calcite content identified the carbonate as calcite. We assume that the main part of Ca is biogenic and that the main contributor to the bioclastic input is foraminiferal and calcareous nannoplankton carbonate, and that gypsum/anhydrite and apatite are very minor constituents in the OSM. This is supported by the XRD analysis. The Al record is attributed to variable dilution of the detrital signal by $CaCO_3$ which constitutes ~25–76 wt. % ($N = 165$) of the bulk sediment, supporting a strong influence of carbonate productivity on the sedimentation. Consequently, we propose a depositional regime dominated by a strong biogenic marine input. This conclusion is consistent with interpretations given to OAE3 black shale deposits from the deep Ivory Basin at ODP Site 959 (Ravizza and Paquay 2008; Schumacher and Lazarus 2004) and OAE2 black shales from the Tarfaya shelf basin off SW Morocco (e.g. Becknamm et al. 2011; Wagner 2002).

Terrigenic/Detrital Input (Al, K, Mg, Ti, Zr)

Aluminium is mainly bonded to detrital silicates in continental marginal at settings and thus represents continental input. Other typical detrital elements include K, Mg, Ti and Zr. The negative correlation between Al and Ca in the PAMA section (Fig. 12.5) might document the dilution effect caused by the delivered detrital input. An extraordinary low content of all detrital elements appears at the base of the OSM (Fig. 12.6), indicating a very low detrital input with an increasing trend along the profile.

Typical elements of light silicates and clay minerals are K, Na and Mg, in plagioclase and K-feldspar. Cross-correlations of K and Na with Al suggest a stronger bond between K, Na and Mg relative to Al, specifically in the UOS (Fig. 12.14). Statistical results reveal good correlations of Al with K in the UOS ($r = 0.65$, $P < 0.05$, $N = 23$), and no correlation in the LOS. In the UOS, a good correlation of Al and Na is observed ($r = 0.69$, $P < 0.05$, $N = 23$). These elements (Na, K, Mg) are also present in seawater, and Mg may be incorporated in carbonates (Mg calcite and dolomite). A statistical non-correlation of Mg with Al is suggested by the analyses obtained by XRD, in which dolomite is detected as a minor component.

The cross plots between Fe, Al and Zr lead to similar conclusions, due to distinct correlation trends along the profile (Fig. 12.14). For example, the correlation between Fe and Al is moderate in the UOS ($r = 0.65$, $P < 0.05$, $N = 137$), high in the LOS II ($r = 0.82$, $P < 0\ 05$, $N = 22$) and non-correlative in the LOS I (Fig. 12.14a). The same pattern is shown by Ti that also correlates well with Al in the UOS ($r = 0.71$, $P < 0.05$, $N = 137$) and LOS II ($r = 0.81$, $P < 0.05$, $N = 22$), whereas no correlation is obtained between the two elements in the LOS I (Fig. 12.14b). A similar pattern is shown by the correlation of Zr with Al (Fig. 12.14c) - in the UOS unit a weaker but significant correlation ($r = 0.58$, $P < 0.05$, $N = 137$) exists, with a non-

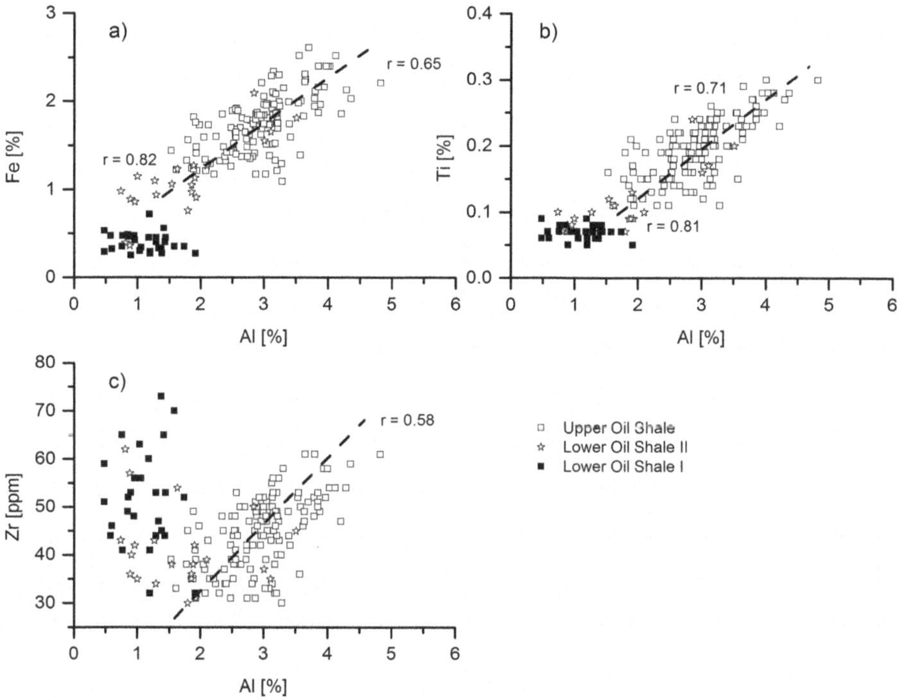

Fig. 12.14 Cross plots show Fe vs. Al (**a**), Ti vs. Al (**b**) and Zr vs. Al (**c**)—the different deposition of heavy minerals (Zr, Ti) and detrital contributors (Fe) and their variation alongside the profile. All plots display a non-correlation in the LOS I. A special status is given by Zr, which also shows a non-correlation with Al in the LOS II. All of these elements tend to correlate in the UOS. This pattern suggests a sharp shift in the composition of delivered sediments

correlation within the LOS (Fig. 12.14c). We will address the question as to what can be the reason for these patterns, after the following section.

Heavy Minerals: Ternary Diagram (Zr–Ti–Al_2O_3)

Changes in the sedimentary input are inferred by the application of a ternary diagram (Fig. 12.15) that yields information on the maturity of the sediments. A preferred enrichment of Zr is known to occur in the coarse-grained fraction, whereas TiO_2 is mostly enriched in the fine-grained fraction, together with Al_2O_3. Generally, after the preliminary settling of heavy minerals they are not regularly added to the transport medium, as is the case with quartz, and tend to remain in the proximal deposition areas in comparison with light minerals (Morton and Hallsworth 1999). Conversely, Al_2O_3 tends to increase with increasing TiO_2/Zr proportions, and vice versa. Thus, the ternary diagram helps to interpret the sorting process, regarding the path travelled by the deposited sediment, and its application as a tool for the chemical identification of shales and sandstones has been well established (e.g. Garcia et al. 1994), indicating for example changes in transport mode and delivery source.

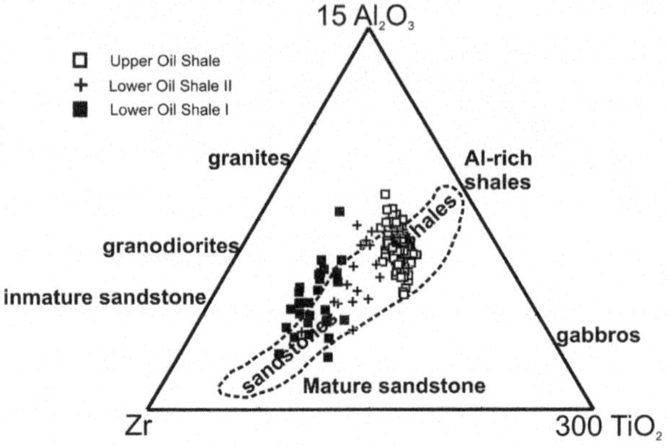

Fig. 12.15 A ternary plot based on the premise that Al, Ti and Zr are immobile during chemical weathering (Garcia et al. 1994). Typical minerals are rutile, zircon and ilmenite. The content of these minerals varies in response to the degree of weathering (leaching process). The ternary diagram displays sorting-related trends in the Al–Ti–Zr relationship: Oxides are expressed as weight percentage and Zr as ppm; Al_2O_3 is presented by a factor of 15, and TiO_2 by a factor of 300 in order to centre the sedimentary cloud. A material with a clastic component that yields a higher level of maturity dominates the base (short transport paths). A gradual change towards the top of the profile, to more immature, clay-rich material, is indicated. A gradual transition horizon starts around 33 m

Additionally, a tentative conclusion might be that in periods of deposition, the composition of the delivered sediment has gradually changed. Plotting the PAMA data on the tertiary diagram (Fig. 12.15) indicates changes in clastic sediment source along the section. A sediment with a clastic component that yields a higher level of maturity dominates the base of the section in the LOS I, indicating a relatively short transport path. A gradual change towards the top of the profile to more immature, clay-rich material is indicated. We concluded that the transport routes became gradually longer and the material came from increasingly farther surrounding landmasses from the LOS I, through the LOS II, with highest values of Al-rich sedimentation towards the UOS (Fig. 12.15). This would indicate that fine-grained sediments/clay minerals should be most dominate within the UOS.

Clay Minerals

Trace elements like Ga, Rb, Ba and Ce as studied through factor analysis plot together with Al in factor 1 (Fig. 12.6), suggesting a possible association to authigenic clay mineral formation (e.g. kaolinite, illite and montmorillonite) or weathered forms of feldspar (biotite and muscovite) (Spiro and Rozenson 1980). Based on XRD analysis, we have detected a high relative abundance of the three most important groups of clay minerals: kaolinite (detrital), smectite and illite (authigenic; Spiro and Rozenson 1980) (Fig. 12.16).

Recent investigations confirm these results and indicate that the dominant clay mineral in the Levant high-productivity sequence is smectitic IS (interstratified

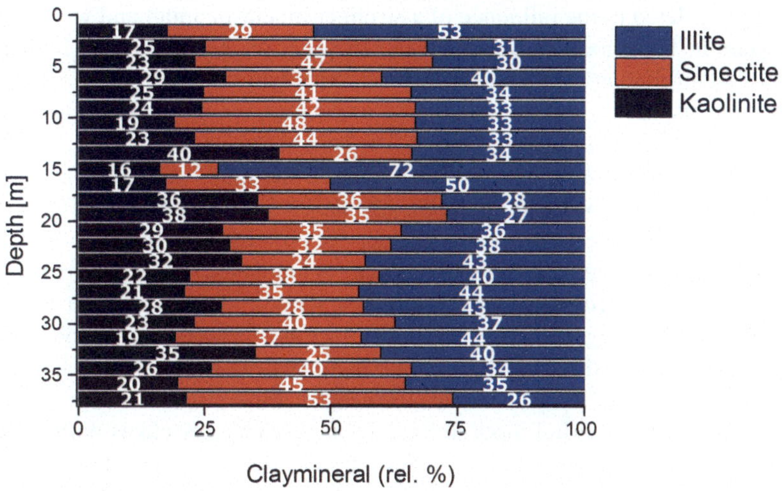

Fig. 12.16 Relative fluctuation in the clay mineral content (OSM at 35–3 m, MM at ~3 m)

illite/smectite) (Shoval 2004a, b). Moreover, it is worth noting that no terrestrial magmatic activity occurred in the region between the Santonian and late Oligocene time (e.g. Almogi-Labin et al. 2012; Shoval 2004a, b). Thus, the occurrence of smectite cannot be a result of weathering of terrestrial basaltic rocks (Shoval 2004b). Thereby argillized submarine volcanic rocks act as the main sources of illite-smectite interlayers. It is thought that these main clay components, especially illite-smectite, formed as part of the detrital contribution to the whole sedimentary record and were transported by upwelling currents across the continental shelf towards the area of deposition (Shoval 2004a, b). Consequently, these clay minerals are transported by upwelled water across the continental slope to the inner shelf area, as was also shown to occur in recent sediments in the Benguela upwelling system (Inthorn et al. 2006). The weathered sandstone reflects a continental source of the Arabo-Nubian Shield, whereby the detrital kaolinite is transported. The occurrence of kaolinite represents a continental weathering product formed during a warm and humid period at the southern Tethys margin (Bolle and Adatte 2001).

Although this is speculative, we suggest that the observed fluctuations in clay mineralogy (Fig. 12.16) are due to changes in the direction and the relative intensity of the different detrital transport mechanisms and authigenic clay formation mechanisms. These analysis-based assumptions advocate a stronger detrital input for the UOS, whereas in the LOS a heavier mineral contribution to the sedimentary record took place, as also suggested by the elemental analysis (Fig. 12.15). This indicates a substantial shift in the depositional regime which might have been caused by changes in the transport mechanisms or sea-level changes along the Upper Cretaceous, cumulating in the UOS.

These conclusions are further validated by the application of a cross plot between factors 1 and 2 (Fig. 12.17). The *y*-axis represents the nature of sedimentary input, with higher factor scores on this axis related to a higher input of detrital material and lower scores related to the influence of a stronger bioclastic input and weak detrital sedimentation. The *x*-axis shows the degree of water oxygenation, with higher factor scores signifying an increasing trend to anoxic conditions.

This plot displays a strong statistical significance for the correlation between calculated factor scores of factors 1 and 2. The base of the OSM displays an excellent correlation ($r = 0.92$, $P < 0.05$, $N = 14$), while the UOS exhibits a very weak correlation while still being statistically significant ($r = 0.3$, $P < 0.05$, $N = 133$). Samples from the lower part of the profile (base of LOS I) correlate to a more intensive bioclastic input or a higher carbonate sedimentation and a low oxic level. Conversely, samples from the top of the OSM show a relatively higher terrigenic input with higher oxygen levels. Scores of factor 2 describe a relatively flatter gradient at the base of the oil shales, indicating relatively lower oxygen levels. The degree of oxygen deficiency tends to decrease gradually up-section to ca. 38 m (LOS I). The shift in detrital input around 35 m (LOS II) correlates with a similar shift towards even lower oxygen levels. The oxygen deficiency of the bottom waters, as reflected by the scores of factor 2, generally tends to correlate positively with the amount of detrital input (factor 1). The slope of the regression line stabilizes for the rest of the section (the UOS—29–3 m; Fig. 12.17).

Lastly, we report on the good correlation identified between the nature of terrigenic input, degree of water oxygenation, and number of benthic foraminiferal species per gram dry sediment (Ashckenazi-Polivoda et al. 2011) (Fig. 12.18).

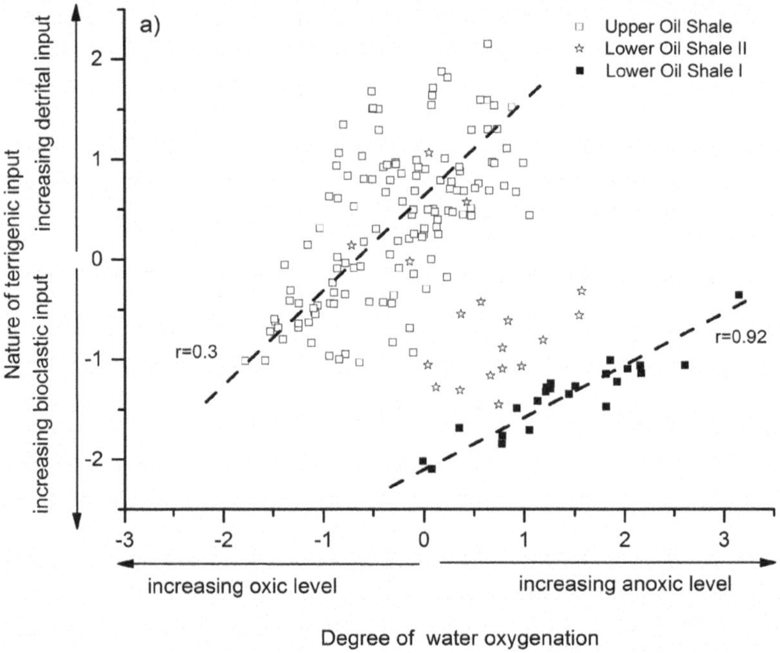

Fig. 12.17 Nature of terrigenic input (factor 2) vs. the degree of water oxygenation (factor 1)

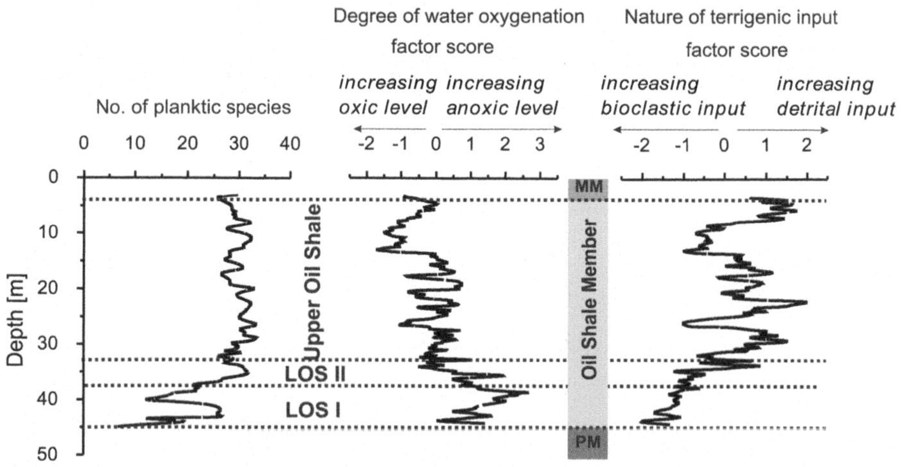

Fig. 12.18 Depth plot displays the adaptation and radiation of benthic foraminifera in relation to shifts in bottom-water oxygen levels and nature of terrigenic input. While benthic numbers increased in response to the change in the environmental conditions at the transition into the UOS at 33 m, they have, nevertheless, adapted to the highly anoxic underlying conditions

Benthic foraminiferal numbers increase from 10–15 specimens per gram dry sediment at the LOS during lower bottom-water oxygen levels to 40–50 in the UOS during times of more elevated bottom-water oxygen levels (Fig. 12.18).

Factor Scores

The calculated factor scores display a variable drift in depth ranking along the studied section and show three distinct depth ranking markers occurring at 33, 13 and 5 m (Fig. 12.19). The base of the OSM starts with high anoxic levels as shown by its high factor scores. Along the profile, a systematic trend is observed, suggesting a shift to more aerated condition at the top of the OSM. In terms of the nature of the terrigenic input, the base of the OSM is characterized by a higher contribution of calcareous sedimentation which continuously decreases up-section. The opposite is true for the terrigenic/detrital related scores, which are low at the base of the OSM and increase up-section. Between 33 and 13 m, the factor scores of the nature of the terrigenic input and corresponding degree of water oxygenation show a similar trend. The next marked change in the depth ranking takes place between 13 and 5 m. According to this analysis, the environment has shifted into continuously more oxic condition accompanied by an increasing detrital input. Significant changes between the interplay of bioclastic and detrital input are observed along the section (28–23–18–5 m, grey-shaded area; Fig. 12.19).

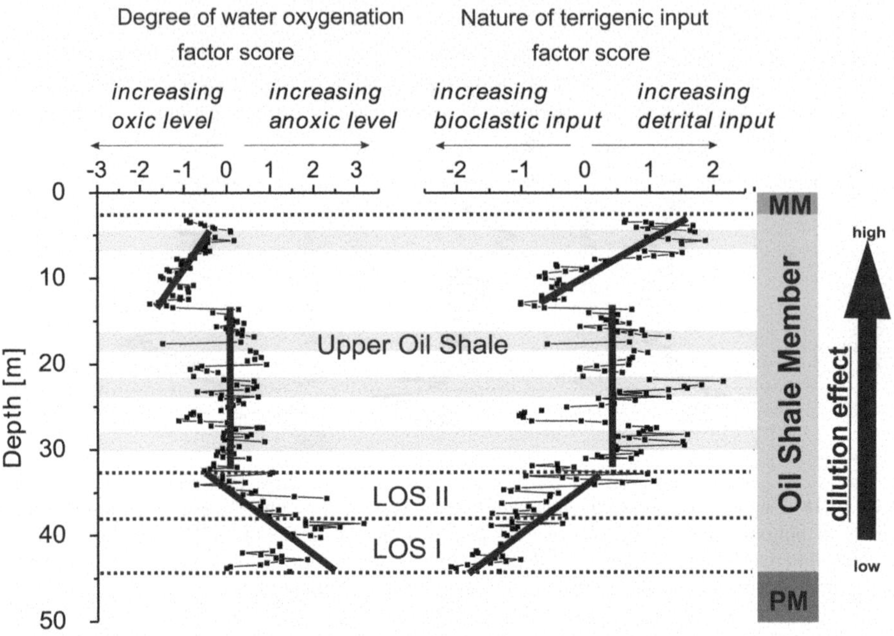

Fig. 12.19 Stratigraphic variation in factor scores for factors 1 and 2. The calculated factor loadings, yielding factor scores plotted along the profile, indicate that the lowest bottom-water oxygen level occurred at the base of the profile, continuously increasing up-section. The highest biomass (calcite-forming organism) production also occurs at this interval. Factor scores of factor 2 (nature of the terrigenic input) show the combined interplay between bioclastic (carbonate) and terrigenic (detrital) input. Obtained factor scores permit to separate the profile into three distinct subunits, which show differences in their geochemical properties. Bold lines may underline the principal tendency for direction towards the environmental evolution of the profile

12.2.2 The Distal Shefela Basin from Central Israel

The same analytical and statistical methods were employed on the samples obtained from the Shefela Basin from central Israel. Additional details on the elemental composition of the Upper Cretaceous sediments from the Shefela Basin are reported in Meilijson et al. (2016). Here we have performed the statistical factor analysis, as presented for the PAMA section above, to further assess not only the degree of bottom-water oxygenation (as done in Meilijson et al. 2016), but also the terrigenic input into the basin. On the basis of the elemental analysis (Fig. 12.20), three factors were determined. Factor 1 is regarded as the combined interplay between the input of bioclastic material (mainly carbonate) and terrigenous detrital input, whereas factor 2 is used for demonstrating oxygen depletion. Factor 3 is mainly controlled by the distribution of CaO, Sr and Mg, relative to that of SiO_2. The former is interpreted to represent a greater abundance of foraminifera and calcareous nannoplankton, and the latter to represent a higher abundance of siliceous diatoms. Figure 12.21 shows the change in the depositional environment as interpreted by the three factor scores.

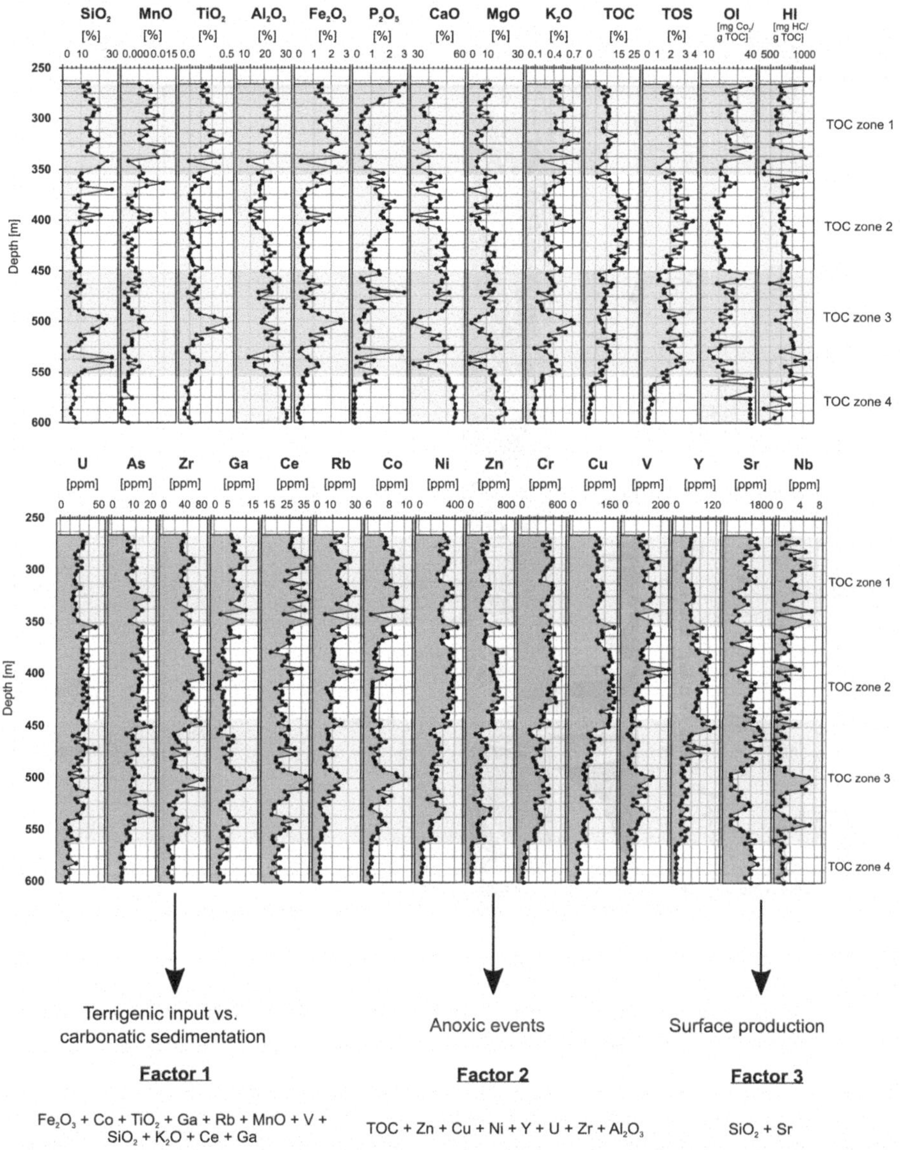

Fig. 12.20 Trace element analyses of the Shefela Basin samples have been interpreted by means of multivariate statistical methods in order to assess the inter-element relationships. The calculated factor loadings are shown in the bottom of the figure. Only factor loadings that show values higher than 0.7 were used. The largest three eigenvalues were found to account for 82% of the total variance. Thereby, three factors have been selected

R-mode cluster A includes biserial and triserial benthic foraminifera identified in the Aderet core in the Shefela Basin (blue-shaded cluster in Fig. 12.21). This is the cluster which includes the benthic foraminifera characterized by morphological adaptations which allowed them to use kleptoplastidy and to better associate with symbiotic bacteria, aiding in their survival within the extremely low bottom-water

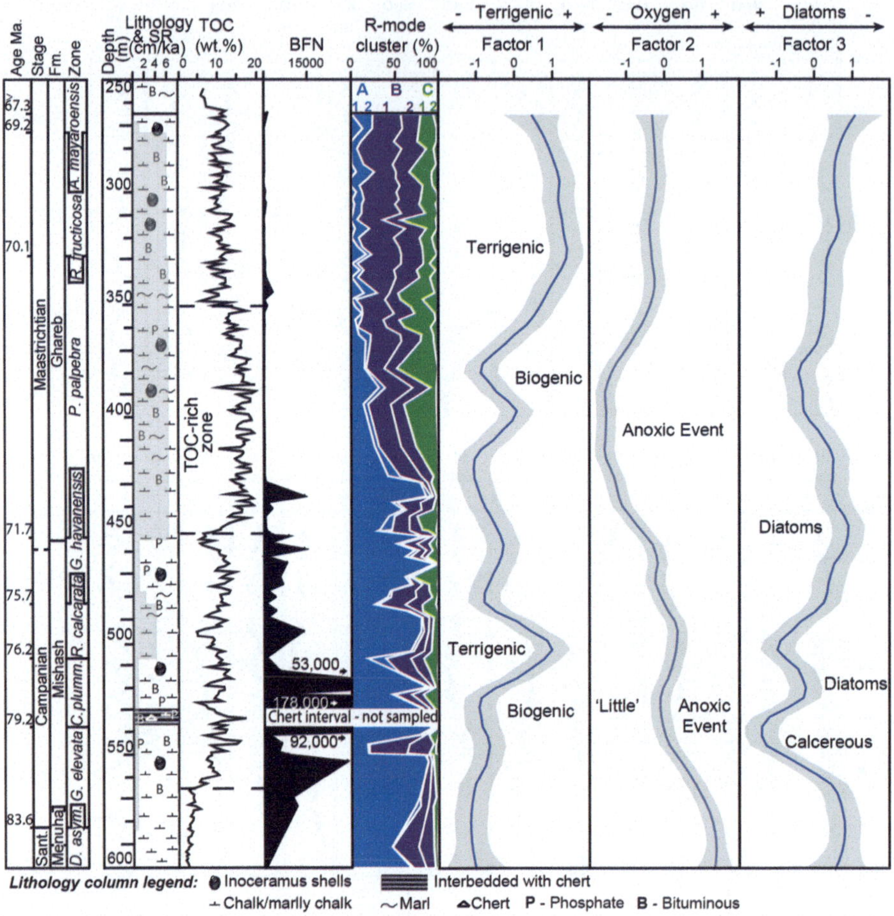

Fig. 12.21 Statistical factor analysis of the trace element distribution in the Aderet core from the Shefela Basin, relative to data from previous studies (nine left columns from Meilijson et al. 2014, 2016) which include age, depth (m), lithology (see legend for symbols), TOC and R-mode benthic foraminiferal clusters. The latter are best described according to their general morphologies: (A) biserial and triserial. (B) Trochospiral; (B1) smooth tests; (B2) perforated by macropores. (C) Mainly uniserial. Factor 1 shows the shifts from biogenic to terrigenous material relative contribution into the basin, with terrigenous peaks at around 535, 500, 450 and 400 m, and a shift into increased terrigenous contribution from 360 m upwards. Factor 2 shows shifts in relative oxygen levels. While the entire core is interpreted as being deposited in oxygen-deficient bottom waters, relative shifts are detected. Factor 3 (F3) shows the interpreted relative abundance of diatom (siliceous) vs. foraminifera/nannoplankton (calcareous) abundance along the section

oxygen levels (see Fig. 12.3 for examples). The abnormally high BFN in the lower Campanian interval, dominated by single species such as *P. prolixa* or *P. Canadensis*, reaching 90,000 and 178,000 specimens per gram dry sediment (Fir. 21) are attributed to their advantageous ability to survive and thrive in the anoxic environment (Meilijson et al. 2016). Based on the factor analysis of the elemental distribution we demonstrate a clear connection between diatom abundance and peaks in the abundance of species thought to have used kleptoplastidy as a morphological adaptation to cope with environmental instability, advocating previous assumptions and hypothesis. Meilijson et al. (2016) presented a theory and suggested that intermittent shifts to a calcareous dominated primary producer population, within a time frame in which diatoms were the dominant algal form, were the cause of the fluctuations between R-mode cluster A (triserial forms using kleptoplastidy) and R-mode cluster B (trochospiral forms using denitrification as an adaptation to life in dysoxic-anoxic bottom-water environments).

Factor analysis of the Aderet elemental composition clearly shows a connection between calcareous sedimentation (factor 3; Fig. 12.21) and R-mode cluster B dominance (purple-shaded abundance; Fig. 12.21), and siliceous diatom sedimentation and peaks in R-mode cluster A, during the lower Campanian from 570 to 500 m where the highest BFN occur. Additionally, while not completely matching in depth, it is evident that along the section in which fluctuations in the relative abundance of primary producers occurred, substantial shifts also transpired in the relative contribution of terrigenic material to the accumulating sediments.

12.3 Discussion

Based on the analysis of the elemental distribution of the studied section, we can now postulate on the cause of the scattered occurrence of both the diatom and the triserial benthic foraminifera during the Upper Cretaceous of the Levant. Factor analysis couples with R-mode cluster analysis; Meilijson et al. (2016) demonstrate that the two coincide, further advocating the possibility of benthic foraminiferal kleptoplastidy from diatoms. This was done through remarkable morphological adaptations of the benthic foraminiferal species. However, if this was the case, why do we not see a continuation of the high BFN and dominance of species such as *P. prolixa* throughout the entire Upper Cretaceous section, as high-productivity conditions seem to have been continuous until the upper part of the Maastrichtian (Almogi-Labin et al. 1993; Ashckenazi-Polivoda 2011; Meilijson et al. 2014, 2018). One possibility is that periodic increases in terrigenous contribution into the basin provided a boost to productivity by increasing the amount of silica. This would enable relatively short-lived intervals in which the diatom population exceeded its usually high numbers, allowing for very fresh and unaltered diatom frustules to reach the deep bottom waters. This in turn would provide fresh chloroplasts for the benthic foraminifera to incorporate into their tests through kleptoplastidy. The correlation between peaks in terrigenous input, diatom blooms, anoxic conditions and matching blooms in R-mode cluster A triserial foraminifera can be seen in Fig. 12.21. The study of modern upwelling systems has clearly shown that episodes of high input of terrestrial material into an area of upwelling (which due to the frequent algal blooms might be nutrient limited at times) bring huge blooms of phytoplankton (Hare et al. 2005; Hutchins and Bruland 1998; Johnson 2001; Takeda 1998). This is usually due to the input of iron and silica from continental sources into marine basins.

Meilijson et al. (2016) attributed the major change in benthic foraminiferal assemblages to a shift in the type of food (i.e. primary producers). Ashckenazi-Polivoda et al. (2018) suggested that these changes were imposed by a deepening of several hundred metres during the latest Campanian global sea-level transgression (Haq 2014), also recognized around the study area (Meilijson et al. 2014). In modern upwelling environments the accumulation of silica due to diatom blooms occurs near shore at water depths below 200 m (Bremner and Willis 1993; Giraudeau et al. 2000). Meilijson et al. (2016) also hypothesized that the shift in primary producers and the consequent turnover in benthic foraminiferal assemblages and life strategies were induced by a southward shift (i.e. towards Egypt and Saudi Arabia) of the main upwelling cells, as a result of a regional deepening that occurred around the Campanian/Maastrichtian boundary. This shift led to the deposition of a homogeneous lithology consisting of organic-rich carbonates throughout the entire Levantine Basin, including the Negev area and accumulation of phosphorite and chert deposits southeastward in Egypt, Saudi Arabia and Iraq (Abed and Sadaqah 2013; Meilijson et al. 2014). In other words, the transition of the upwelling foci shifted and with it the huge diatom blooms, resulting in the deposition of siliceous and phosphatic lithologies in those distant southern localities. This study demonstrates two main attributes to advocate this claim. The first is the similarities observed in the trace element distribution, or rather their factor score-based interpretation, between the two basins (Negev and Shefela). These similarities strengthen the notion that sea levels during the Upper Cretaceous in the region were high enough above the seals of the Syrian Arc basins to create one connected system, as opposed to a series of disconnected basins. This was proposed by Meilijson et al. (2014, 2016) due to similar trends in organic carbon deposition, lithological and faunal changes occurring in the same time across the region. Here we demonstrate this to be true also for the distribution of terrigenous material within the two basins, trends in bottom-water oxygen content, deepwater renewal time and finally relative abundance of siliceous vs. calcareous primary producer sedimentation. This is important as without oceanic connectivity between the basins any high-productivity episodes would be local, and a portrait of gradually shifting upwelling cells would be difficult to draw. The second finding supporting the shift in upwelling foci is the general increase in the detrital component recorded following the Campanian/Maastrichtian boundary sea-level rise in both basins, coupled with a decrease in the maturity of the detritus, indicating longer transport of the terrigenous material.

Meilijson et al. (2016, 2018) argue that changes in bathymetry, which impose shifts in benthic foraminiferal assemblages in normal marine conditions, would not have such an effect on benthic assemblages in upwelling-induced high-productivity environments. There, the massive amount of organic production and consequent bottom-water oxygen deficiency are the main factors controlling benthic populations. There is no doubt that the extremely low oxygen values, determined for the bottom-water environments of the Late Cretaceous in the Levant, would have a vast effect on benthic communities. This is a classic Darwinistic setting for the survival of the fittest. Those who can adapt to these extreme environments will survive, and the rest will not. However, surviving in low-oxygen environments and flourishing in them are not the same thing. The positive relation between the lowest oxygen levels and terrigenous input, providing iron and silica into an already high-nutrient surface-water environment, triggers blooms of diatoms. The synchronous occurrence of abnormally high numbers of low-diversity benthic foraminifera demonstrates the existence and success of functional adaptations. Our identification of morphological adaptations in *P. prolixa*, which are identical to those recognized in modern diatom-based kleptoplastidy of benthic foraminifera, acts as the missing link for understanding this complex system. It does so by tying between productivity, oceanography, continental-marine interactions and remarkable biochemical reciprocity and adaptiveness of present and deep-time ecosystems.

Acknowledgements We would like to thank the 'Levantine Upper Cretaceous group' for their support and valued conversations and comments: S. Abramovich, A. Almogi-Labin, S. Feinstein, Z. Berner, W. Püttmann, R. Speijer and H. Alsenz. We wish to express our gratitude to S. Volin from Rotem Amfert Negev Company for his cooperation in the sampling process of the PAMA section. We wish to express our gratitude to Israel Energy Initiatives Ltd. for the use of the Aderet core material and laboratory equipment. We would like to thank the Israeli Ministry of Science and Technology (MOST) for their continued support of the ADSSC. This study was supported by The German-Israeli Foundation for Scientific Research and Development (GIF) grant no. 956-38.8/2007, by the Israeli Ministry of Infrastructure grant no. 277-17-018 and 27-17-005, and the Helmholtz Association (GRACE) in Karlsruhe.

References

Abed AM, Sadaqah RM (2013) Enrichment of uranium in the uppermost Al-Hisa phosphorite formation, Eshidiyya basin, southern Jordan. J Afr Earth Sci 77:31–40. https://doi.org/10.1016/j.jafrearsci.2012.09.009

Algeo TJ, Lyons TW (2006) Mo-total organic carbon covariation in modern anoxic marine environments: implications for analysis of paleoredox and paleohydrographic conditions. Paleoceanography. https://doi.org/10.1029/2004PA001112

Algeo TJ, Maynard JB (2004) Trace-element behavior and redox facies in core shales of Upper Pennsylvanian Kansas-type cyclothems. Chem Geol. https://doi.org/10.1016/j.chemgeo.2003.12.009

Algeo TJ, Rowe H (2012) Paleoceanographic applications of trace-metal concentration data. Chem Geol. https://doi.org/10.1016/j.chemgeo.2011.09.002

Almogi-Labin A, Bein A, Sass E (1993) Late Cretaceous upwelling system along the Southern Tethys Margin (Israel): interrelationship between productivity, bottom water environments, and organic matter preservation. Paleoceanography 8:671–690. https://doi.org/10.1029/93PA02197

Almogi-Labin A, Ashckenazi-Polivoda S, Edelman-Furstenberg Y, Benjamini C (2012) Anoxia-dysoxia at the sediment-water interface of the Southern Tethys in the Late Cretaceous: Mishash Formation, Southern Israel. In: Altenbach A, Bernhard JM, Seckbach J (eds) Anoxia: evidence for eukaryote survival and paleontological strategies. Springer, Dordrecht, pp 553–572. https://doi.org/10.1007/978-94-007-1896-8_29

Alsenz H, Illner P, Ashckenazi-Polivoda S, Meilijson A, Abramovich S, Feinstein S, Almogi-Labin A, Berner Z, Püttmann W (2015) Geochemical evidence for the link between sulfate reduction, sulfide oxidation and phosphate accumulation in a late cretaceous upwelling system. Geochem Trans 16:2. https://doi.org/10.1186/s12932-015-0017-1

Altenbach AVA, Bernhard JMJ, Seckbach J (2011) Anoxia: evidence for eukaryote survival and paleontological strategies. In: Altenbach AV, Bernhard JM, Seckbach J (eds) Cellular origin, life in extreme habitats and astrobiology. Vol. 21, Cellular origin, life in extreme habitats and astrobiology. Springer, Dordrecht, p 480. https://doi.org/10.1007/978-94-007-1896-8

Arias YM, Tebo BM (2003) Cr(VI) reduction by sulfidogenic and nonsulfidogenic microbial consortia. Appl Environ Microbiol. https://doi.org/10.1128/AEM.69.3.1847-1853.2003

Ashckenazi-Polivoda S (2011) The Late Cretaceous Southern Tethyan upwelling system: a case study from the high productivity sequence, Negev, Israel

Ashckenazi-Polivoda S, Abramovich S, Almogi-Labin A, Schneider-Mor A, Feinstein S, Puttmann W, Berner Z (2011) Paleoenvironments of the latest Cretaceous oil shale sequence, Southern Tethys, Israel, as an integral part of the prevailing upwelling system. Palaeogeogr Palaeoclimatol Palaeoecol 305:93–108. https://doi.org/10.1016/j.palaeo.2011.02.018

Ashckenazi-Polivoda S, Titelboim D, Meilijson A, Almogi-Labin A, Abramovich S (2018) Bathymetric trend of Late Cretaceous southern Tethys upwelling regime based on benthic foraminifera. Cretac Res 82:40–55. https://doi.org/10.1016/j.cretres.2017.10.014

Austin HA, Austin WEN, Paterson DM (2005) Extracellular cracking and content removal of the benthic diatom Pleurosigma angulatum (Quekett) by the benthic foraminifera Haynesina germanica (Ehrenberg). Mar Micropaleontol. https://doi.org/10.1016/j.marmicro.2005.07.002

Barnes CE, Cochran JK (1993) Uranium geochemistry in estuarine sediments: controls on removal and release processes. Geochim Cosmochim Acta. https://doi.org/10.1016/0016-7037(93)90367-6

Barwise AJG (1990) Role of nickel and vanadium in petroleum classification. Energy Fuels. https://doi.org/10.1021/ef00024a005

Becknamm B, Wagner T, Hofmann P (2011) Linking Coniacian–Santonian (OAE3) black-shale deposition to African climate variability: a reference section from the Eastern Tropical Atlantic at orbital time scales (Odp Site 959, Off Ivory Coast and Ghana). In: Deposition of organic-carbon-rich sediments: models. pp 125–143. https://doi.org/10.2110/pec.05.82.0125

Bein A, Amit O (1982) Depositional-environments of the Senonian Chert, phosphorite and oil-shale sequence in Israel as deduced from their organic-matter composition. Sedimentology 29:81–90

Bein A, Almogi-Labin A, Sass E (1990) Sulfur sinks and organic carbon relationships in Cretaceous organic-rich carbonates: implications for evaluation of oxygen-poor depositional environments. Am J Sci. https://doi.org/10.2475/ajs.290.8.882

Bernhard JM (1993) Experimental and field evidence of Antarctic foraminiferal tolerance to anoxia and hydrogen sulfide. Mar Micropaleontol. https://doi.org/10.1016/0377-8398(93)90033-T

Bernhard JM (2003) Potential symbionts in bathyal foraminifera. Science 299(5608):861. https://doi.org/10.1126/science.1077314

Bernhard JM, Bowser SS (1999) Benthic foraminifera of dysoxic sediments: chloroplast sequestration and functional morphology. Earth Sci Rev 46:149–165

Bernhard JM, Reimers CE (1991) Benthic foraminiferal population fluctuations related to anoxia: Santa Barbara Basin. Biogeochemistry. https://doi.org/10.1007/BF00003221

Bernhard JM, Visscher PT, Bowser SS (2003) Submillimeter life positions of bacteria, protists, and metazoans in laminated sediments of the Santa Barbara Basin. Limnol Oceanogr. https://doi.org/10.4319/lo.2003.48.2.0813

Bernhard JM, Habura A, Bowser SS, (2006) An endobiont-bearing allogromiid from the Santa Barbara Basin: implications for the early diversification of foraminifera. J Geophys Res Biogeosci. https://doi.org/10.1029/2005JG000158

Bernhard JM, Goldstein ST, Bowser SS (2010) An ectobiont-bearing foraminiferan, Bolivina pacifica, that inhabits microxic pore waters: cell-biological and paleoceanographic insights. Environ Microbiol. https://doi.org/10.1111/j.1462-2920.2009.02073.x

Bernhard JM, Casciotti KL, McIlvin MR, Beaudoin DJ, Visscher PT, Edgcomb VP (2012a) Potential importance of physiologically diverse benthic foraminifera in sedimentary nitrate storage and respiration. J Geophys Res Biogeosci. https://doi.org/10.1029/2012JG001949

Bernhard JM, Edgcomb VP, Casciotti KL, McIlvin MR, Beaudoin DJ (2012b) Denitrification likely catalyzed by endobionts in an allogromiid foraminifer. ISME J. https://doi.org/10.1038/ismej.2011.171

Bolle M-P, Adatte T (2001) Palaeocene-early Eocene climatic evolution in the Tethyan realm: clay mineral evidence. Clay Miner. https://doi.org/10.1180/000985501750177979

Böning P, Brumsack HJ, Schnetger B, Grunwald M (2009) Trace element signatures of Chilean upwelling sediments at ~36°S. Mar Geol. https://doi.org/10.1016/j.margeo.2009.01.004

Borchers SL, Schnetger B, Böning P, Brumsack HJ (2005) Geochemical signatures of the Namibian diatom belt: perennial upwelling and intermittent anoxia. Geochem Geophys Geosyst. https://doi.org/10.1029/2004GC000886

Bremner JM, Willis JP (1993) Mineralogy and geochemistry of the clay fraction of sediments from the Namibian continental margin and the adjacent hinterland. Mar Geol. https://doi.org/10.1016/0025-3227(93)90076-8

Bruland KW (1980) Oceanographic distributions of cadmium, zinc, nickel, and copper in the North Pacific. Earth Planet Sci Lett. https://doi.org/10.1016/0012-821X(80)90035-7

Bruland KW (1983) Trace elements in sea-water. Chem Oceanogr. https://doi.org/10.1016/b978-0-12-588608-6.50009-2

Calvert SE, Pedersen TF (1993) Marine sediments, burial, pore water chemistry, microbiology and diagenesis geochemistry of recent oxic and anoxic marine sediments: implications for the geological record. Mar Geol. https://doi.org/10.1016/0025-3227(93)90150-T

Corliss BH (1991) Morphology and microhabitat preferences of benthic foraminifera from the northwest Atlantic Ocean. Current 17:195–236

Corliss BH, Chen C (1988) Morphotype patterns of Norwegian Sea deep-sea benthic foraminifera and ecological implications. Geology:716–719

Cranston RE (1983) Chromium in Cascadia Basin, northeast Pacific Ocean. Mar Chem. https://doi.org/10.1016/0304-4203(83)90020-8

Cutter GA, Cutter LS (1995) Behavior of dissolved antimony, arsenic, and selenium in the Atlantic Ocean. Mar Chem. https://doi.org/10.1016/0304-4203(95)00019-N

Dybek J (1963) Zur Geochemie und Lagerstättenkunde des Urans. Clausthaler Heft zur Lagerstättenkunde und Geochemie der mineralische Rohstoffe, vol I. Borchert H (ed). Borntraeger, Berlin. Mineralogical Magazine - Mineralogical Society. https://doi.org/10.1180/minmag.1963.033.262.13

Eshet Y, Almogi-Labin A (1996) Calcareous nannofossils as paleoproductivity indicators in upper cretaceous organic-rich sequences in Israel. Mar Micropaleontol. https://doi.org/10.1016/0377-8398(96)00006-0

Eshet Y, Almogi-Labin A, Bein A (1994) Dinoflagellate cysts, paleoproductivity and upwelling systems: a Late Cretaceous example from Israel. Mar Micropaleontol 23:231–240. https://doi.org/10.1016/0377-8398(94)90014-0

Fleischer L, Gafsou R (2003) Top Judea Group digital structural map of Israel (1:200.000). Geophysical Institute of Israel Report 753/312/03

Friedrich O (2010) Benthic foraminifera and their role to decipher paleoenvironment during mid-Cretaceous Oceanic Anoxic Events—the "anoxic benthic foraminifera" paradox. Rev Micropaléontologie 53:175–192. https://doi.org/10.1016/j.revmic.2009.06.001

Garcia D, Fonteilles M, Moutte J (1994) Sedimentary fractionations between Al, Ti, and Zr and the genesis of strongly peraluminous granites. J. Geol. https://doi.org/10.1086/629683

Garfunkel Z, Zak I, Freund R (1981) Active faulting in the dead sea rift. Tectonophysics 80:1–26. https://doi.org/10.1016/0040-1951(81)90139-6

Geslin E, Risgaard-Petersen N, Lombard F, Metzger E, Langlet D, Jorissen F (2011) Oxygen respiration rates of benthic foraminifera as measured with oxygen microsensors. J Exp Mar Biol Ecol. https://doi.org/10.1016/j.jembe.2010.10.011

Geslin E, Barras C, Langlet D, Nardelli MP, Kim JH, Bonnin J, Metzger E, Jorissen FJ (2014) Survival, reproduction and calcification of three benthic foraminiferal species in response to experimentally induced hypoxia. In: Kitazato H, Bernhard JM (eds) Approaches to study living foraminifera. Environmental science and engineering. Springer, Tokyo. https://doi.org/10.1007/978-4-431-54388-6_10

Giraudeau J, Bailey GW, Pujol C (2000) A high-resolution time-series analyses of particle fluxes in the Northern Benguela coastal upwelling system: carbonate record of changes in biogenic production and particle transfer processes. Deep Res Part II Top Stud Oceanogr. https://doi.org/10.1016/S0967-0645(00)00014-X

Gooday AJ, Bernhard JM, Levin LA, Suhr SB (2000) Foraminifera in the Arabian Sea oxygen minimum zone and other oxygen-deficient settings: taxonomic composition, diversity, and relation to metazoan faunas. Deep Res Part II Top Stud Oceanogr. https://doi.org/10.1016/S0967-0645(99)00099-5

Haq BU (2014) Cretaceous eustasy revisited. Glob Planet Change 113:44–58. https://doi.org/10.1016/j.gloplacha.2013.12.007

Hare CE, DiTullio GR, Trick CG, Wilhelm SW, Bruland KW, Rue EL, Hutchins DA (2005) Phytoplankton community structure changes following simulated upwelled iron inputs in the Peru upwelling region. Aquat Microb Ecol. https://doi.org/10.3354/ame038269

Henderson P, Henderson G (2010) The Cambridge handbook of earth science data, Choice Reviews Online. Cambridge University Press, Cambridge, U.K, New York. https://doi.org/10.5860/choice.47-2354

Høgslund S, Revsbech NP, Cedhagen T, Nielsen LP, Gallardo VA (2008) Denitrification, nitrate turnover, and aerobic respiration by benthic foraminiferans in the oxygen minimum zone off Chile. J Exp Mar Biol Ecol 359:85–91. https://doi.org/10.1016/j.jembe.2008.02.015

Huerta-Diaz MA, Morse JW (1990) A quantitative method for determination of trace metal concentrations in sedimentary pyrite. Mar Chem. https://doi.org/10.1016/0304-4203(90)90009-2

Hutchins DA, Bruland KW (1998) Iron-limited diatom growth and Si:N uptake ratios in a coastal upwelling regime. Nature. https://doi.org/10.1038/31203

Inthorn M, Mohrholz V, Zabel M (2006) Nepheloid layer distribution in the Benguela upwelling area offshore Namibia. Deep Sea Res Part I Oceanogr Res Pap 53:1423–1438. https://doi.org/10.1016/j.dsr.2006.06.004

Johnson KS (2001) Iron supply and demand in the upper ocean: is extraterrestrial dust a significant source of bioavailable iron? Global Biogeochem Cycles. https://doi.org/10.1029/2000GB001295

Jorissen FJ, De Stigter HC, Widmark JGV (1995) A conceptual model explaining benthic foraminiferal microhabitats. Mar Micropaleontol 26(1–4):3–15

Jorissen FJ, Fontanier C, Thomas E (2007) Chapter seven paleoceanographical proxies based on deep-sea benthic foraminiferal assemblage characteristics. Dev Mar Geol. https://doi.org/10.1016/S1572-5480(07)01012-3

Klinkhammer GP, Palmer MR (1991) Uranium in the oceans: where it goes and why. Geochim Cosmochim Acta. https://doi.org/10.1016/0016-7037(91)90024-Y

Kniewald G, Branica M (1988) Role of uranium(V) in marine sedimentary environments: a geochemical possibility. Mar Chem. https://doi.org/10.1016/0304-4203(88)90002-3

Koho KA, Piña-Ochoa E (2012) Benthic foraminifera: inhabitants of low-oxygen environments. In: Altenbach AV, Bernhard JM, Seckbach J (eds) Anoxia: evidence for eukaryote survival and paleontological strategies. Springer, Dordrecht. https://doi.org/10.1007/978-94-007-1896-8_14

Krenkel E (1924) Der Syrische Bogen. Cent Miner Geol Palaeontol 9:274–281

Kuhnt T, Friedrich O, Schmiedl G, Milker Y, Mackensen A, Lückge A (2013) Relationship between pore density in benthic foraminifera and bottom-water oxygen content. Deep Res Part I Oceanogr Res Pap. https://doi.org/10.1016/j.dsr.2012.11.013

Langmuir D (1978) Uranium solution-mineral equilibria at low temperatures with applications to sedimentary ore deposits. Geochim Cosmochim Acta. https://doi.org/10.1016/0016-7037(78)90001-7

Leiter C, Altenbach AV (2010) Benthic foraminifera from the diatomaceous mud belt off Namibia: characteristic species for severe anoxia. Palaeontol Electron 13:19

Lewan MD (1984) Factors controlling the proportionality of vanadium to nickel in crude oils. Geochim Cosmochim Acta. https://doi.org/10.1016/0016-7037(84)90219-9

Lewan MD, Maynard JB (1982) Factors controlling enrichment of vanadium and nickel in the bitumen of organic sedimentary rocks. Geochim Cosmochim Acta. https://doi.org/10.1016/0016-7037(82)90377-5

Lovley DR (1995) Microbial reduction of iron, manganese, and other metals. Adv Agron 54:175–231. https://doi.org/10.1016/S0065-2113(08)60900-1

McArthur JM, Algeo TJ, Van De Schootbrugge B, Li Q, Howarth RJ (2008) Basinal restriction, black shales, Re-Os dating, and the Early Toarcian (Jurassic) oceanic anoxic event. Paleoceanography. https://doi.org/10.1029/2008PA001607

McManus J, Berelson WM, Klinkhammer GP, Hammond DE, Holm C (2005) Authigenic uranium: relationship to oxygen penetration depth and organic carbon rain. Geochim Cosmochim Acta. https://doi.org/10.1016/j.gca.2004.06.023

Meilijson A, Ashckenazi-Polivoda S, Ron-Yankovich L, Illner P, Alsenz H, Speijer RP, Almogi-Labin A, Feinstein S, Berner Z, Püttmann W, Abramovich S (2014) Chronostratigraphy of the Upper Cretaceous high productivity sequence of the southern Tethys, Israel. Cretac Res 50:187–213. https://doi.org/10.1016/j.cretres.2014.04.006

Meilijson A, Ashckenazi-Polivoda S, Illner P, Alsenz H, Speijer RP, Almogi-Labin A, Feinstein S, Püttmann W, Abramovich S (2015) Evidence for specific adaptations of fossil benthic foraminifera to anoxic-dysoxic environments. Paleobiology 42. https://doi.org/10.1017/pab.2015.31

Meilijson A, Ashckenazi-Polivoda S, Illner P, Alsenz H, Speijer RP, Almogi-Labin A, Feinstein S, Püttmann W, Abramovich S (2016) Evidence for specific adaptations of fossil benthic foraminifera to anoxic-dysoxic environments. Paleobiology 42:77–97. https://doi.org/10.1017/pab.2015.31

Meilijson A, Ashckenazi-Polivoda S, Illner P, Speijer RP, Almogi-Labin A, Feinstein S, Püttmann W, Abramovich S (2018) From phytoplankton to oil shale reservoirs: a 19-million-year record of the Late Cretaceous Tethyan upwelling regime in the Levant Basin. Mar Petrol Geol 95:188–205. https://doi.org/10.1016/j.marpetgeo.2018.04.012

Moodley L, Heip CHR, Middelburg JJ (1998) Benthic activity in sediments of the northwestern Adriatic Sea: sediment oxygen consumption, macro- and meiofauna dynamics. J Sea Res 40:263–280. https://doi.org/10.1016/S1385-1101(98)00026-4

Morford JL, Martin WR, Carney CM (2009) Uranium diagenesis in sediments underlying bottom waters with high oxygen content. Geochim Cosmochim Acta. https://doi.org/10.1016/j.gca.2009.02.014

Morton AC, Hallsworth CR (1999) Processes controlling the composition of heavy mineral assemblages in sandstones. Sediment Geol. https://doi.org/10.1016/S0037-0738(98)00118-3

Okrusch M, Matthes S (2005) Eine Einführung in die spezielle Mineralogie, Petrologie und Lagerstättenkunde. Springer, Berlin

Pan Y, Fleet ME (2019) Compositions of the apatite-group minerals: substitution mechanisms and controlling factors. Phosphates Geochem Geobiol Mater Importance. https://doi.org/10.2138/rmg.2002.48.2

Pillet L, de Vargas C, Pawlowski J (2011) Molecular identification of sequestered diatom chloroplasts and kleptoplastidy in foraminifera. Protist. https://doi.org/10.1016/j.protis.2010.10.001

Piña-Ochoa E, Høgslund S, Geslin E, Cedhagen T, Revsbech NP, Nielsen LP, Schweizer M, Jorissen F, Rysgaard S, Risgaard-Petersen N (2010) Widespread occurrence of nitrate storage and denitrification among Foraminifera and Gromiida. Proc Natl Acad Sci USA. https://doi.org/10.1073/pnas.0908440107

Piper DZ, Perkins RB (2004) A modern vs. Permian black shale—the hydrography, primary productivity, and water-column chemistry of deposition. Chem Geol. https://doi.org/10.1016/j.chemgeo.2003.12.006
Pucci F, Geslin E, Barras C, Morigi C, Sabbatini A, Negri A, Jorissen FJ (2009) Survival of benthic foraminifera under hypoxic conditions: results of an experimental study using the CellTracker Green method. Mar Pollut Bull 59:336–351. https://doi.org/10.1016/j.marpolbul.2009.08.015
Ravizza G, Paquay F (2008) Os isotope chemostratigraphy applied to organic-rich marine sediments from the Eocene-Oligocene transition on the West African margin (ODP Site 959). Paleoceanography. https://doi.org/10.1029/2007PA001460
Risgaard-Petersen N, Langezaal AM, Ingvardsen S, Schmid MC, Jetten MSM, Op Den Camp HJM, Derksen JWM, Piña-Ochoa E, Eriksson SP, Nielsen LP, Revsbech NP, Cedhagen T, Van Der Zwaan GJ (2006) Evidence for complete denitrification in a benthic foraminifer. Nature. https://doi.org/10.1038/nature05070
Ruokolainen M, Pantsar-Kallio M, Haapa A, Kairesalo T (2000) Leaching, runoff and speciation of arsenic in a laboratory mesocosm. Sci Total Environ. https://doi.org/10.1016/S0048-9697(00)00521-0
Schneider-Mor A, Alsenz H, Ashckenazi-Polivoda S, Illner P, Abramovich S, Feinstein S, Almogi-Labin A, Berner Z, Püttmann W (2012) Paleoceanographic reconstruction of the late Cretaceous oil shale of the Negev, Israel: integration of geochemical, and stable isotope records of the organic matter. Palaeogeogr Palaeoclimatol Palaeoecol 319–320:46–57. https://doi.org/10.1016/j.palaeo.2012.01.003
Schumacher S, Lazarus D (2004) Regional differences in pelagic productivity in the late Eocene to early Oligocene—a comparison of southern high latitudes and lower latitudes. Palaeogeogr Palaeoclimatol Palaeoecol. https://doi.org/10.1029/2002PA000804
Schwartz W (1976) Zeitschrift für allgemeine Mikrobiologie. In Wedepohl KH (ed) Handbook of geochemistry, vol II/4. 898 S, 113 Abb. Springer, Berlin, DM 238,00 (bei Subskription DM 238,40). https://doi.org/10.1002/jobm.3630160626
Sen Gupta BK (1999) Systematics of modern foraminifera. In: Modern foraminifera. Kluwer, Dordrecht. https://doi.org/10.1007/0-306-48104-9
Shoval S (2004a) Clay sedimentation along the southeastern Neo-Tethys margin during the oceanic convergence stage. Appl Clay Sci 24:287–298. https://doi.org/10.1016/j.clay.2003.08.010
Shoval S (2004b) Deposition of volcanogenic smectite along the southeastern Neo-Tethys margin during the oceanic convergence stage. Appl Clay Sci 24:299–311. https://doi.org/10.1016/j.clay.2003.08.009
Sinninghe Damsté JS, Kohnen MEL, De Leeuw JW (1990) Thiophenic biomarkers for palaeoenvironmental assessment and molecular stratigraphy. Nature 345:609–611. https://doi.org/10.1038/345609a0
Smillie RH, Hunter K, Loutit M (1981) Reduction of chromium(VI) by bacterially produced hydrogen sulphide in a marine environment. Water Res. https://doi.org/10.1016/0043-1354(81)90007-5
Sohrin Y, Matsui M, Kawashima M, Hojo M, Hasegawa H (1997) Arsenic biogeochemistry affected by eutrophication in lake Biwa, Japan. Environ Sci Technol. https://doi.org/10.1021/es960846w
Soudry D, Glenn CR, Nathan Y, Segal I, VonderHaar D (2006) Evolution of Tethyan phosphogenesis along the northern edges of the Arabian–African shield during the Cretaceous–Eocene as deduced from temporal variations of Ca and Nd isotopes and rates of P accumulation. Earth Sci Rev 78:27–57. https://doi.org/10.1016/j.earscirev.2006.03.005
Spiro B, Rozenson I (1980) Distribution of iron species in some "oil shales" of the Judea Desert, Israel. Chem Geol. https://doi.org/10.1016/0009-2541(80)90034-0
Spiro B, Dinur D, Aizenshtat Z (1983) Evaluation of source, environments of deposition and diagenesis of some Israeli "oil shales"—N-Alkanes, fatty acids, tetrapyrroles and kerogen. Chem Geol. https://doi.org/10.1016/0009-2541(83)90015-3
Takeda S (1998) Influence of iron availability on nutrient consumption ratio of diatoms in oceanic waters. Nature. https://doi.org/10.1038/31674

Thomas E (1990) Late Cretaceous through Neogene deep-sea benthic foraminifers (Maud Rise, Weddell Sea, Antarctica). Proc Ocean Drill Program Sci Results 113:571–594

Tribovillard N, Algeo TJ, Lyons T, Riboulleau A (2006) Trace metals as paleoredox and paleo-productivity proxies: an update. Chem Geol. https://doi.org/10.1016/j.chemgeo.2006.02.012

Tribovillard N, Bout-Roumazeilles V, Algeo T, Lyons TW, Sionneau T, Montero-Serrano JC, Riboulleau A, Baudin F (2008) Paleodepositional conditions in the Orca Basin as inferred from organic matter and trace metal contents. Mar Geol. https://doi.org/10.1016/j.margeo.2008.04.016

Turekian KK (1977) The fate of metals in the oceans. Geochim Cosmochim Acta. https://doi.org/10.1016/0016-7037(77)90109-0

Turekian KK (2010) Marine Chemistry and Geochemistry, 2. ed. ed, Carbon. Academic Press, Amsterdam u.a. ISBN:9780080964836

Wagner T (2002) Late Cretaceous to early Quaternary organic sedimentation in the eastern Equatorial Atlantic. Palaeogeogr Palaeoclimatol Palaeoecol. https://doi.org/10.1016/S0031-0182(01)00415-1

Widmark JGV, Malmgren BA (1992) Biogeography of terminal Cretaceous deep-sea benthic foraminifera from the Atlantic and Pacific Oceans. Palaeogeogr Palaeoclimatol Palaeoecol 92:375–405. https://doi.org/10.1016/0031-0182(92)90092-J

Yao W, Millero FJ (1995) The chemistry of the anoxic waters in the Framvaren Fjord, Norway. Aquat Geochem. https://doi.org/10.1007/BF01025231

Yokoyama Y, Tanaka K, Takahashi Y (2012) Differences in the immobilization of arsenite and arsenate by calcite. Geochim Cosmochim Acta. https://doi.org/10.1016/j.gca.2012.05.022

Chapter 13
Evolution as a Timeless Continuum

John S. Torday and William B. Miller Jr.

Keywords Endosymbiosis · Exaptation · Niche construction · Reverse evolution · Parathyroid hormone-related protein

13.1 Introduction

In "On the Origin of Species," Darwin describes the terrain of Patagonia at great length during his voyage on the HMS Beagle. From that experience, he inferred the importance of the environment in understanding the adaptive evolution of flora and fauna. To this day, and despite the passage of time, observations of the interrelationship between the complexity of environmental stresses and their relevance to evolution remain inferential rather than proven through rigorous scientific measurement. In the Galapagos, the Grants have attempted a distinct longitudinal study of the interrelationship between the environment and adaptive phenotype. However, their scrutiny of finch beaks and size is confounded with substantial observational bias, included no rigorous species definitions, and answered no substantive evolutionary questions since no novelty was observed (Grant, P. R., Grant, B. R. How and Why Species Multiply: The Radiation of Darwin's Finches. Princeton University Press, 2011).

Even as it is acknowledged that evolution is adaptation to an ever-changing environment, the specific connections between biologic and topographical phenomena remain in the realm of metaphor. However, that gap can be resolved by properly viewing the organism within its environment as a continuum. In so doing, the reasons for reverse evolution are clarified.

J. S. Torday (✉)
Department of Pediatrics, Obstetrics and Gynecology, Evolutionary Medicine Program, David Geffen School of Medicine, University of California, Los Angeles, CA, USA
e-mail: jtorday@ucla.edu

W. B. Miller Jr.
Department of Medicine, Banner Health System, Paradise Valley, AZ, USA

J. Guex et al. (eds.), *Morphogenesis, Environmental Stress and Reverse Evolution*, https://doi.org/10.1007/978-3-030-47279-5_13

13.2 Endosymbiosis, or How the Inanimate Is Animated

Torday and Rehan (Evolutionary Biology, Cell-Cell Communication and Complex Disease. Wiley, 2012) have asserted that cell–cell signaling mechanisms permit embryogenesis and result in consistent phylogeny to effect patterns of evolution, though leaving open how the intermediary steps came about. Throughout evolution, all existential have been accommodated, including heavy metals, ions, gases, bacteria, and viruses. When viewed on the basis of cell–cell interactions and cellular problem-solving in response to cell-based threats, evolution can be understood differently from being a result of simple selection. Instead, evolution can only be understood as the continuous process of self-referential and self-organizing cellular networks coping with their environmental circumstances (Miller, W.B. Cognition, Information Fields and Hologenomic Entanglement: Evolution in Light and Shadow. Biology (Basel) 2016, 5, 21.). Of those means, endogenization and compartmentation are primary.

Unicellular organisms have dominated the Earth for over 3.5 billion years. Even within their own limitations, their cellular narrative was learning how to cope with environmental stresses. The result was biofilms, sustained through quorum sensing. When Eukaryota (cells with nuclei) began approximately 500 million years ago, the basic unicellular patterns of cell–cell signaling were exapted to develop means by which collections of these cells could cooperate with one another metabolically, giving rise to multicellular eukaryotes. Their modifications in structure and function were responses to stresses generated by the environment.

13.3 Cell-Molecular Basis for Environmental Adaptation

In earlier publications, the causal interrelationships between environmental change and cell-molecular adaptation have been elucidated. Global changes in water salinity, water-land transition due to greenhouse effect of CO_2, and atmospheric oxygen increases and subsequent decreases have all been shown to have given rise to cell-molecular adaptations for gas exchange as the evolution of the vertebrate lung from the fish swim bladder. It has been demonstrated that those mechanistic, causal interrelationships were based on developmental and phylogenetic changes in lung alveolar structure and function, having been corroborated by genetic gain- and loss-of-function experiments consistent with that hypothesis.

Moreover, by tracing similar interlinked cellular-molecular changes backwards from the present-day form of the mammalian lung alveolus based on preadaptive traits, or exaptations, insight is obtained into how and why the lung and other physiologic traits evolved, all of which can be traced as having been first initiated within the unicellular state.

13.4 Niche Construction Theory Instructs How Biology and Ecology Merge

Niche construction has been defined as the process by which organisms modify their environments in order to effect optimal adaptation. The first documented example of niche construction was reported by Darwin himself, noting that earthworms retain their aquatic kidneys on land by engineering the surrounding soil. Likewise, beavers build dams and humans build homes, towns, states, and nations. Perhaps of greatest importance has been the insight that the cell was actually the first niche construction. When this is properly understood as the starting point for the perpetual dynamic of niche construction, then it becomes clear that there is a necessary continuum from cell physiology to the surrounding environment that extends across the entirety of evolution. Importantly, it is this same cellular impulse towards niche construction that enables all of the cellular ecosystems that define multicellular life (Miller cognition). It is contended that it is this exact relationship, scalable across evolution, that explains the continuum from the unicell to Gaia.

13.5 Stress-Induced Adaptation Can Seem Like Reverse Evolutionary Simplifications

The greenhouse effect is believed to have been caused by the accumulation of carbon dioxide in the atmosphere, beginning some 500 million years ago, as described by Romer. That event raised the air temperature by x degrees, drying up the ocean that covered the Earth, offering adaptive opportunities for the water-to-land transition. There are two types of bony fish, physoclistous and physostomous. It is the latter which are best suited for air breathing since they have a trachea-like tube that connects the esophagus with the swim bladder. The gene necessary for alveolar formation, parathyroid hormone-related protein (PTHrP), is expressed in the swim bladder of fish, as well as their bony skeleton. Under the increased gravitational force on land (compared to buoyancy in water), there was an adaptive cellular impulse and resultant filtering selection towards the remodeling of bone in order to support the body of land-adapted fish. It is known that physiologic stress generates radical oxygen species, which cause gene mutations and duplications. Among those consequences, the PTHrP receptor gene duplicated during the water-land transition, amplifying PTHrP signaling in the lung and kidney, giving rise to the alveoli and glomeruli necessary for adaptation to air breathing and salt/water economy on land.

As has been previously described in prior publications, chronic diseases of the bone, lung, and kidney result in reversal of PTHrP signaling, leading to a regression of these structures/functions back to their earlier forms in reptiles and amphibians. Correctly viewed, this can be viewed as a form of reverse evolution.

13.6 Berner Oxygen Fluctuations, Hypoxic Stress, and Warm-Bloodedness

The present-day concentration of oxygen in the atmosphere is 21%, but it did not arise gradually. Rather, it fluctuated substantially over the last 500 million years, estimated between 15 and 35%. This fluctuation had profound, lasting effects, including "giantism" as the best known of its consequences. However, the effects of low oxygen tension subsequent to prior episodes of high atmospheric oxygen levels also have other profound ramifications from a physiologic perspective, despite having not been previously addressed in the evolution literature. The reason for this lapse is that these phenomena entail cell physiology, which has been overarched by genetics. It has been generally ignored since hypoxia is not a function of the random mutations that evolutionary biologists have long presumed are the primary source of evolutionary variation. Yet hypoxia is the most potent of all physiologic agonists, causing great stress. As a result, during those hypoxic episodes, the hypothalamic-pituitary-adrenal axis was highly activated, causing the production of catecholamines that acutely alleviated the hypoxia at the level of the lung parenchyma by stimulating lung surfactant secretion, allowing the alveoli to distend further. This stretch-mediated mechanism also enhanced PTHrP secretion by the epithelial type II cells, promoting the formation of additional alveoli. That combination alleviated hypoxia on a long-term basis.

In response, adaptive cells deployed increased PTHrP signaling in the skeleton and lung as cellular based niche constructions, promoting the evolution of the kidney glomerulus. PTHrP signaling between the podocytes lining the renal glomerulus and the mesangial cells lining the renal tubules regulates water and salt balance in the systemic circulation. Moreover, PTHrP began to be expressed in the anterior pituitary and adrenal cortex, amplifying the production of catecholamines which further alleviated both the short- and long-term stress of hypoxia.

The combination, in tandem with the effect of catecholamines on lung and kidney evolution, fostered warm-bloodedness in mammals and birds. Catecholamines stimulate lipolysis of fat cells, causing secretion of free fatty acids into the circulation. Free fatty acids are the most efficient substrate for aerobic metabolism, increasing body temperature. In the course of consistent reactive adaptation, it was oxytocin production by the posterior pituitary that became the constitutive regulator of body temperature. In reciprocating support of this series of cell-based land adaptations, body temperature increased from 25° centigrade to 37°. In turn, lung surfactant composition changed in order for its phase transition temperature to function optimally. Again, this was due to loss of homeostatic control in the alveoli, resulting in cellular remodeling to re-establish homeostasis. It is clear from the above that adaptation is a cellular based process and evolutionary adaptation cannot be understood outside of an understanding of a constant reciprocation between the environment and the cells that constitute all organisms.

13.7 Time as an Artifact in Biology

It has recently been advanced that any understanding of evolution with respect to the eukaryotic life cycle requires a completely new perspective that centers on the obligatory recapitulation of the zygote, whose adjudicating role in epigenetic inheritance has only recently been elucidated (Torday JS, Miller WB. The Unicellular State as a Point Source in a Quantum Biological System. Biology (Basel) 2016, 5(2):25). When this is understood, the proper narrative of biology is revealed. It is always in support of the three essential cellular domains (Prokaryota, Archaea, Eukaryota) despite any outward appearances (Miller Jr, W.B. and Torday, J.S., 2018. Four Domains: The Fundamental Unicell and Post-Darwinian Cognition-Based Evolution. *Progress in Biophysics and Molecular Biology*). When these realities are fully considered, further central ramifications emerge. Time can essentially be removed from biology after the origin of those three perpetual cellular lines. That may seem counterintuitive to our living perspective, but it is consistent with the work of Einstein and Feynman, both of whom thought that time was irrelevant to physics. Einstein's theory of relativity obviates time from consideration of the Cosmos, as does Lee Smolin's conception of the Cosmos based on Darwinian natural selection. Feynman's "random walks" also factor time out of quantum mechanics. And because biology and physics both derive from the singularity/big bang, they thereby necessarily comply with the same laws of nature, and the elimination of time from physics can also be thought of as applying to biology.

This conjecture is strengthened by considering biology as analogous with the organizational structure of the periodic table of elements. Mendeleev's "table of elements" succeeded in creating a productive ordering of chemistry where others had failed because he used atomic number as a common denominator to organize the elements. His crucial insight was that the number of protons is balanced by the number of electrons. Thus, the electron configurations in the outermost shell of the atom, or valence, can be identified as the basis for the periodic chemistry as reconfiguration of energy and mass. Similarly, cell–cell interactions are founded on communication mediated by growth factors and their receptors, constituting flows of energy along phylogenetic trees. Importantly too, Mendeleev also considered the chemical reactivity of the elements in his table, which permits an analogy with biological form and function.

From the foregoing, it can be seen that all of the seemingly complex properties of biology are epiphenomena of cell–cell interactions that follow an ordered series in the continuous perpetuation of the unicellular domain. In Eukaryota, this is exhibited by the unicellular adjudication and utilization of acquired epigenetic marks to serve the continuation of the eukaryotic unicellular form despite consistent environmental assaults. By understanding that this is the actual narrative across the entire course of evolution, and from which all physiology devolves, time can be factored out of the equation. The artifactual nature of time in biological evolution as we conventionally measure it through geologic history becomes superfluous, bringing into focus the true nature of the process of evolution as a continuous approximation of the singularity/big bang, which can be considered to be epitomized within unicellular forms. Thus, biology changes quite sharply from its conventional underpinnings, akin to the transition to heliocentrism in the late eighteenth century that placed the Sun at the center of the solar system and ushered in the Enlightenment and the Age of Reason. A proper understanding of biology permits a complete reconceptualization of our interrelationships with our fellow man, other organisms, communities, and the planet.

Since time can now be removed as a prime factor in biology, it is no longer surprising that biological processes might be fluid in direction and reversible. Previously, there was the improper assumption that evolution necessarily moved towards greater complexity. Yet, this is not observed biologically, as enormous complexity has been apparent at the unicellular level for billions of years. Jean Guex has shown that under physiologic stress conditions invertebrate ammonites exhibit properties of reverse evolution. Others in this volume have demonstrated the same. Their consummate painstaking biological observations are not merely theoretical. Instead, they are a natural consequence of the actual biological order. It should be pointed out that there is an important further evolutionary codicil to removing the time component of biology. Examples of reverse evolution vitiate the concept of evolution as a series of random mutations. If it were, then there should be no retrievable pathway to reverse changes once they have occurred in the fluid manner that the chapters of this book enunciate. It would be impossible to recapitulate ontogeny and phylogeny with the required fidelity. And further yet, it would be expected that removing the time aspect of biology would also relate to our own lives. Indeed it does, as under physiologic stress conditions, chronic diseases can be characterized as biological simplifications that can be seen to correspond to reversal of evolutionary paths (Torday J.S., Rehan V.K. Evolutionary Biology, Cell-Cell Communication and Complex Disease. Wiley, 2012).

13.8 Predictions of Cell–Cell Communication Not Available to Descriptive Biology

It has been well documented in this volume that under conditions of physiologic stress, biologic systems can resolve towards patterns consistent with reverse evolution. This is due to the principle of terminal addition, which indicates that new traits are appended at the end of a series of evolved traits due to the nature of cellular-molecular evolution, which is based on cell–cell communications mediated by ligand-receptor interactions. The downstream effect of ligand binding to its cognate receptor is the production of so-called second messengers, such as cyclic AMP and inositol phosphate, which ultimately interact with nuclear DNA to affect cell proliferation or differentiation. Changing such a pathway other than at the end of the sequence would be highly inefficient, placing the organism at risk of extinction. Conversely, adding a new trait on to the end of a set of evolved traits is highly efficient and more easily permits cells to adhere to the first principles of physiology, which are the ground-state requirements to maintain their homeostatic equipoise.

13.9 Dictyostelium

The slime mold *Dictyostelium discoideum* exists in one of the two states, a free-swimming amoeboid form, or a colonial form. Which state the organism exists within relates to the nutrients in the environment. Abundant food yields the amoeboid form. The colonial form occurs under conditions of low-food abundance. Nicole King has shown that the amoeboid form has the complete "toolkit" for forming the colonial form, demonstrating that the unicellular state gave rise to the metazoan sponge.

Dictyostelium has the capacity for epigenetic inheritance, inferring that the colonial form is a derivative means for acquiring information from the environment under stressful conditions. At the molecular level, Dictyostelium expresses the target of rapamycin (TOR) gene, which senses the levels of nutrients in the environment, and regulates signaling for cytoskeletal polarization. TOR integrates a vast number of unicellular pathways that are critical for eukaryotic development and metabolism. For example, an important group of mTOR factors are involved in regulating the use of lipids for energy in the cell. Therefore, it should not be surprising that in human biology, mTOR dysfunction has been linked to a diverse set of disorders such as diabetes, cancer, tumors, epilepsy, degenerative brain disorders, depression, and autism. This clearly illustrates the unicellular developmental influences that resonate across the vastly differentiated tissues that compromise all complex organisms.

Such reciprocations are mirrored by the interactions between physical phenomena and biologic systems. For example, microgravity profoundly affects both polarization and reproduction in yeast under experimental conditions. Both traits are impaired in near 0 × g conditions. Obviously, eukaryotes have adapted to gravity, likely due to the interaction between TOR and AKT signaling for cytoskeletal form as the ultimate determinant of the cellular "state" as either homeostatic, meiotic, or mitotic. This fundamental integration of calcium flux, cell division, and adaptation to gravity extends to all organisms through its role in reproduction. Reproduction is a critical means of monitoring the environment that allows for adaptation by sorting epigenetic impacts through reciprocation with basic cellular mechanisms via the unicell as the primary focus for environmental adaptations that can be disciplined within filtering selection.

13.10 *Turritopsis dohrnii*

The jellyfish, *Turritopsis dohrnii*, is considered immortal because under environmental stress conditions, it reverts to its immature polyp phenotype, seemingly rediscovering the "fountain of youth." However, this is another example of the phenotypic agent acting to acquire epigenetic marks. It has been argued that this phenotype has been misunderstood as an endpoint, when in fact, its purpose is directed towards the exploration of the environment relevant to an organism that is perpetuated iteratively through the unicellular form. Therefore, the immature state of the jellyfish and the other stages of its life cycle should be considered according to that context. Each stage is a means of optimizing its epigenetic inheritance. Its actual "immortality" is through its link to the perpetual unicellular form, from which it continuously renews its exploration of the environment through a variety of life cycle phenotypes, including reversion to its polyploid state.

13.11 Maternal Food Restriction, Truncation of the Life Cycle, and Metabolic Syndrome

Similarly, the observation that maternal nutrient restriction leads to metabolic syndrome as obesity, hypertension, and diabetes in the next generation is a misinterpretation of its ultimate importance. It is also known that this phenomenon causes precocious puberty due to accelerated adrenarche. A shortening of the reproductive cycle enables timing the next adult recapitulation into what is likely to be nutrient rich since all biological processes including food availability follow largely similar cycles. Indeed, metabolic syndrome can be viewed as a maladaptation since it is associated with premature senescence and aging.

13.12 Hibernation as Reverse Evolution of Endothermy/Homeothermy

The causal interrelationships between physiologic stress, catecholamines, and being warm-blooded are supported by the effects of hibernation on lung surfactant lipid composition in the alveoli, and on membrane fatty acid in peripheral cells. Such low-stress conditions decrease catecholamine production, resulting in increased surfactant cholesterol content, decreased lung surfactant surface activity, and decreased unsaturated fatty acid content of cell membranes, adaptively reducing oxygen uptake in the periphery.

There is experimental evidence to support the causal effect of environmental temperature on lung surfactant composition. In a study by Lau and Keogh, map turtles maintained at different ambient temperatures promptly adapted the lipid composition of their lung surfactant. This capacity to optimize lung alveolar physiology in response to various environmental temperatures can be seen as a precursor to endothermy/homeothermy.

Moreover, the causal relationship between land adaptation, neuroendocrine system, respiratory system, and endothermy/homeothermy may account for the observation by Weibel et al. that the lung is physiologically "overengineered." While trying to determine whether biological organisms are structurally evolved to match their functional demands, which is termed "symmorphosis," it was discovered that this same principle held for all of the components of the respiratory system—blood, heart, capillaries, and mitochondria—except the lung itself. Based on these observations, they concluded that for some unknown reason the capacity of the lung exceeds its physical requirements. Based on Darwinian evolution, that would have been seen as an epiphenomenon of land adaptation. But from a mechanistic cellular frame, based on the exaptation of atavistic cellular-molecular traits, it can be viewed as reflecting reactive cellular adaptation in support of homeostasis through PTHrP-PTHrP receptor signaling, facilitating the adaptation of multiple organs for life on land, namely the lung, kidney, skin, and brain. Those organisms with the most robust PTHrP-PTHrP signaling capacity would have been the most flexibly adaptive. Since duplication of the PTHrP receptor amplifies that signaling pathway during the water-land transition, the result of that flexibility led to the seeming "overengineering" of the lung.

13.13 Senescence and Aging Exemplify "Reverse Evolution"

As pointed out above for the effects of maternal food restriction, what is thought of as metabolic syndrome may actually be premature aging, as a consequence of premature menarche to accelerate the next generation's entry into a more food-abundant environment. Yet the loss of physiologic homeostasis is characteristic of the aging process. This is due to loss of cell–cell signaling capacities. Because bioenergetics is finite, and skewed towards the beginning of the life cycle in order to ensure successful reproduction, it wanes during the post-reproductive years. As such, it recapitulates the evolutionary strategy, but in reverse. This perspective on aging would suggest that by prolonging the reproductive phase of the life cycle, longevity would be extended as well.

13.14 Discussion

As indicated in this chapter, there is a close, causal relationship between the environment and evolution, mediated by cell–cell communication. In fact, what is referred to as "reverse evolution" is actually the means by which organisms remain in synchrony with their environments in order to function as efficiently in the prevailing conditions as possible. Once it is realized that time is largely a biologic artifact, documented changes in cell–cell signaling that can be traced towards a recapitulation of the evolutionary course of the organism justify the conclusions based on the many elegant biological observations enumerated in this volume. If it were not for such structural-functional plasticity in support of perpetual unicellular forms, life would not be feasible. Reverse evolution is one example of that fluid plasticity.

Chapter 14
Chronic Disease as Reverse Evolution

John S. Torday and William B. Miller Jr.

Keywords Just So Stories · Cell–cell signaling · Environmental stress · Terminal addition · Reverse evolution

14.1 Introduction

Dobzhansky famously said, "Evolution is all of biology." This resolute dictum raised the bar by challenging the field to exactly explain what he meant. The crux in biology is that it is easier to describe the relationship between ontogeny and phylogeny than it is to determine the mechanistic interrelationships between them. Retrospective reasoning has resulted in the generation of "Just So" teleologies instead of causal relationships. In the spirit of Dobzhansky, the following is a direct means of exploiting developmental biologic mechanisms of cell–cell signaling to understand how and why chronic diseases are a manifestation of "reverse evolution."

14.2 Cell–Cell Signaling Determines Embryologic Growth and Differentiation

Understanding the interrelationship between evolution, reverse evolution, and chronic disease requires an understanding of the mechanisms of cell–cell signaling that determine pattern formation during embryogenesis. Starting with the penultimate cell–cell communication between the sperm and egg, culminating in the consonant zygote, the

J. S. Torday (✉)
Department of Pediatrics, Obstetrics and Gynecology, Evolutionary Medicine Program, David Geffen School of Medicine, University of California, Los Angeles, CA, USA
e-mail: jtorday@ucla.edu

W. B. Miller Jr.
Department of Medicine, Banner Health System, Paradise Valley, AZ, USA

J. Guex et al. (eds.), *Morphogenesis, Environmental Stress and Reverse Evolution*, https://doi.org/10.1007/978-3-030-47279-5_14

subsequent steps in embryo formation are mediated by soluble growth factors produced by one cell type stimulating the growth and differentiation of other cell types through ligand-receptor interactions. These "combination lock and key" reactions trigger cascades, referred to as second messengers, that produce a series of intermediate chemical signals that ultimately affect DNA transcription for either mitosis or differentiation of the target cell. Many of these are under epigenetic influences that serve to maintain all cells in consistent resonance with environmental circumstances.

These serial cell–cell signaling processes terminate in homeostatic control of the intermediate physiologic steps at birth, ultimately giving rise to the offspring. More importantly, they give rise to the homeostatic regulation of physiology from the cell to the organism, and everything in between. The entire process is predicated on the first principles of physiology—negentropy, chemiosmosis, and homeostasis. In this manner, the formation of the first cell is continually referencing its environment so that its reciprocal interactions govern the course of the evolution of the organism. Therefore, it is this consistent interaction, expressed through epigenetic inheritance, that allows the phenotype to be properly assessed as the agent for monitoring and modifying the offspring of the next generation. It is through this exact cell–cell narrative that the homeostatic mechanisms of physiology and agency act to tether the organism to its origins mechanistically.

14.3 Environmental Stress Disrupts Cell–Cell Signaling, Causing Remodeling of Physiologic Homeostasis

It can be maintained that "experiments of nature" reveal the true properties of the evolutionary process. Major examples are the water-land transition and the increases and decreases in oxygen levels in the atmosphere that have occurred over the course of the last 500 million years, documented by Berner et al. (2007).

In the case of the water-land transition there were specific terrestrial adaptations that were necessary in order for fish to transition from water to land. Up until 2004 this association was considered merely hypothetical. However, it was from within this transitional period that Neil Shubin and his colleagues discovered Tiktaalik, the fossil remains of a transitional quadruped. What has been missing is the fossilized evidence for the concomitant remodeling of the visceral organs to accommodate this major transition.

This illuminating data came about serendipitously from the observation that there were three gene duplications that occurred during this period of time, namely the parathyroid hormone-related protein (PTHrP) receptor, β-adrenergic receptor (βAR), and glucocorticoid receptor (GR). All three are receptors, and are not coincidentally conjoined, as will be explained.

As had been hypothesized by Romer, bony fish self-selected to transition from water to land because of the "greenhouse effect" caused by the buildup of carbon dioxide in the atmosphere some 500 million years ago. Because gravitational force is greater on land than in water, these fish "sensed" the increased gravitational force, causing physiologic stress, primarily on the skeleton, but also on the internal organs. The primary effect of this stress was directly on the skeleton, causing remodeling based on Wolff's law, the capacity of bone to remodel itself in response to environmental factors. The molecular mechanism for this process is due to mechanical mediation by parathyroid hormone-related protein (PTHrP), the decrease in its activity causing decalcification of the bone, allowing the bone to deform and reform in compliance with gravitational forces on land. The necessary re-establishment of homeostasis was then mediated by increased PTHrP signaling for calcification of the bone, generating bony structures in support of terrestrial life.

Within the visceral organs, increased physiologic stress caused increased blood flow, "shearing" the microvasculature, generating radical oxygen species as a consequence, which further caused gene mutations and duplications.

Within the swim bladder, duplication of the PTHrP receptor gene would have facilitated the remodeling of the swim bladder to form alveoli, increasing the surface area-to-blood volume ratio, increasing oxygenation. Similarly, the kidney evolved from the simpler circulation of the glomus to the more complex vascular structure of the glomerulus by means of PTHrP signaling, which is angiogenic, i.e., promotes blood vessel formation. This enhanced vasculature surface area, in turn, facilitated PTHrP signaling between the podocytes lining the glomerulus and the mesangial cells lining the kidney tubules, enhancing the homeostatic control of water and electrolytes for land adaptation.

Such coordinated amplification of PTHrP signaling also facilitated the evolution of the integument since skin evolution is also contingent on PTHrP signaling. And because it is now thought that the skin gave rise to the brain, known as the "skin-brain hypothesis," the evolution of the central nervous system would have also been advanced by the concerted effects of PTHrP signalings on bone, lung, kidney, and skin in adaptation to land.

In each case—skeleton, lung, kidney, and skin—physiologic stress reversed the cell–cell communications of fish physiologic structures, allowing for the remodeling of structure and function over the course of the water-land transition. Importantly, in the case of earthworms their aquatic kidneys have been retained on land due to the refashioning of their soil surroundings, first observed by Darwin as an exception to the rule cited above. In modern terms, this is considered as an example of niche construction, whereby an organism can interact with its environment to engineer a set of changes to suit its own physiologic requirements. For example, organisms such as beavers build dams, whereas humans build homes, villages, towns, cities, and nations.

It is clear from research data that niche construction is a consequential part of cellular life, across evolutionary space-time. The reasoning is direct: the cell is the first example of niche construction. Niche construction is the cellular means of sustaining a causally linked relationship between itself and its environment.

14.4 Terminal Addition as the Underlying Principle for Reverse Evolution

Terminal addition describes the serial accrual of physiologic traits over the course of ontogeny and phylogeny. The underlying mechanism of cell–cell signaling provides an explanation for the "how and why" of this process. The binding of a ligand produced by one cell type to a cell of a different germ line origin to affect its growth or differentiation is mediated by the production of what are called second messengers, such as cyclic adenosine monophosphate and inositol phosphate. Such messengers cascade through the cell until they affect DNA transcription for either the further growth or the differentiation of the target cell. In turn, the affected cell subsequently produces other ligands that influence other cells during embryogenesis to yield the consecutive formation of differential features in a reproducible pattern recognized as terminal addition. Such terminal additions, mediated by second messengers, are highly evolved pathways for structure and function.

There is a direct biological logic behind this pathway. Adding new features other than at the end would undermine the intricate, integrated evolutionary process, and would undercut the problem-solving linkages that cells must make to successfully respond to successive cycles of environmental stress.

This precise point is also the key toward understanding the capacity for cells to collaborate and coordinate in complex tissues to provide "reverse evolutionary" phenotypes. It is much easier to deconvolute phenotypic expression to an earlier form if the most recent additions are those that are no longer being expressed. The other basics of tissue ecological expression remain firm. When faced with the effects of environmental stress, terminal addition affords the option of "reversing" the process of evolution in order to re-establish a previous state of homeostasis, sustaining itself under such suboptimal conditions. The cell–cell interactions that formed the most recent level of the homeostatic structure and function can be broken down under conditions of environmental stress.

For example, the increased blood flow caused by the physiologic stress "shears" the walls of the microvessels, producing radical oxygen species that cause gene mutations and duplications and a variety of epigenetic impacts. The communicating cells will default to an earlier stage of evolution mediated by atavistic growth factor/receptor signaling, reconstituting the earlier ontogenetic and phylogenetic structure and function, i.e., reverse evolution.

This is the most efficient way to survive under stress conditions until the organism can reproduce. Furthermore, if conditions were to improve, the organism might recapitulate the prior terminal addition through the process of regeneration.

Alternatively, if the environmental conditions are different from those under which the initial terminal addition occurred evolutionarily, the process of heterochrony might prevail, described as "a developmental change in the timing or rate of embryologic events, leading to changes in size and shape of organs and features over evolutionary time scales." At the cellular-molecular level, heterochrony allows for changes in cell–cell communication in order to maintain homeostasis under adverse conditions.

14.5 Self-Referential Self-Organization and Evolution

Life is often referred to as self-referential and self-organized. Where these characteristics originated is still unknown. It has been argued that the Cosmos is the product of the Big Bang, which exploded the putative singularity that existed before it, giving rise to dualities throughout. However, "for every action there is an equal and opposite reaction" based on Newton's third law of motion. That equal and opposite reaction can be reasonably equated with what is descriptively referred to as homeostasis—the force that stabilizes the mass and energy in a balanced chemical reaction, or the cell–cell interactions that generate a physiologic trait. If it were not for this homeostasis requirement based on the third law of motion, the Cosmos would solely be comprised of chaotic free energy traveling through space.

14.6 The Big Picture of $E = mc^2$ as the Basis for "Reverse Evolution"

Up until Einstein equated mass and energy in his famous equation, $E = mc^2$, the idea that there is unity in the Cosmos was just speculation. Beginning with the ancient Greeks like Parmenides and Anaximander, and more recently L.L. Whyte and E.O. Wilson, there has been conjecture for the universal interconnectedness that we all sense but cannot prove. However, with the advent of our knowing of mass-energy equivalence which accounts for the status of all existence, there is new opportunity to identify the organizing principle of the Cosmos, matter, energy, life, and all.

As indicated above, the explosive expansion of the Cosmos was counterbalanced by Newton's third law of motion, as reflected by the equivalence of reactants and products in a chemical equation, or the homeostatic balance formed by cell–cell interactions that generate physiologic form and function. Importantly, the presence of obligatory counterbalances in the Cosmos that began with and overlay the expansion implies that evolution, too, must have its own counterbalancing force, which can be seen as "reverse evolution" under stressful conditions that undermine homeostasis. It would therefore be expected that just as reverse evolution can be identified among the phenotypic features of diverse organisms, such as ammonites and others that have been discussed earlier in this book, reverse evolution would also be an expected manifestation of the cellular response to the stress of disease.

14.7 Examples of Reverse Evolution

14.7.1 Emphysema

Emphysema is characterized by a breakdown in the alveolar structure of the lung due to a variety of factors. Because the surface area-to-blood volume ratio is inefficient for gas exchange, the patient must use supplemental oxygen. The breakdown in the structure of the lung follows a course that is the literal step-by-step reverse of lung development, during which the larger saccules are subdivided into smaller and smaller compartments in order to increase the surface area-to-blood volume ratio in preparation for air breathing at birth. In addition to the breakdown in the compartmentalization of the alveolar bed, the fibroblasts produce connective tissue that physically maintains the integrity of the alveoli, referred to as pulmonary fibrosis.

At the cellular-molecular level the fibroblasts revert from their differentiated state as lipofibroblasts to their more primitive state as smooth muscle cells, more like the cytoarchitecture of the frog lung, having large faveoli with muscle-lined partitions between the air sacs.

14.7.2 Nephrosis

The number of kidney tubules varies from 30,000 to 300,000. The mechanism by which systemic salt and water are regulated by the kidney glomerulus resembles that of parathyroid hormone-related protein regulation of surfactant in the alveolus distension of the podocyte that stimulates PTHrP production; the secreted PTHrP binds to its receptor on the mesangial cells that line the kidney tubule, regulating the secretion and reuptake of fluid and electrolytes. If the glomerulus is damaged, there is loss of tubules. It has been demonstrated that one of the prime manifestations of nephrotic syndrome is the formation of epithelial cell foot process lesions, which correlate with the extent of proteinuria (Vernier et al. 1961).

In mature renal nephron cells, epithelial cell foot processes demonstrate tight interdigitation along the basement membrane. The nephrotic epithelium demonstrates foot processes that are reduced in number and abnormally broad, closely resembling less developed forms. These have been characterized as representing a form of "dedifferentiation" to a more primitive organization (Reeves et al. 1978).

14.7.3 Osteoporosis

The calcification and decalcification of bone are also regulated by PTHrP. Bone can remodel itself, referred to as Wolff's law. Pressure on the bone causes decreased expression of PTHrP, resulting in loss of calcium. Once the bone re-establishes homeostatic control at its newly established set point, there is increased PTHrP expression resulting in recalcification of the remodeled bone. This mechanism is the way in which the fish skeleton adapted to living on land, with the increase in gravitational force on land compared to water placing physiologic stress on the bone. This decalcification-recalcification to facilitate remodeling of bone is homologous with Jean Geux's experience with ammonites, changing their shell conformations under environmental stress.

14.7.4 Effect of Cholesterol-Deficient Lung Surfactant

The lung has evolved to increase oxygenation by decreasing the size of the alveoli, thus increasing the surface area-to-blood volume ratio to improve oxygen transfer. But in order to do so, lung surfactant reduction of surface tension within the alveoli had to evolve, beginning with cholesterol in the swim bladder of fish, preventing the walls of the bladder from adhering to one another. As the gas exchanger evolved from amphibians to reptiles, birds, and mammals, the surface tension-reducing activity of the lung surfactant increased in tandem.

This causal relationship was demonstrated by deleting the SCAP gene in rats. The SCAP gene governs the rate-limiting step in cholesterol synthesis, from the lung alveolar type II cells in which surfactant is produced. The alveoli compensated for the resulting suboptimal rise in surface tension by increasing the capacity of the lipofibroblasts adjacent to the type II cells to provide substrate for surfactant phospholipid synthesis. This homeostatic compensatory mechanism is actually reverse evolution since the lipofibroblasts initially evolved to protect the alveoli against rising levels of oxygen in the atmosphere, and then later evolved the increased capacity to produce lung surfactant by actively "trafficking" fatty acids to the alveolar type II cells for enhanced surfactant phospholipid production.

14.7.5 PTHrP-KO Phenotypes

Reverse evolution is epitomized by deleting the PTHrP gene during mouse embryogenesis, spanning effects on the lung, kidney, skin, bone, and brain, all of which "simplify" due to loss of PTHrP signaling for morphogenesis. The key to understanding how and why that occurred is in the duplication of the PTHrP receptor during the water-land transition. This was likely due to selection pressure for adapting to the increased force of gravity on land, causing the remodeling of the skeleton under the auspices of PTHrP, which regulates the calcification and decalcification of bone. Positive selection for PTHrP signaling would, in turn, have increased alveolar lung evolution, glomerular evolution of the kidney, barrier function of the skin, and brain neuronal evolution. These evolutionary changes represent the evolution of those physiologic traits expressed by fish, possessing swim bladders and kidney glomus instead of terrestrially adapted lung alveoli and kidney glomeruli. In order to evolve land-adaptive phenotypes, the fish cell–cell communications for homeostasis had to have been disrupted by physiologic stress, leading to newly established cell–cell communications for homeostasis that constituted the evolved structures and functions in adaptation to land. To be clear, the evolution of structure and function under environmental stress requires a breakdown in prior normative homeostasis through physiologic stress causing the production of radical oxygen species, causing a loss of cell–cell communication before reconstruction of cell–cell communication for a new homeostatic set point can occur.

14.7.6 Brain Pathology as "Reverse Evolution"; General Anesthesia Recovery as Brain Evolution

Hughlings Jackson, the British neurologist, hypothesized that chronic brain illness was reverse evolution. He observed that the higher centers of the brain inhibited the lower centers, and loss of such relationships was consistent with brain diseases. Consistent with these structural hierarchical interrelationships in the brain, Mashour (Mashour and Alkire 2013) has observed that as patients come out from general anesthesia they recapitulate the phylogeny of the brain.

14.7.7 Liver Function

The liver is one of the few organs that are capable of significant self-repair. Hepatocytes have the distinct ability to dedifferentiate to a progenitor phase as part of their regenerative mechanism, and can do so without any requisite genetic and epigenetic manipulations (Chen et al. 2012).

For example, the liver of a rat can regenerate by 75% in 1 week after partial resection. The stem cell/progenitor population that permits this regenerative repair derives in part from existing pluripotent stem cells, but is also produced by the dedifferentiation of a portion of the remaining hepatocytes in a reverse evolution process. Research in the treatment of acute liver failure is being directed to an enhancement of hepatocyte dedifferentiation in vivo by attempting the conversion of mature hepatocytes into progenitor cells to activate fetal programming through Hippo/Yes-associated protein signaling pathways (Cubero and Martinez-Chantar 2019).

This phenomenon of reverse evolution in the liver is not confined to liver regeneration. For example, in a model of tyrosinemia-precipitated stress in mice, the liver can overcome this dysfunction by inducing the formation of aneuploid hepatocytes that lack chromosome 16. The capacity for reverse evolution therefore may be seen as a feature of the cellular repertoire that persists in differing degrees among cellular types and populations across evolutionary space-time.

14.7.8 Neoplasia

It has been previously suggested that cancer is connected in some way to ancient roots, perhaps extending as far back as the origin of life (Vincent 2010). Specific stem cells in tumor tissues appear to exist within a progenitor state of development as a connection to the prokaryotic cellular homolog toolbox (Fernandes et al. 2012).

High rates of glycolysis, chemoresistance, and radioresistance are all shared between cancers and unicellular realm. These traits arose early in evolution. Therefore, it has been proposed that one of the keys to the aggressive proliferation of cancer is a backward connection to a primordial toolkit, i.e., a form of neoplastic reverse evolution (Miller and Torday 2017). Cancer phenotypes based on a reverse evolutionary backwardization toward unicellular behaviors permit neoplasia the flexibility to profit in local tissue ecologies at the expense of other ecological partners. One suggested pathway is through triggered dysregulation of mitochondria (Davila and Zamorano 2013). It is hypothesized that stress from the resulting cumulative oxidative damage causes a previously differentiated cell to revert to the phenotype of a facultative anaerobic heterotrophic cell optimized for survival and proliferation. It has even been proposed that the cancer phenotype is a mirror image of the phenotype of the last eukaryotic common ancestor (LECA) (Davila and Zamorano 2013).

It is through this fluid dedifferentiation back to primitive pluripotent forms that the chromosomal instabilities seen in neoplasia can rapidly express as a full range of phenotypes and still escape normal cellular checkpoints (Kastan and Bartek 2004).

Hyper-adaptabilities are a well-known feature of neoplasia (Vineis 2003). This adaptability can now be understood as tumor cell utilization of prior evolutionary tools to succeed in both local and distant (metastatic) tissue ecological environments. In this manner, neoplasia outcompetes typical differentiated cells.

14.8 Reverse Evolution and the Commutative Law

Reverse evolution is a robust aspect of evolution in part because it satisfies the mathematical commutative law that any finite sum or product is unaffected by the order of the multiplicative terms, whereas Darwinian evolution is noncommutative, i.e., unidirectional. That is to say, evolution based on cell–cell communication allows the operation to go either forward or backward along a string of developmental phenotypes, each a product of terminal addition, depending upon the environmental conditions. This demonstrates the power of placing evolutionary development in the framework of cellular interactions and cell–cell signaling. Conversely, neo-Darwinian evolution due to random mutation is unidirectional, as there is no consistent way to reverse a mutation except through its complete silencing, which need not relate to contemporaneous environmental stress.

Cell–cell interactions are a continuous cascade of energy transfer from one cell to another, determining the structure and function of the embryo, ending in homeostatic regulation. The process is characterized by a linked sequence of high-energy phosphates compounded with either adenosine or inositol that mediate communication between cells. In contrast to that, gene mutations are discrete events within cell nuclei that regulate cellular activities under the control of homeostatic mechanisms.

Rowlands (2015) teaches us that commutative quantities are one-dimensional, whereas anti-commutative quantities are three-dimensional, consistent with the difference between cell–cell communication and gene mutations as commutative and noncommutative, respectively.

The cell–cell signaling mechanism for evolution is consistent with other examples of reverse evolution cited in this book, whereas evolution based within genetic mutation cannot accommodate such reverse evolutionary events with the same facility. For example, it is specifically emphasized that under environmental stress, organisms exhibit loss of traits consistent with terminal addition. Torday and Miller Jr. (2018) have explained this phenomenon as the result of cell–cell signaling via second messengers that determine the growth and differentiation of tissues, organs, and organisms. Adding new traits other than at the end of a series of evolutionary adaptations would be highly inefficient, risking extinction of the organism, as previously emphasized.

The term terminal addition was first used by Haeckel to explain his biogenetic law, or "ontogeny recapitulates phylogeny." Of course, Haeckel's theory was dismissed in favor of genetics as the basis for the modern synthesis back in the mid-nineteenth century. That seemed expeditious at the time, but has become unfortunate as it displaced cell biology from evolution theory for decades. At this contemporary juncture, it is understood that embryologic development and phylogeny occur as a result of cell–cell communication. In turn, this elucidates the process of terminal addition as serial signaling mechanisms for form and function that are fluid enough to also account for the observed reversions to earlier states of ontogeny and phylogeny under stress within Cope's rule and Dollo's law. Each is a differing facet of cell–cell signaling pathways that ratchet cellular evolution within genomic constraints. When placed under sufficient stress, flexible cells can use their toolkit to meet contemporary environmental problems through a fluid reversal to ancient evolved pathways.

References

Berner RA, Vandenbrooks JM, Ward PD (2007) Evolution. Oxygen and evolution. Science 316(5824):557–558

Chen Y, Wong PP, Sjeklocha L, Steer CJ, Sahin MB (2012) Mature hepatocytes exhibit unexpected plasticity by direct dedifferentiation into liver progenitor cells in culture. Hepatology 55(2):563–574

Cubero FJ, Martinez-Chantar ML (2019) Plasticity of adult hepatocytes and readjustment of cell fate: a novel dogma in liver disease. Gut 68(6):954–956

Davila AF, Zamorano P (2013) Mitochondria and the evolutionary roots of cancer. Phys Biol 10(2):026008

Fernandes J, Guedes PG, Lage CL, Rodrigues JC (2012) Claudia de Alencar SL. Tumor malignancy is engaged to prokaryotic homolog toolbox. Med Hypotheses 78(4):435–441

Kastan MB, Bartek J (2004) Cell-cycle checkpoints and cancer. Nature 432(7015):316–323

Mashour GA, Alkire MT (2013) Evolution of consciousness: phylogeny, ontogeny, and emergence from general anesthesia. Proc Natl Acad Sci U S A 110(Suppl 2):10357–10364. https://doi.org/10.1073/pnas.1301188110

Miller WB, Torday JS (2017) A systematic approach to cancer: evolution beyond selection. Clin Transl Med 6(1):2

Reeves W, Caulfield JP, Farquhar MG (1978) Differentiation of epithelial foot processes and filtration slits: sequential appearance of occluding junctions, epithelial polyanion, and slit membranes in developing glomeruli. Lab Investig 39:90–100

Rowlands P (2015) The foundations of physical law. World Scientific, Singapore

Torday JS, Miller WB Jr (2018) Terminal addition in a cellular world. Prog Biophys Mol Biol 135:1–10

Vernier RL, Worthen HG, Good RA (1961) The pathology of the nephrotic syndrome. J Pediatr 58(5):620–639

Vincent MD (2010) The animal within: carcinogenesis and the clonal evolution of cancer cells are speciation events sensu stricto. Evolution 64(4):1173–1183

Vineis P (2003) Cancer as an evolutionary process at the cell level: an epidemiological perspective. Carcinogenesis 24(1):1–6

Index

A
Abiotic events
anoxic, 151
environmental changes, 151
evolutionary diversification, 152
internal microstructures, discoidal tests, 145, 153, 154
mesoendothyrids, 154–159
Mesozoic isomorphic morphotypes, 152
oligotrophic conditions, 151
orbitopsellids, 158–159
pfenderinids, 159–163
shape and coiling, 142, 143, 152, 153
valvulinids, 158–159
Adaptive phenotype, 289
Ad hoc functional explanations, 15
Ad hoc morphofunctional origination, 15
Aganane formation, 165
Alkanes, 192
Aluminium, 270
Alveolae, 138, 140–141
Alveoli, 305
Ammonia tepida, 189, 192
Ammonites, 14
Ammonoid assemblages, 224
Ammonoids, 63
Amoco Cadiz black tides
anomalies observations, 183, 184
CNEXO, 178
environmental stress, 183
flatworms egg capsules, 187
method and materials, 180, 183
morphological abnormalities, 186
oil spills, 178
pre- and post-spill observations, 180
proximity, 179
rocks of Portsall, 179
weather conditions, 179
Amoco Cadiz vs. Erika spills
chemical composition, 189, 192
foraminiferal sampling, 193
oil concentration, 192, 193
oil pollution sediments, 193
parameter, 194
physical oil properties, 192, 193
resilience threshold, 192
weathering processes, 193
Anaticinella, 7
Anaticinella multiloculata, 8
Anchignathodontidae, 62, 75
Anisian, 70
Anoxia, 248
Anoxic crises, 152
Anoxic environments, 262, 265
Anoxic events, 2, 151
Apertures, 139
Aromatic hydrocarbons, 186
Arsenate, 266
Arsenic (As), 266
Artificial scheme, 62
Atavic morphology, 133
Atavic persistent morphotypes, 147
Atavism, 49
Atavistic development, 7
Atavistic reversals, 19, 74–76, 79

B
Bacterial filaments, 140–141
Benthic foraminifera, 14, 15
Berner oxygen fluctuations, 292
Bioclastic input, 270, 274

J. Guex et al. (eds.), *Morphogenesis, Environmental Stress and Reverse Evolution*, https://doi.org/10.1007/978-3-030-47279-5

Bio-environments, 60
Biogenetic law, 17
Biological evolution, 293
Bioseries, 134
Biostratigraphical and palaeogeographical data, 225
Biostratigraphic units, 198
Bivalve species, 210
Black tide effects
  Amoco Cadiz vs. Erika spills, 189–191
  oil pollution, 189
Black tides, Brittany (NW France)
  Amoco Cadiz, 177
  oil spills, 177, 178
  Torrey Canyon, 177
Bleaching, 242
Boreal, 83, 84
Boreal Russian, 63
Boreal Superrealm, 222, 224
Borinella (Dienerian-Spathian), 74, 76, 78, 79, 83, 86, 87
Bottom-water oxygenation, factor analysis
  Cu, 262–264
  nickel, 260–262
  oxyanions (*see* Oxyanions)
Brahmanian and Jakutian stages, 62
Brain pathology, 306
Brown eye pathway, 98
Brown eye pigment, 98
Bunter facies, 61

C

Calcification and decalcification, 304
Cale du Dourduff, 180, 187
Cancer phenotypes, 307
C- and N-isotope records, 217
Cantabriconus, 148, 159
Capture, 73
Carbon isotopes, 72
Carbonate successions, 64
Carnian Pluvial Episode, 65
Carnian Pluvial Event (CPE), 41
Catastrophe theory, 6
Catastrophic oil spills, 194
Catecholamines, 292, 296
Cell-based land adaptations, 292
Cell–cell communication, 294, 297, 308
Cell–cell interactions, 290, 293, 302, 308
Cell–cell narrative, 300
Cell–cell signaling, 297
  capacities, 297
  embryologic growth and differentiation, 299, 300
  environmental stress, 300–301
  mechanisms, 290, 299, 308
  pathways, 308
Cell-molecular adaptations, 290
Cell-molecular basis, 290
Cell proliferation/differentiation, 294
Cellular domains, 293
Cellular environment
  comb orientation
    in single rotating combs, 103–106
    in VCWR, 106
  flexibility, 106
  sex comb during development, 103
Cellular environment bias comb evolution, 103
Central Atlantic Magmatic Province (CAMP), 4, 5, 60
Centre National pour l'Exploitation des Océans (CNEXO), 178
Ceratitid ammonoids, 221
Chamberlets, 135, 138
Changhsingian, 219
Chelif Basin of Northwestern Algeria, 116
Chert gap, 71
Chocolate mousse, 193
Cholesterol-deficient lung surfactant, 305
Chromium (Cr), 265, 266
Chronic diseases, 294
Chrysalidinids, 160–161
$\delta^{13}C$ isotope, 62
cis-regulatory elements, 98
Cladistic approach, 143
Clarkina, 76
Clay minerals, 262, 272–275
Climatic changes, 2
Clonal animals, 235
Coal gap, 72
Coevolutive genetical processes, 143
Colonial animals
  assembly-line animals, 235
  clonal and colonial organisms, 235
  divergent biology, 235
  environmental impacts, 235
  symbiotic associations, 235
Colonial hydroid *Myrionema*, 237
Colony development, 234
  cluster feeding polyps, 236
  feeding triggers contractions, 236
  food-rich environment, 236
  gastrovascular system, 237
  genes recruitment, 237
  hydroids, 237
  phenotypically plastic growth forms, 236
  photosynthesis, 237
  sheets and runners, 236
  Symbiodinium spp., 237

Compartmentation, 290
Conflict mediation, 242
Conodont apatite ($\delta^{18}O_{phos}$), 65
Conodont radiation
  Anchignathodontidae, 75
  Ellisoniidae, 75
  Gladigondolellidae, 80, 81
  Gondolellidae (*see* Gondolellidae)
Conodonts
  depth stratification model, 38
  Dovško succession, 43, 45, 47, 48, 55
  jawless vertebrate animal, 38
  juvenile mortality, 40, 50, 54, 55
  lateral segregation model, 38
  Middle and Late Triassic, 40–42
  Middle Norian juvenile assemblages, 42–54
  paleoecology, 38
  paleogeographic distribution, 38
  pectiniform elements, 41, 44
  Permian-Triassic extinction, 39, 40
  temperature dependence, 38
Cope's rule, 13, 14, 16, 19, 20, 308
Copper (Cu), 262–264
Coral-dinoflagellate symbiosis, 235
Coral reef communities
  bleaching crux, 242
  colonial animals, 235
  colony development, 234
  component modules, 234
  conflict mediation, 242
  coral-symbiont cross talk, 239–241
  cryptic species, 241
  environmental signalling, 236–238
  environmental stress, 241
  evolutionary adaptation, 241
  phenotypic plasticity, 235, 241
  polymorphism/hybridization, 242
  population structure, 242
  scleractinian hexacorals, 234
  water currents, 238
Coral species, 241
Coral-symbiont cross talk
  conflict mediation, 240
  cooperation, 240
  defector symbiont, 240
  evolutionary conflict, 240
  export mechanisms, 240
  host-symbiont community, 240
  photosynthate, 241
  replicatory advantage, 240
  Symbiodinium dinoflagellates, 239
Cosmos, 303
Cretaceous planktonic foraminiferal anagenetic lineage, 7
Cusp catastrophe, 6, 7
Cycloclypeus, 137

D

Darwinian anagenesis, 74, 76, 79
Darwinian evolution, 17, 297
Darwinian natural selection, 293
Dedifferentiation, 304, 307
Deep-sea microbes, 73
Deepwater Horizon pollution, 177
Deepwater renewal time (τ_{DW}), 268, 269
Defector symbiont, 240
Developmental plasticity
  description, 97
  externally induced, 98
  nutritional plasticity, 98
  phenotypic, 102
  reverse evolution, 102
  single-gene mutation, 102
Dictyostelium, 295
Dienerian-Smithian boundary, 66, 68
Discharged oils, 194
DNA transcription, 302
Dollo's law, 19, 20, 308
Dollo's point, 19
Drosophila insulin-like factor (dilp6), 100

E

Early Jurassic abiotic events, 155
Early Middle Triassic Anisian, 70
Early Triassic conodonts
  environmental changes (*see* Environmental changes)
  environmental instabilities, 60, 61
  gaps (*see* Environmental gaps)
  Griesbachian zonation, 81, 82
  Late Permian-Early Triassic extinction, 60
  methane, 73
  morphogenesis, 82, 83
  ocean water, 70
  Period of Time, 61–63
  phylogenetic reconstructions, 73–75
  radiation (*see* Conodont radiation)
  stages, 63, 64
  warmest periods, 60
Early Triassic environmental changes, 85
Ecological survey, 183

Ellisonia teicherti, 75
Ellisoniidae, 75
Emphysema, 304
Endogenization, 290
Endosymbiosis, 290
Endothermy/homeothermy, 296, 297
End Triassic extinction (ETE), 4, 5
Energy-dispersive X-ray fluorescence (ED-XRF), 255
Energy-rich compounds, 73
Energy-rich dead bodies, 73
Environmental changes
  Anisian, 70
  conodont apatite ($\delta^{18}O_{phos}$), 65
  cooling periods, 66
  Dienerian-Smithian boundary, 68
  geochemical tools, 64
  Late Permian-Early Triassic negative shifts, 68
  middle-late Spathian boundary, 70
  $\delta^{18}O$ composite profile, 66–68
  $\delta^{18}O_{carb}$ excursions, 65
  Paleozoic atmosphere, 64
  Phanerozoic carbon isotopic excursions, 65
  SSB, 65, 68–70
  strontium isotope, 64
  terrestrial material, 66
  uppermost Spathian, 70
  warming events, 65
  warming periods, 66
  water temperature, 64
Environmental conditions, 60
Environmental gaps
  carbon isotopes, 72
  chert gap, 71
  coal gap, 72
  reefs, 71, 72
Environmental instabilities, 60, 61, 279
Environmental nature, 63
Environmental parameters, 40
Environmental perturbations, 1
Environmental stresses, 123
  catastrophic climate change, 25
  development and evolution, 1
  evolutionary/taphonomic processes, 30
  geology, 1, 2
  gradual changes, 6
  on late Permian pollen grains, 25–30
  moderate, 6
  origins, 1
Epigenetic inheritance, 300
Epiparasitized specimens, 183
Episodical, 135
Erika black tides, 189
  Ammonia tepida, 189
  aromatic hydrocarbons, 192
  field samples, 188
  Maltese oil tanker, 188
  morphological abnormalities, 189
  oil slick ashore, 188
  turbellarian egg capsules, 189
Erika oil-treated mesocosms, 189
Eukaryota, 290, 293
Eukaryotic clades, 132
Eukaryotic life cycle, 293
Eukaryotic symbiosis, 242
Eurygnathodus, 77, 78
Eusociality in ants, 99
Everticyclamminids, 147
Evolution, 112
Evolutionary diversification, 152
Evolutionary lineages, 14
Evolutionary sciences principles
  classical, 132
  genetic proximity, 132
  microstructures morphogenesis, 135
  ontogenesis/heterochrony, 133
  phyletic arborescences reconstruction, 132
  phyletic relations, 132
  sedimentary sequential evolution, 133
  stratigraphic/paleogeographic repartition, 132
Evolutionary trends, 14
External abiotic factors, morphogenetical trend
  abiotic events, 151
  bioseries, 149
  early Jurassic extinction event, 149
  extinctions, 149
  K/T crisis, 151
  LIP eruptions, 149
  P/T extinction, 149
Extinctions
  catastrophic evolutionary changes, 1
  Cretaceous Paleogene boundary, 3
  end-Pliensbachian, 5
  ETE, 4
  and giant volcanism, 3
  in Greenland, 5
  modes of evolution, 1
  origins, 2
  and periods of volcanism, 2
  Permian-Triassic boundary, 3
  post-extinction recovery, 3
  Rhaetian, 5

sublethal environmental stress, 1
Toarcian OAE, 6
Triassic ammonoids, 4
Extraterrestrial impacts, 2
Extreme warming, 68

F
Factor analysis, Mishor Rotem basin
bioclastic input, 270
bottom-water oxygenation (*see* Bottom-water oxygenation, factor analysis)
factor scores, 275, 276
OSM, 258
terrigenic/detrital input
Al, 270
clay minerals, 272–275
heavy minerals, 271, 272
Factor scores, 275, 276
Feeding triggers contractions, 236
Flatworms, 187
Fluorescence microscopy, 99
Food-abundant environment, 297
Foraminifera
abnormal deformed, 176, 177
adaptive advantage, 253
Amoco Cadiz (*see Amoco Cadiz* black tides)
anomalous, 194
anoxic environments, 248
bulminid assemblages, 253
cultures, 193
Deepwater Horizon pollution, 177
diatom chloroplast sequestration, 253
diatom-related kleptoplastidy, 253
epi-parasitized, 187
Erika black tides, 188, 189
and geochemical proxies, 248
kleptoplastidy, 247, 251, 253, 279
late Cretaceous upwelling regime, 251
marine ecosystems, 247
micropaleontological records, 251
morphologic features, 251, 252
morphological adaptations, 248
N. canadensis, 253
nitrate, 247
palaeoceanography, 247, 248
physiological adaptations, 247, 248
phytoplankton, 247
P. prolixa, 252
Shefela basin, 253, 254
sulphidic conditions, 247
sulphidic environments, 253
TROX model, 253
type of food, 280
Foraminiferal fauna, 186–188
Foraminiferal morphology, 194
Foraminiferal test deformations, 177
Fusiform-to-globular alveolinids, 153

G
Gastrovascular system, 237
Gene capture, 73
Genetic memory, 168
Genetic proximity, 132
Genus *Chablaisia*, 161
Genus *Episageceras*, 221
Genus *Otoceras*, 225
Genus *Planisepta*, 167
Geochemical analyses, 255
Geochemical tools, 64
Giant volcanism, 2, 3
Giantism, 292
Gladigondolella, 80
Gladigondolellidae
Gladigondolella, 80
Wapitiodus, 80
Global Stratotype Section and Point (GSSP), 62
Globular alveolinids, 153
Golden Spikes, 62
Gondolellidae, 62
incertae sedis (*see* Incertae sedis)
multi-element apparatus, 75
Neogondollelinae (*see* Neogondollelinae)
Gondolellids, 73
Green autofluorescent kynurenine accumulation, 99
Greenhouse effect, 291
Griesbachian environment, 62
Griesbachian zonation, 81, 82
Growth stages, conodonts, 44, 47, 50, 51
Gulf of Oman, 120, 124

H
Haeckel's law, 17, 18
Haeckel's theory, 308
Haynesina germanica, 184, 189
Heavy minerals, 271, 272
Heliocentrism, 294
Hepatocytes, 306
Hermatypic coral reefs, 131

Heterochrony, 82, 302
Hibernation, 296
Himalayas, 61
Hindeoodus typicalis, 202
Homeostasis, 292
Homeostatic regulation, 300
Hydrocarbon evolution, 194
Hydroids, 237
Hyper-adaptabilities, 307
Hypermorphosis, 153
Hypodermic network, 138, 140–141

I

Icriospathodus, 80, 86
Immortality, 296
Imperforate larger foraminifera (ILF), 131
Imtachan formation, 202
Incertae sedis in Gondolellids
  Icriospathodus, 80
  morphotypes, 79
  Neogondolella, 80
  Paullella, 80
  rediversifications, 79
  Siberigondolella, 79
Induan, 64
Induan ammonoids, 221
Inessin Klyuch section, 204
Insulin, 99–101
Insulin-TOR nutritional sensing pathways, 99
Internal microstructures
  apertures, 139, 141
  coiling modes, 138, 139
  hypodermic network, 138
  miliolids, 143
  orbitolinids, 141
  pillars, 139
  vertical radial partitions, 139
  wall, 137
Internal structural innovations, 154
International N-air standard for nitrogen, 200
Intertropical Convergence Zone (ITCZ), 66
Irreversibility of evolution, Dollo's law, 19, 20

J

Just So stories, 299
Juvenile mortality, conodonts, 40, 50, 54, 55

K

Kashmirella, 77
Kleptoplastidy, 247, 251, 253, 277, 279, 281
Knötchenstadium of *P. spelae*, 15
K/T crisis, 151
K/T extinction, 149
Kynurenine, 98, 99, 101
Kynurenine accumulation, 99, 100
Kynurenine ex vivo, 99

L

Larger foraminifera
  adult-shape, 144
  algal symbiosis, 131
  bioseries, 165
  Cretaceous innovation, 145
  diversification, 152
  ILF, 131
  impact, 145
  microgranular, 152
  sensitive, 151
  trochospiral, 145
Last eukaryotic common ancestor (LECA), 307
Last Glacial Maximum sediments, 120, 122
Late Cenomanian anoxic crisis, 169
Late Cenomanian events, 152
Late Liassic ammonites, 18
Late Permian
  gymnosperm bisaccate pollen grains, 23
  Jugasporites genus, 24
  miospore assemblages, Poland, 29
  Zechstein Basin, 25–27
Late Permian–Early Triassic Boundary, 76
Late Permian–Early Triassic extinction, 60
Late Permian–Early Triassic interval, 66
Late Permian–Early Triassic negative shifts, 68
Late Permian marine environmental crises
  accumulations, 218
  biotic crisis, 214
  brachiopod fossils, 215
  Changhsingian, 219
  C-isotope excursions, 214
  correlation, 217
  extinction, 214
  interpretation, 218
  Kolyma–Omolon region, 215
  Lower Triassic, 215
  N-isotope data, 218
  N-isotope interval, 218
  N-isotope superintervals, 217, 218
  N-isotopic composition, 215
  nitrogen, 215
  $\delta^{15}N_{sed}$ values, 215
  O- and N-isotope records, 215
  palaeolatitudes, 214, 219
  Pravyi Suol-1 section, 218
  P–T crises, Arctic Canada, 219

P–T marine environmental/biotic crises, 219
Siberian flood basalt volcanism, 214
Triassic strata, 214
Verkhoyansk region, 219
Late Permian mass extinction, 61
Late Pleistocene sediment, 120
Late Spathian (LSE-SAB), 66–68
Late Smithian event, 61
Late Smithian Thermal Maximum, 66
Late Triassic, conodonts, 40, 41, 43, 54
Lee Smolin's conception, 293
Levant region, 66–68
Ligand-receptor interaction, 294
Lindström formation, 219
Lituolids, 131, 137, 143, 147
Lituoliporidac, 138
Lituosepta-Orbitopsella bioseries, 165
Liver function, 306
Living fauna, 183
Lower Triassic conodonts, 61, 62, 64, 70, 74
Lung surfactant composition, 296

M

Maastrichtian warming, 145
Macrofossil, 199
Marine regressions, 2
Marl member (MM), 257
Maternal nutrient restriction, 296
Meishanensis zone, 76
Mendeleev's "table of elements", 293
Mesoendothyrids
characteristics, 155
Cretaceous, 158
diversification, 157
everticyclamminids, 147
genera, 154
Jurassic abiotic events, 155
Jurassic chambers, 155
orbitopsellids, 155
"Planisepta" flat morphotype, 155
pseudokeriothecal-to-alveolar, 146
pseudokeriotheca-to-alveolae, 155–157
Tethyan realm, 155
Triassic endothyrids, 155–157
uncoiled-cylindrical, 155
Mesozoic carbonate platforms, 131
Mesozoic interval, 135
Mesozoic isomorphic morphotypes, 152
Mesozoic Neotethys, 165
Messinian salinity crisis (MSC), 116, 124
Metabolic syndrome, 296, 297
Metal-reducing bacteria, 265
Metazoa, 132
Methane, 73
Methanogens, 73
Microfauna, 143
Microfossil-rich intervals, 136
Microgranular/agglutinant imperforate larger foraminifera (ILF)
biseriality, 143
extinction, 154
fossil, 131
iterative evolution, 142
Mesozoic clades, 138
morphogenetical history, 135, 136
morphological characters, 168
morphology, 131
sedimentological setting, 136
unkeeled, 168
Microgranular calcitic rhombs, 137
Mid-Cretaceous OAE 1b-d events, 153
Middle-late Spathian boundary, 70
Middle-to-Late Jurassic valvulinids, 159
Middle-Upper Toarcian crisis, 19
Milankovitch's cycles, 60
Minor genetic mutations, 133
Miospores, 26, 27, 29, 30
Mishor Rotem basin, Upper Cretaceous
factor analysis (*see* Factor analysis, Mishor Rotem basin)
major and minor elements, 256
MM, 257
OSM, 257
PM, 257
trace elements, 258
Mitochondrion-rich myoepithelial cells, 237
Moderate Canadian, 63
Molluscs, 201, 202, 208, 219
Monocyclic aromatic hydrocarbons, 191
Morphogenesis
in biology, 86
bios, 61
conodont, 82, 83
palaeontology, 86
Morphogenetic innovations, 133
Morphogroups, 147
Morphological abnormalities, 186, 189
Morphological deformations, 176
Morphological plasticity, 123
Morphological retrogradation, 123
Morphological variations, 133
Morphotypes, 131, 138
mTOR dysfunction, 295
Multicellular life, 291

Multi-element apparatus, 60
Multivariate analysis, 255
Mutations, 294
Mytilus beds, 131

N
Natural environmental stress, 1
Nautilus, 199
Nekuchan formation
    ammonoid zones, 201
    Changhsingian–Induan, 219
    fossils, 208
    Otoceras boreale beds, 198
    Ustupny, 210
    Verkhoyansk region, 198
    zones, 198
Neobulimina canadensis, 253
Neo-Darwinian evolution, 307
Neogondolella, 79, 80, 86
Neogondollelinae
    Borinella, 78
    Clarkina, 76
    Eurygnathodus, 77, 78
    Kashmirella, 77
    meso- and jinogondolellin, 76
    Neospathodus, 76–78
    Paragondolella, 79
    Platyvillosus, 78
    Scythogondolella, 79
    Smithodus, 78
Neoplasia, 307
Neospathodus, 76–78
Neospathodus collinsoni, 80
Neotethyan successions, 68
Nephrosis, 304
Newell's regression hypothesis, 151
Niche construction, 291
Nickel, 260–262
Nicoraella budaensis, 42
Nikolkin Klyuch section, 200
    conodonts *H. typicalis*, 202
    fossils, 204
    Nekuchan formations, 201
    O. concavum Tozer, 202
    Otoceras boreale zone, 202
    Otoceras cf. *gracile*, 201
    sandstone, 201, 202
    T. morpheous epibole zone, 203
    T. pascoei epibole zone, 203
N-isotope ratio, 199
Nonionella, 253
Non-platform *Sweetospathodus* (Dienerian), 74
Nutrient deprivation, 98
Nutrition, 98, 99
Nutritional plasticity, *vermilion* (v) mutants, 98

O
OAE2 (environmental perturbations), 7
Oceanic anoxic events (OAE), 248
Ocean water, 70
$\delta^{18}O$ composite profile, 66–68
Octocorals, 234, 237
Oil-polluted areas, 194
Oil shale member (OSM), 255, 257
Oil-treated mesocosms, 194
Olenekian, 64
Ontogeny, 16, 17, 294
Orange pigment pathway, 98
Orbitolinids, 141
Orbitopsellids, 136
    Aganane formation, 136, 164, 165
    genus *Planisepta*, 167
    growth and demise, 166
    Liassic carbonate platform, 165
    lithiotids, 164
    Lituosepta-Orbitopsella bioseries, 165
    Mesozoic Neotethys, 165
    microfaunal assemblage, 167
    microfaunal faunal turnover, 167
    parasequences, 165
    rapid disappearance, 167
Orbulina-Praeorbulina-like assemblage, 118–121
Orbulina suturalis, 115
Orbulina universa, 112, 114, 115, 120
Orientation, *Drosophila* sex combs, 103–104
Osteoporosis, 304
Otoceras-bearing deposits, 211
Otoceras boreale Spath, 198, 199
Otoceras boreale zone, 202
Otoceras cf. *gracile* Tozer, 205
Otoceras concavum Tozer, 202, 208
Otoceras concavum zone, 208
Otoceratid and dzhulfitid ammonoids
    bipolar distribution, 223
    Boreal species, 221
    ceratitid, 221
    Changhsingian, 223
    genus *Episageceras*, 221
    genus *Otoceras*, 221
    O. concavum Tozer, 222
    onthogenetic development, 222
    palaeolatitudes, 224
    Permian Otoceratoidea, 223

succession lineages, 220
Tompophiceras, 220
Wuchiapingian Anderssonoceratidae, 221
Oxyanions
As *vs*. P_2O_5, 266, 267
bottom-water oxygenation, 267, 269
Cr, 265, 266
deepwater renewal time, 267, 269
U, 264, 265
Oxygen deficiency, 274

P
Pacific Terranes, 85
Paedomorphosis, 82
Palaeobiogeography, 64
Palaeoceanography, 247, 248
Palaeogeographical map, 224
Palaeogeography
Boreal, 83, 84
early Triassic, 83, 84
Pacific Terranes, 85
Tethys, 83
Werfen, 84, 85
Palaeolatitudes, 214
Paleotectonics, 136
Paleozoic atmosphere, 64
Paleozoic-Mesozoic transition, 68
Paleozoic Oxygenation Event, 64
Pancake-like tar globs, 193
Paragondolella, 79
Parallel isomorphic clades, 132
Parathyroid hormone-related protein (PTHrP), 291, 292, 300
receptor gene, 301
signaling, 301
Paris Biota, 61
Paullella, 80
Peat-forming plants, 72
Pelagic ammonoid zones, 62
Penglai oil spill, 192
Peramorphic spherisation lineage, 122
Period of Time, 61–63
Permian mass extinction, 64
Permian reefs, 71
Permian-Griesbachian boundary, 86
Permian-Triassic boundary (PTB), 62, 64, 65, 72
biostratigraphy, 198
conodonts, 39, 40
Imtachan, 200
mass extinction, 86
otoceratid and dzhulfitid ammonoids, 199
sequences, 198
Setorym River basin, 199
See also Nekuchan Formation
Pfenderinids
Cretaceous, 163
genus *Chablaisia*, 161
hypodermic network, 151
Middle Jurassic, 159
morphological innovations, 150
Neokilianina, 163
Nezzazatinella type, 164, 165
pillars/calcitic material, 159
polyphyletic origin, 159
Praekurnubia ancestor, 161
pseudokeriothecal wall, 163
Siphovalvulinidae, 159, 162–163
Triassic, 162–163
trochospiral LIF, 161
Turriglomina-Turritellella type, 159
vertical adaxial, 159
Phanerozoic carbon isotopic excursions, 65
Phenotypic variation, 235
Phosphorite member (PM), 257
Photosynthate export, 242
Photosynthesis, 237
Photo-synthesisers, 72
Photosynthetic *Symbiodinium* symbionts, 237
Phylogenetically non-related genera, 154
Phylogenetic reconstructions, 73–75
Phylogeny, 16, 17
Physiologic homeostasis
alveoli, 301
blood flow, 301
greenhouse effect, 301
molecular mechanism, 301
PTHrP, 300, 301
water-land transition, 300, 301
Physiologic stress, 294, 302
Physiology principle, 300
Physoclistous, 291
Physostomous, 291
Phytoplankton, 247, 254, 264, 266, 279
Planches Inédites, 176
Planispiral, 153
Planispiral-discoid LIF, 154
Planispiral Jurassic mesoendothyrids, 153
Planispiral mesoendothyrids, 139
Planispiral-to-discoid genera, 153
Planktic foraminifera, 112, 115, 118, 120, 123
Planktonic rotaliporiids, 147
Planktonics, 168
Platyvillosus, 78
Polish Zechstein deposits, 24, 26

Pollen grains
abnormal grains, 28
bladderlike sacci enlarge volatile surface, 23
contemporary teratological, 27
features, 27
Jugasporites delasaucei, 28
Klausipollenites schaubergeri, 25
late Permian, 25
Lunatisporites genus and polysaccate, 28
malformed, 24, 28
sacci, 28
teratological forms, 30
UV-B radiation, 30
Polycyclic hydrocarbons (PAHs), 191
Polymorphism/hybridization, 242
Population biodiversity, 134
Praebulimina prolixa, 252
Praeorbulina, 114, 115
Praeorbulina glomerosa circularis, 112, 113, 115, 118–121
Praeorbulina glomerosa glomerosa, 116–117
Praeorbulina-like morphotypes, 116, 122–125
Praevalvulinid, 160–161
Principal component analysis (PCA), 255
Proteromorphosis, 3, 70, 71, 75, 83, 123
Proterozoic–Cenozoic age, 216
Protoelphidium paralium (Tintant), 180
Protozoa, 132
Provincial nature, 63
Pseudofurnishius murcianus, 40
Pseudokeriotheca, 138, 140–141, 154, 162–163
Pseudokeriothecal-to-alveolar microstructure, 154
PTHrP-KO phenotypes, 305, 306
PTHrP-PTHrP receptor signaling, 297

R
Radial vertical partitions, 140–141
Radiolarian oozes, 71
Random walks, 293
Rediversification, 74, 79, 82
Reef-building corals, 235
Reef gap, 71, 72
Reef organisms, 64
Reproduction, 295
Reservoir, 147
Respiratory system, 297
Retrogradation, 19, 20
Retrograde evolution, 123, 135
Retrospective reasoning, 299
Reverse evolution, 14, 102, 123, 297
brain pathology, 306
cholesterol-deficient lung surfactant, 305
commutative law, 307, 308
Cosmos, 303
emphysema, 304
liver function, 306
neoplasia, 307
nephrosis, 304
osteoporosis, 304
PTHrP-KO phenotypes, 305, 306
Reverse evolutionary backwardization, 307
Reverse evolutionary phenotypes, 302
R-mode cluster A, 277
r-strategist planktonics, 135
Rugose corals, 71

S
Scleractinian hexacorals, 234
Scythian, 61
Scythogondolella, 79
Second messengers, 294, 300
Sedimentary sequential evolution, 133
Self-referential self-organization, 303
Seryogin Creek-1 and -2 and -3, 211
Setorym River basin
Inessin Klyuch, 204
Nikolkin Klyuch section, 201–204
nitrogen isotopes, 211
Seryogin Creek-1 and -2 and -3, 211
Siberian platform, 200
Suol Creek and Pravyi Suol-1, 201–204
Ustupny, 210
Sex combs
cellular environment, 103–105
in *D. melanogaster*, 103
Drosophila species, 102
leg bristles, 102
ontogeny and evolution, 106
orientation, *Drosophila*, 104
vertical comb orientation, 106
Shefela Basin, 276–279
Siberian Traps volcanism, 64
Siberigondolella, 79
Sicani basin, 65
Skin-brain hypothesis, 301
Smithian-Spathian boundary (SSB), 65, 68–70
Smithian-Spathian Thaynes/Werfenian environments, 86
Smithian-Spathian transition, 85

Smithian-Spathian warming event, 68–70
Smithodus, 78
Soil nutrients, 72
South Verkhoyansk region
  biostratigraphic purposes, 199
  Changhsingian, 224
  earliest Induan, 218
  Episageceras shells, 221
  high-boreal fauna, 219
  Imtachan Formation, 210
  Nekuchan Formation, 198
  O. boreale Spath, 198, 221
  Otoceras-bearing sequences, 201
  P–T boundary sections, 199
  P–T boundary transition, 218
  Verkhoyansk–Kolyma folded area, 200
Spathian-Anisian environmental bottleneck, 86
Spathian-Julian octomembrate apparatus, 60
Species-wide plasticity, 98
Spiculidendron, 131
Stratigraphical and paleoecological data, 143
Stratigraphical calibration, 136
Stratigraphic/paleogeographic repartition, 132
Stress-induced adaptation, 291
Strigatus zone, 76
Strontium isotope, 64
Structural-functional plasticity, 297
Sublethal environmental stress, 1
Sulphate-reducing bacteria (SRB), 264
Sulphide-forming elements
  Cu, 262–264
  nickel, 260–262
  oxyanions (*see* Oxyanions)
Sulphidic environments, 253
Sulphurization, 262
Suol Creek and Pravyi Suol-1
  benthic fossils, 208
  bivalve species, 210
  Imtachan formation, 208
  Nekuchan formation, 208
  Nizhny Suol sections, 207
  O. boreale Spath, 208
  O. concavum Tozer, 208
  O. concavum zone, 208
  powerful storm event, 207
  P–T boundary transition, 206
  stratigraphical position, 210
  W. decipiens zone, 210
Surface sediments, 186
Symbiodinium dinoflagellates, 235, 239
Symbiodinium spp., 237, 242
Symbiosis, 132
Symbiotic algae, 134
Symbiotic cnidarians, 235
Symbiotic corals, 241
Symbiotic microorganisms, 131, 132
Symmorphosis, 296

T
Tabulate corals, 71
Target of rapamycin (TOR), 295
Tempestites/turbidites, 133
Temporal fluctuations, 186
Terminal addition, 17, 18, 20, 302, 308
Terrigenic/detrital input
  Al, 270
  clay minerals, 272–275
  Fe *vs.* Al, 271
  heavy minerals, 271, 272
Tethyan realm, 155
Tethys, 70, 83
Tetrapyrrole, 260
Thalmanninella greenhornensis, 7, 8
Thalmanninella multiloculata, 7, 8
Thanatocoenosis, 183
Ticinella-Thalmanninella lineage, 8
Time-oriented morphological
    transformations, 122
Tithonian warming event, 145
T/J extinction events, 153
Tompophiceras pascoei epibole zone, 203
TOR and AKT signaling, 295
Torrey Canyon, 177
Toxicity, 192
Trace elements, 255, 258, 259, 262, 264, 267,
    272, 277
Trafficking, 305
Triassic diversification, 70
Triassic endothyrids, 155–157
Triassic Period, 61–63
Triassic strata, 214
Trilobatus–Praeorbulina–Orbulina
    evolutionary lineage, 112–115
Trisaccate pollen grains, 28–30
Trochospiral cylindrical-conical complex
    lituolids, 153
Trochospiral valvulinids/pfenderinids, 139
Tropical Tethyan, 63
Tryptophan 2,3-dioxygenase (TDO), 98–101
Tukey's test, 99
Turritopsis dohrnii, 295, 296
Tyrosinemia-precipitated stress, 306

U
Ultraviolet-B (UV-B) irradiation, 30
Unicellular organisms, 290

Unitarian association zones (UAZ), 81, 82
Upper Cretaceous
  factor loadings, 255
  Levant Basin, 247
  Mishor Rotem basin (*see* Mishor Rotem basin, Upper Cretaceous)
  OSM, 255
  PCA, 255
  Shefela Basin, 276–279
  Southern Tethyan, upwelling system, 248, 249, 251
  statistical analyses, 255
  trace elements, 255, 280
  upwelling systems, 279
  Varimax method, 255
Uppermost Spathian, 70
Upwelling system, 248, 249, 251, 279
Uranium (U), 264, 265
U.S. National Oceanic and Atmospheric Administration (NOAA), 178–179
Ustupny section, 210

V
Valvulammina picardi, 164
Valvulinids
  apertural face, 158
  epigenesis, 158
  genus *Cantabriconus*, 159
  Jurassic praevalvulinids*, 159
  Middle-to-Late Jurassic, 159
  multiple porous apertures, 160–161
  pfenderinids, 159
  siphovalvulinids-pfenderinids, 158
  trematophore-like plate, 158
  Triassic and Liassic carbonate platforms, 158
  trochospiral morphotypes, 148, 158
  valvular tooth plate, 158
Varimax method, 255
Vermilion gene (v^l), 98
Vertical radial partitions, 139, 141
Vertical sex combs without rotation (VCWR), 106
Virtual phylogenetical model, 134
Vitamin D signaling, 102
Volcanism, 61

W
Wall, 137
Wapitiodus, 80
Warm-bloodedness, 292
Warming events, 65
Water currents, 238
Water temperature, 64
Wavelength-dispersive (WD)-XRF, 255
Werfen, 84, 85
Western Tethys climate trends, 67
Wordieoceras decipiens zone, 210
Wuchiapingian–Changhsingian boundary transition, 225

X
X-ray diffraction (XRD), 255, 272

Z
Zechstein Basin, 25
Zooplankton, 238

The manufacturer's authorised representative in the EU is Springer Nature Customer Service Centre GmbH, Europaplatz 3, 69115 Heidelberg, Germany. If you have any concerns regarding our products, please contact ProductSafety@springernature.com

Printed and bound by CPI Group (UK) Ltd, Croydon, CR0 4YY
30/09/2026
02225503-0008